Environmental Science, Engineering and Technology

Mine Drainage and Related Problems

ENVIRONMENTAL SCIENCE, ENGINEERING AND TECHNOLOGY

Additional books in this series can be found on Nova's website under the Series tab.

Additional E-books in this series can be found on Nova's website under the E-book tab.

ENVIRONMENTAL SCIENCE, ENGINEERING AND TECHNOLOGY

MINE DRAINAGE AND RELATED PROBLEMS

BROCK C. ROBINSON
EDITOR

Nova Science Publishers, Inc.
New York

Copyright © 2010 by Nova Science Publishers, Inc.

All rights reserved. No part of this book may be reproduced, stored in a retrieval system or transmitted in any form or by any means: electronic, electrostatic, magnetic, tape, mechanical photocopying, recording or otherwise without the written permission of the Publisher.

For permission to use material from this book please contact us:
Telephone 631-231-7269; Fax 631-231-8175
Web Site: http://www.novapublishers.com

NOTICE TO THE READER

The Publisher has taken reasonable care in the preparation of this book, but makes no expressed or implied warranty of any kind and assumes no responsibility for any errors or omissions. No liability is assumed for incidental or consequential damages in connection with or arising out of information contained in this book. The Publisher shall not be liable for any special, consequential, or exemplary damages resulting, in whole or in part, from the readers' use of, or reliance upon, this material. Any parts of this book based on government reports are so indicated and copyright is claimed for those parts to the extent applicable to compilations of such works.

Independent verification should be sought for any data, advice or recommendations contained in this book. In addition, no responsibility is assumed by the publisher for any injury and/or damage to persons or property arising from any methods, products, instructions, ideas or otherwise contained in this publication.

This publication is designed to provide accurate and authoritative information with regard to the subject matter covered herein. It is sold with the clear understanding that the Publisher is not engaged in rendering legal or any other professional services. If legal or any other expert assistance is required, the services of a competent person should be sought. FROM A DECLARATION OF PARTICIPANTS JOINTLY ADOPTED BY A COMMITTEE OF THE AMERICAN BAR ASSOCIATION AND A COMMITTEE OF PUBLISHERS.

Additional color graphics may be available in the e-book version of this book.

LIBRARY OF CONGRESS CATALOGING-IN-PUBLICATION DATA

Mine drainage and related problems / editor, Brock C. Robinson.
    p. cm.
  Includes index.
  ISBN 978-1-60741-285-4 (hardcover)
  1. Mine drainage--Environmental aspects. I. Robinson, Brock C.
  TD428.M56M563 2010
  622'.5--dc22
                            2010004375

Published by Nova Science Publishers, Inc. ✦ New York

CONTENTS

Preface vii

Chapter 1 Effectiveness and Sustainability of Active and Passive Systems for the Treatment of Acid Mine Drainage and the Recovery of Metals: A Review 1
T. A. Hughes and N. F. Gray

Chapter 2 Acid Mine Drainage in the Portuguese Iberian Pyrite Belt 71
Maria Manuela Abreu, Maria João Batista, Maria Clara F. Magalhães and João Xavier Matos

Chapter 3 Biogeochemistry of Acid Mine Drainage Formation: A Review 119
V. Sheoran, A. S. Sheoran and R. P. Choudhary

Chapter 4 Status of Acid Mine Drainage and Metal Leaching Studies in Ghana 155
Thomas M. Akabzaa, Bernard Kortatsi and Justine Seyire

Chapter 5 Characterization, Dewatering, Disposal and Beneficial Use of Acid Mine Drainage Sludge 189
Xinchao Wei

Chapter 6 Neutralisation of Acid Mine Drainage with Fly Ash in South Africa 211
Vernon Somerset, Michael Klink, Leslie Petrik and Emmanuel Iwuoha

Chapter 7 How Many Data Are Enough? Estimation of an Optimum Sampling Density in Mining and Environmental Applications 227
Konstantinos Modis and Konstantinos Komnitsas

Chapter 8 Treatment of Acid Mine Drainage by a Combined Chemical/Biological Column Apparatus: Mechanisms of Heavy Metal Removal 239
Francesca Pagnanelli, Ida De Michelis, Michele Di Tommaso, Francesco Ferella, Luigi Toro, and Francesco Vegliò

Index 265

PREFACE

Acid mine drainage refers to the outflow of acidic water from (usually abandoned) metal mines or coal mines. However, other areas where the earth has been disturbed (e.g. construction sites, subdivisions, transportation corridors, etc.) may also contribute acid rock drainage to the environment. This book discusses the effectiveness and sustainability of active and passive systems for the treatment of acid mine drainage; acid mine drainage in the Portuguese Iberian pyrite belt; the geochemistry of acid mine drainage and it's environmental pollution problem; the status of acid mine drainage and metal leaching in Ghana; the disposal and beneficial use of acid mine drainage sludge and a host of other related topics.

Chapter 1 - There has been enormous interest in the past two decades in the use of active and passive treatment processes for acid mine drainage (AMD) remediation. Many systems have been successfully employed, and new technologies are being continually developed and tested. This review evaluates both existing and new technologies by examining the underlying chemical and biological principles of each process, their design, and their operation. Treatment systems are compared using data from both full- and pilot-scale systems, with special attention paid to long-term performance and secondary waste products. The advantages and disadvantages of currently used treatment technologies are discussed in the light of new AMD treatment technologies.

Chapter 2 - The Iberian Pyrite Belt (IPB), a major Variscan basement unit, located in the SW sector of the Iberian Peninsula, is one of the oldest mining districts in Europe. This world-class volcanic-hosted massive sulfide province extends for nearly 250 km. Pyrite ore exploitation in the Portuguese IPB sector was developed intensively in the XIXth and XXth centuries at São Domingos, Caveira, Lousal, Aljustrel, Montinho and Chança mines which are responsible for significant acid mine drainage (AMD). Outcropping Aljustrel, São Domingos and Caveira deposits were mined during Roman times, especially the gossan and supergene enrichment zones, up to 118 m depth at Aljustrel, promoting two thousand years of AMD. At present, AMD is related with unprotected open pit and underground mining, exposed mining wastes and host mineralization volcanic and sedimentary rocks of Devonian and Carboniferous ages. The São Domingos, Aljustrel, Lousal and Caveira main mining areas present lagoons, dams and stream waters with very low pH and high electric conductivity. The minimum most recent (July 2008) measured pH was 1.04 in Caveira main stream, and the maximum conductivity of 59.5 mS cm^{-1} was measured in the São Domingos open pit in July 2000. Electric reduction potential also reflects the occurrence of AMD, being values as high as 677 mV measured in the Caveira main stream. At São Domingos and Aljustrel, different materials deposited along the main streams and in the dump areas show the ore processing, while seepage waters indicate different pH-E stabilities. From these AMD waters different mineral precipitations occur, such as copiapite, jarosite, melanterite, and iron oxides. Dissolved elements such as copper, zinc, molybdenum, selenium, cadmium, iron and arsenic attained high concentrations in Lousal and Caveira mines. In Aljustrel and São Domingos mines the dissolved metals and metalloids in AMD areas are mainly iron, aluminium, manganese, cobalt, copper, arsenic, antimony and nickel.

In the Portuguese IPB, especially in São Domingos, Aljustrel, Lousal and Caveira mines, extensive areas are affected by AMD, where different mining practices (mineral processing, period of exploitation, abandon conditions) create different impacts on superficial environments. These situations need different approaches and requirements in case of decommissioning and rehabilitation. Modern mining is implemented in Neves Corvo mine where no significant AMD impact is witnessed.

Chapter 3 - Acid Mine Drainage (AMD) and the contaminants associated with it is the most persistent environmental pollution problem, which occurs worldwide in the coal/metal mining regions. It occurs as a result of natural oxidation of sulphide minerals contained in the mining wastes at operating/closed/decommissioned mine sites. Once it develops at a mine, its control can be difficult and expensive. Treatment of AMD usually costs more than control of AMD and may be required for many years after mining activity is ceased. Thus the early diagnosis of the problem would reduce the potential role of AMD formation and controlling the factors responsible for generation. The focus of this paper is on the review of basic biogeochemistry involved in the formation of AMD and presents the various factors, which control the rate and extent of AMD formation.

Chapter 4 - This chapter reviews the status of acid mine drainage studies and management in Ghana. Acid drainage, and the contaminants associated with it, has been acknowledged as the single largest environmental problem facing the mining industry worldwide. Metals mobilised from acid rock/mine drainage conditions invade the environment with disastrous consequences. Fortunately, knowledge of both local and regional geology and geochemical processes can lead to a better understanding of the phenomenon and its environmental impacts. This knowledge has constituted the bases of management strategies, defining practices in acid drainage prediction, prevention, control, treatment and monitoring, developed to reduce risk of formation of acid drainage from mine wastes in most developed countries. These management approaches rely on the correct identification of mine rock, both in the pre-mining phase and as an ongoing part of mine planning. The Birimian and Tarkwaian Rocks Systems are the main source of gold in Ghana. Gold mineralisation in the Birimian occur in the form of quartz-veined gold in mainly carbonaceous phyllites while mineralisation in disseminated sulphides is in the metavolcanics. Base metal sulphides and sulphosalts including arsenopyrite, pyrite, sphalerite, pyrrohorite, and terahedrites dominate ore mineralogy. Although mineralisation in the Tarkwaian is not directly associated with sulphide-bearing rocks, dolerite intrusions among the Tarkwaian country rocks have been reported to contain considerable sulphide minerals. These sulphide-bearing lithologies present potential acid generating sites. Various studies of stream drainage emanating from some mine sites have recorded pH values as low as 2.4 with associated high doses of As (23. mg/L), Mn (15 mg/L) and Fe (103mg/L). Low pH (<3.4) and high trace metals have been reported from some boreholes in mining areas. Areas outside mining centres with similar geology have water bodies with similar low pH and high metal content. Unfortunately, acid mine drainage and metal leaching management practices are still a very gray area in Ghana. Mining environmental management in itself is a novelty in Ghana. The guidelines for mining environmental management and Environmental Regulations are two key tools for mining environmental management in Ghana. However, these do not have definite benchmarks, protocols and methodologies to guide companies in the evaluation of rocks for acid generating potential in Environmental Impact Assessment (EIA) processes. Limited studies on the subject matter has been carried out by mining companies on the mine scale and largely restricted to the mine site, as part of the project's EIA.

Chapter 5 - The formation of acid mine drainage (AMD) has been an environmental problem for decades in coal mining and hard rock mining regions throughout the world. Typical AMD treatment processes are designed to raise the pH of the effluent to within the range of 6.0-9.0 and to precipitate various metals. A major problem in the AMD treatment is the generation of large volumes of sludge through pH neutralization and subsequent metal precipitation. The sludge typically contains a low concentration of solids, resulting in associated handling and disposal difficulties. Good sludge properties can increase the effective throughput rate of AMD treatment facilities, reduce the capital costs of settling ponds, alleviate the task of sludge dewatering and lower the transportation cost of sludge to ultimate disposal sites. In this chapter, detailed physical and chemical characterization is provided for AMD sludge generated from different neutralization processes. Factors affecting sludge properties include chemical composition of AMD, chemical agents applied for active AMD treatment, passive treatment systems, and AMD treatment operations. AMD sludge treatment can include conditioning, thickening, and dewatering. Sludge disposal can be achieved by deep mine disposal, retained-in-pond disposal, or disposal at coal refuse areas. Finally, the potential beneficial uses of AMD sludge are discussed.

Chapter 6 - This paper reports the results obtained when FA is neutralised with AMD, using different FA:AMD ratios. XRF analysis of South African FA has shown that it consists of three main phases of SiO_2, Al_2O_3 and Fe_2O_3. These three phases form more than 70% of the composition of the FA source, while the other major components are CaO and MgO representing the relative soluble bases present in the FA and related materials. The neutralisation process was investigated by adding different volumes of AMD to a constant mass of FA, using ratios of 1:1; 1:2; 1:3; 1:4; 1:5; and 1:10. The results obtained have shown that the FA in FA:AMD ratios of 1:10; 1:5; 1:4 and 1:3 were able to raise the pH of the AMD from pH 2.4 to pH 6.5. The use of FA:AMD ratios of 1:1 and 1:2 have shown that the FA was able to raise the pH to 11.7 and 11.2 respectively. For ratios 1:10, 1:5 and 1:1 a good plateau was observed between 60 and 210 minutes, indicating that a limited alkalinity was available after this period to influence the pH. Analysis and determination of the sulphate, iron and aluminium concentrations for the co-disposal reactions have shown clear increasing and decreasing trends for the different FA:AMD ratios investigated. Investigation of the B and As concentrations have shown no clear increasing or decreasing trends, while for the Co and Cu concentrations evaluated, it was observed that a clear increasing trend was observed as the FA:AMD ratio increased and the pH of the co-disposal reaction dropped from pH 12 to 5.

Chapter 7 - In mining, environmental and geochemical applications sampling is considered as an extremely important factor to accurately evaluate reserves, spatial heterogeneity of trace elements and contaminants and reliably assess risk as well as environmental impacts. Various sampling campaigns are designed and huge numbers of samples are often analyzed resulting thus in high costs and delays. It would have been therefore useful if a critical sampling density was established for the analysis of earth related space-distributed natural variables.

In the present chapter a case study that predicts the net neutralization potential (NNP) and therefore the capacity of acid generation of the rock formations in a mixed sulphide mine in northern Greece is discussed. A critical sampling grid that extracts maximum information is proposed, so that sampling above this threshold provides limited improvement in the mapping results. The study also revealed that in some parts of the irregular ore body, the initial sampling grid used was denser than required.

Chapter 8 - Natural oxidation of sulphide minerals, exposed to the combined action of oxygen and water, results in the worst environmental problem associated with mining activities, i.e. acid mine drainage (AMD). Waters polluted by AMD are often characterised by low pH, elevated concentrations of iron, sulphates and toxic metals.

Biological remediation options in passive systems (permeable reactive barriers, PRB) usually exploit sulphur production by sulphate reducing bacteria, SRB.

In this report a combined chemical-biological treatment was tested for decontamination of synthetic AMD containing iron, arsenic, copper, manganese and zinc.

Particular attention was paid to the investigation of the mechanisms involved in pollutant removal (chemical precipitation, sorption, bioprecipitation and biosorption) as a fundamental preliminary step for permeable reactive barrier design and long term performance estimation.

Experimental tests were performed both in batch reactors and in a two-column apparatus for sequential treatment by chemical precipitation (first column filled with natural limestone) followed by bioprecipitation/biosorption (second column filled with a natural organic mixture inoculated by sulphate reducing bacteria).

Distinct mechanisms of removal for each metal were identified by combining theoretical data of metal solution chemistry, and results obtained from independent experimental tests: batch and column tests, blank tests using natural organic mixture as biosorbing materials, acid digestions, and selective extractions of metals using solid samples of filling material after column dismantlement.

This analysis allowed isolating metal-specific mechanism of abatement and denoted the relevant contribution of biosorption phenomena in metal removal in biological column. This contribution, generally neglected in biological PRB design with respect to bioprecipitation, should be taken into account in order to avoid misleading estimation of SRB performance and also to better estimate PRB duration.

In: Mine Drainage and Related Problems
Editor: Brock C. Robinson, pp. 1-69

ISBN: 978-1-60741-285-4
© 2010 Nova Science Publishers, Inc.

Chapter 1

EFFECTIVENESS AND SUSTAINABILITY OF ACTIVE AND PASSIVE SYSTEMS FOR THE TREATMENT OF ACID MINE DRAINAGE AND THE RECOVERY OF METALS: A REVIEW

T. A. Hughes[1] and N. F. Gray[2]*
Centre for the Environment, Trinity College, University of Dublin, Dublin 2, Ireland

ABSTRACT

There has been enormous interest in the past two decades in the use of active and passive treatment processes for acid mine drainage (AMD) remediation. Many systems have been successfully employed, and new technologies are being continually developed and tested. This review evaluates both existing and new technologies by examining the underlying chemical and biological principles of each process, their design, and their operation. Treatment systems are compared using data from both full- and pilot-scale systems, with special attention paid to long-term performance and secondary waste products. The advantages and disadvantages of currently used treatment technologies are discussed in the light of new AMD treatment technologies.

Keywords: Acid mine drainage; Active treatment systems; Bioremediation; Passive treatment systems; Metal recovery; Treatment options

---

* Email addresses: [1]hughesta@tcd.ie; [2]nfgray@tcd.ie.

1. INTRODUCTION

1.1. Scope of Review

Mining processes have long-term detrimental impacts on freshwater systems (Kelly, 1988). In particular, the generation of water contaminated with dissolved metals, sulphate, and acidity presents a serious threat to the stability and maintenance of freshwater communities, often causing complete elimination of invertebrate and vertebrate species (Gray and Delaney, 2008). The contaminated waters, known as acid mine drainage (AMD) (or acid rock drainage (ARD)) may emanate from mine workings, drain from waste rock heaps, or seep from tailings dams. The chemistry of AMD varies with site characteristics, as does the longevity of AMD release, which ranges from decades to centuries. The duration of AMD generation depends on the size of the mine, length of associated shafts and tunnels, volume of waste generated, hydraulic parameters, and geological characteristics (Younger et al., 2002; Lottermoser, 2007).

Many treatment systems for AMD exist. The minerals sector and government agencies charged with environmental protection continue to invest in research and development of increasingly efficient and cost effective treatment systems. This chapter reviews the current technologies used to treat AMD, and presents an overview of the technical application of various active and passive treatment systems. Emphasis is placed on the long-term effectiveness of different systems.

1.2. Chemistry of Acid Mine Drainage

Mining and processing of sulphide-bearing ores and coal deposits lead to increased exposure of mineral surfaces to water and oxygen. Water flowing over the exposed mineral surfaces becomes laden with dissolved metals and sulphate as the minerals are oxidized and the acidity (hydrogen ion concentration, $[H^+]$) of the water is increased. This metal- and sulphate-laden, low-pH water, known as acid mine drainage (AMD), may drain from flooded mine workings, surface tailings deposits or tailings dams, and waste rock heaps (Salomons, 1995; Feng et al., 2000; Younger et al., 2002; Kaksonen and Puhakka, 2007; Mendez-Ortiz et al., 2007), and contaminate surface waters or underlying aquifers. Carbonates, oxides, and aluminosilicate minerals (e.g. clays) may be dissolved by the acidic waters. Acid mine drainage may contain iron (Fe), copper (Cu), zinc (Zn), nickel (Ni), cadmium (Cd), arsenic (As), aluminum (Al), manganese (Mn), magnesium (Mg), calcium (Ca), lead (Pb), silver (Ag), mercury (Hg), uranium (U), chromium (Cr), sulphate (SO_4), and other solutes, in concentrations ranging from 10^{-4} to 10^3 g/L (Table 1). Each metal contributes specific metal acidity (Kalin et al., 2006).

High concentrations of alkalinity (CO_3^{2-} + HCO_3^- + OH^-) (e.g. from groundwater infiltration) may neutralize the acidity of AMD. Sulphide-rich, carbonate-poor sites tend to produce acidic drainage, whereas carbonate-rich sites (even with significant sulphide concentrations) produce neutral- to alkaline drainage (Skousen et al., 2002).

Table 1. Comparison of typical chemical characteristics of AMD from coal and hard rock mines

Parameter (metal concentrations in mg/L)	Coal mine AMD	Hard rock mine AMD
pH	3-7	2-6
Fe	0.1-300	10-5,000
Al	0.1-60	1-1,000
Zn	0.005-2	0.1-1,000
As	0.01-10	1-50
Cd	0.001-0.002	0.1-0.25
Cu	0.001-0.2	0.01-350
Mn	0.1-100	0.01-100
Ni	0.01-0.3	0.5-500
Pb	0.01-0.03	0.1-100
SO_4	200-2000	100-10,000

Oxidation of iron sulphide

The oxidation of sulphides, particularly pyrite (iron sulphide, FeS_2), releases dissolved metals, sulphate, and hydrogen ions (Eq. 1) (Stumm and Morgan, 1981). Dissolved ferrous iron (Fe^{2+}) may then be oxidized to the ferric form (Fe^{3+}) (Eq. 2). After oxidation, ferric iron may either be hydrolyzed, forming insoluble ferric hydroxides (Eq. 3) (Nordstrom, 1982), or it may oxidize pyrite (Eq. 4) (Stumm and Morgan, 1981). The hydrolysis reaction (Eq. 3) is abiotic and depends on both the pH and the concentration of dissolved ferric iron (Gazea et al., 1996). Under equilibrium conditions, concentrations of dissolved ferric iron at pH >3 are negligible (Stumm and Morgan, 1981), because ferric oxyhydroxides and oxyhydroxysulphates (e.g. schwertmannite) precipitate above pH 2.3-3.5 (Woulds and Ngwenya, 2004).

$$2\ FeS_2 + 7\ O_2 + 2\ H_2O \rightarrow 2\ Fe^{2+} + 4\ SO_4^{2-} + 4\ H^+ \quad (1)$$

$$4\ Fe^{2+} + 4\ H^+ + O_2 \rightarrow 4\ Fe^{3+} + 2\ H_2O \quad (2)$$

$$Fe^{3+} + 3\ H_2O \rightarrow Fe(OH)_3 + 3\ H^+ \quad (3)$$

$$FeS_2 + 14\ Fe^{3+} + 8\ H_2O \rightarrow 15\ Fe^{2+} + 2\ SO_4^{2-} + 16\ H^+ \quad (4)$$

The process of pyrite oxidation may be considered as three sequential stages (Kleinmann et al., 1981; Nordstrom, 1982). In the first stage, while pH is circum-neutral or slightly acidic, pyrite oxidation (Eq. 1) is both biotic and abiotic, and ferrous iron oxidation is primarily abiotic. In the pH range 5-8, the rate of ferrous iron oxidation (Eq. 5) is relatively fast and strongly dependent on pH, with a unit increase in pH causing a 100-fold increase in the oxidation rate (Sung and Morgan, 1980; Stumm and Morgan, 1981; Wehrli, 1990).

$$-d[Fe^{2+}]/dt = k_0[Fe^{2+}][O_2][H^+]^{-2} \quad (5)$$

In the second stage, when pH is 3-4.5, pyrite oxidation (Eq. 1, Eq. 4) is biotic and abiotic, and ferrous iron oxidation is primarily biotic. In the pH range 2-5, the rate of ferrous iron oxidation is given by Eq. 6.

$$-d[Fe^{2+}]/dt = k_1[Fe^{2+}][O_2][H^+]^{-1} \qquad (6)$$

In the third stage, as the pH drops below 3, formation of ferric hydroxides is hindered and ferric iron activity in solution increases (Evangelou, 1995). Thus, oxidation of pyrite by ferric iron (Eq. 4) rather than by molecular oxygen becomes the primary mechanism at low pH, with reoxidation of ferrous iron to the ferric form (Eq. 2) continuing the process. Because ferric iron is reduced to the ferrous form in Eq. 4, reoxidation of ferrous iron is often the rate-determining step in pyrite oxidation (Singer and Stumm, 1970). At pH <2, the rate of ferrous iron oxidation is given by Eq. 7.

$$-d[Fe^{2+}]/dt = k_2[Fe^{2+}][O_2] \qquad (7)$$

At pH <3.5, the oxidation rate becomes independent of pH, and the activity of acidophilic bacteria such as *Thiobacillus ferrooxidans* becomes increasingly more important in catalyzing the reaction (Singer and Stumm, 1970; Kleinmann and Crerar, 1979; Kleinman et al., 1981; Stumm and Morgan, 1981; Cohen, 2006). In Eqs. 5-7, the rate constants k_0, k_1, and k_2 depend on oxygen concentration and have been calculated in Wehrli (1990). The overall reaction sequence of pyrite oxidation is acid-generating (Eq. 8) (Banks et al., 1997).

$$4\ FeS_2 + 14\ H_2O + 15\ O_2 \rightarrow 4\ Fe(OH)_3 + 8\ SO_4^{2-} + 16\ H^+ \qquad (8)$$

Although sulphides other than pyrite may also contribute to AMD formation (e.g. galena, PbS; sphalerite, ZnS; millerite, NiS; greenockite, CdS; and covellite, CuS), the oxidation of these sulphides (Eq. 9) does not release hydrogen ions (Kaksonen and Puhakka, 2007). For example, hydrolysis of dissolved zinc may release hydrogen ions without causing a significant decrease in pH below 5.5 (Banks et al., 1997).

$$ZnS + 2\ O_2 \rightarrow Zn^{2+} + SO_4^{2-} \qquad (9)$$

As weathering reactions dissolve sulphides and aluminosilicates, a crust of water-soluble sulphates may form on evaporation. Metals such as cadmium, lead, cobalt, and nickel may be held in this crust temporarily until being washed into the receiving waters (Acero et al., 2007).

2. KEY CONSIDERATIONS FOR TREATMENT SYSTEM DESIGN

The ultimate objective for AMD treatment systems is to restore natural background water quality, a process that typically requires neutralization of acidity and the removal of metals and sulphate. Many characteristics of a mine site affect the temporal and spatial pattern of AMD formation. Geological, hydrological, and topographical parameters must all be considered, as well as the particular extraction processes used (e.g. open pit or deep shaft). The first step in any mitigation strategy must be the temporal and spatial characterization of the variability in site-generated AMD chemistry. Other important factors relating to the site include land and energy availability and accessibility for construction and maintenance (Gray, 1997). Acid mine drainage generation is self-perpetuating because of its leaching capacity and bacterial activity (Kalin et al., 2006), and may continue for decades or even centuries, depending on site characteristics. Therefore, sustainability, maintenance requirements, and cost are also key factors in treatment selection.

2.1. Source Control versus Migration Control

The feasibility of treatment methods depends on mine site characteristics, which are extremely variable and possibly in some situations unique (Gray, 1997). Isolating the contaminant source, suppressing the reactions that release contaminants (e.g. sealing the mine from air to prevent oxidation), and/or installing active (Section 3) or passive (Section 4) treatment systems may minimize the problems associated with AMD (Banks et al., 1997). In some cases, AMD formation may be prevented by a "source control" method. Because of the key role played by ferric iron in pyrite oxidation, strategies to minimize or prevent the generation of AMD typically aim to immobilize ferric iron by complexation or precipitation, inhibit ferric iron production, and/or limit the ingress of oxygen (Evangelou, 1995).

Source control methods aim to minimize AMD generation by preventing air and water from coming into contact with exposed sulphides. If the locations of all mine shafts and adits are known, it may be feasible to flood and seal the mine, preventing oxygen ingress and sulphide weathering. This method is difficult to implement in practice, because the material used for sealing adits and any underground fractures must be able to withstand high concentrations of acidity and sulphate as well as continuous and possibly increasing hydraulic pressure.

Management of mine tailings and wastes is also a key factor in minimizing AMD formation (Bussière, 2007). A common method is backfilling, where tailings are returned to the mine shafts after processing, thus minimizing air and water contact with the tailings. Backfilling is often the most cost-effective option for mine owners, because it reduces the volume of tailings for surface management facilities. Impermeable caps made of plastic, compacted mine spoil, or organic material (e.g. clay) may also be used to cover tailings and prevent contact with air and water. Geosynthetic clay liners may limit oxygen diffusion into mine wastes and attenuate metals from AMD (Bouazza and Rahman, 2007; Lange et al., 2007; Liang and Thomson, 2008). Pyrite oxidation may be minimized by mixing coal-combustion by-products into grout for encapsulating pyritic mine wastes (Bulusu et al., 2007). Anionic surfactants or natural organic wastes may be applied to tailings piles to decrease bacterial activity and retard pyrite oxidation (Kleinman, 1989; Evangelou, 1995; Paganelli et al., 2007; Sand et al., 2007).

Tailings impoundments may be effective in minimizing AMD formation. Romero et al. (2007) studied an abandoned lead and zinc flotation tailings impoundment, and observed that a series of precipitation and sorption reactions was effectively retaining toxic metals from the tailings within cemented layers of secondary precipitates such as gypsum, geothite and hematite. Another method of preventing AMD from forming in mine tailings is underwater disposal, although this method is inappropriate in areas where evaporation is likely to occur. In addition, seepage may be a problem. For example, compacted clay liners have been found to be ineffective at preventing leaching of iron, arsenic, nickel, sulphate, and acidity from evaporation ponds into groundwater (Mapanda et al., 2007). Tailings may be treated with lime (CaO) and soluble phosphate (PO_4) to reduce leaching of lead, copper, cadmium, and zinc, by converting these metals into insoluble metal phosphate precipitates (Eusden et al., 2002). Evangelou (1998) described a method of forming an iron oxide/silica coating on pyrite surfaces, preventing further oxidation. Alternatively, if soluble phosphate and hydrogen peroxide (H_2O_2) are added to pyritic mine waste, the hydrogen peroxide oxidizes the pyrite, producing ferric iron, which reacts with the phosphate to form a surface coating of acid-resistant ferric phosphate (Evangelou, 1995; Ueshima et al., 2004). Low-permeability membranes have also been successfully used to isolate tailings piles (Arnesen and Iversen, 1995). Nehdi and Tariq (2007) reviewed the use of cementitious materials and other material such as fly ash (a residue of coal combustion) to prevent AMD forming in mine tailings. Finally, AMD from tailings may be recirculated through the mine workings. In Løkken Mine in Norway, AMD from spoil tips is recirculated through the excavated and flooded sulphide mine, which serves as a giant copper adsorption basin (Banks et al., 1997), effectively reducing copper concentrations from 75-200 mg/L to 0.5-10 mg/L (Arnesen et al., 1994; Arnesen and Iversen, 1995). Sulphate concentrations in the AMD are not decreased; however, pH is increased from 2 to 4, and the total metal toxicity of the leachate is reduced.

It is difficult to prevent oxidation processes in waste rock, because the particle size is larger than that of tailings, making the rock more free-draining, with the larger interstices resulting in greater oxygen exchange rates and creating ideal environments for AMD formation (Salomons, 1995). In addition, waste rock piles may be high and sloping, resulting in high erosion and runoff potential. Covers may reduce oxygen infiltration into the waste, but they are unlikely to minimize the overall load of AMD contamination.

It is often the case that AMD generation cannot be prevented "at source", and an active or passive treatment system must be installed, i.e. "migration control" (Johnson and Hallberg, 2005). A short-term migration control option is to continue to pump water from the mine, preventing flooding of the mine workings and subsequent AMD formation (Younger and Harbourne, 1995; Younger et al., 1995). For long-term treatment, active or passive systems must be installed to intercept and treat AMD before it enters ground- or surface waters.

2.2. Temporal Changes in AMD

The longevity of AMD formation is an important factor in designing a treatment system. For example, passive treatment systems are often designed with a 20-year lifespan (Ziemkiewicz et al., 2003); thus, predicted longevity of AMD formation at the mine site must be carefully considered to determine if such a system would be appropriate.

Long-term and seasonal changes

Contamination by AMD may persist for decades or centuries after a mine has been abandoned (Kalin et al., 2006). In surface coal mines, over 90% of the coal is typically removed (Demchak et al., 2004), and because the coal usually contains most of the sulphide content (and acid-generating potential), little AMD-generating waste rock remains once mining has ceased. Meek (1996) suggested that within 20 years, acid products would be leached from the pyrite-bearing coal and rocks. At some coal mine sites, the AMD may be neutralized by naturally-occurring carbonates or added carbonates (e.g. limestone dust) used as fire retardants during mining (Wood et al., 1999).

Acid mine drainage from underground mines may also improve on a decadal time scale (Jones et al., 1994; Wood et al., 1999), although this depends on whether or not the mine is below the water table and permanently flooded (Younger, 1997; Demchak et al., 2004). The water quality of AMD from mines above the water table is harder to predict. Demchak et al. (2004) compared the water quality of underground coal mine drainages from two regions of West Virginia in 1968, and again 31 years later. The mine discharges were above the water table, and therefore were not flooded after abandonment. Acidity was significantly improved at 34 out of 44 (77%) sites. Over the three decades, average total acidity decreased by 79% in one region and 56% in another. The acidity of drainage from 10 sites showed no improvement, and at three sites, the acidity had become significantly worse. Across all sites, iron concentration decreased from an average of 400 to 72 mg/L (approximately 80%), and sulphate concentration decreased by between 50% and 75%.

Seasonal changes in precipitation and temperature have been observed to affect AMD chemistry. Kumpulainen et al. (2007) found that after spring snowmelt, precipitates in drainage ponds around tailings and waste rock piles at abandoned sulphide mines in Finland changed from mainly geothite (prevalent during warm summer months) to schwertmannite. Søndergaard et al. (2007) observed seasonal increases in metal and sulphuric acid concentrations and decreases in pH after thaw in high arctic coal mine waste rock, and concluded that waste products were being released as a seasonal flush of pollutants. Snowmelt in spring and heavy rain in autumn have been reported to cause dissolution of metal-sulphate salt precipitates in the soil and affect concentrations of iron, zinc, aluminum, sulphate, and acidity in AMD from an abandoned mine with ores containing pyrite, chalcopyrite, pyrrhotite, sphalerite, and galena (Bloom et al., 2007). Smuda et al. (2007) observed secondary mineralization during the dry season and metal release during the wet season from waste rock dumps in Peru, a phenomenon also recorded elsewhere (Alpers et al., 1994; Gray, 1998)

Thus, AMD chemistry changes over time, and although acidity and metal concentrations typically decrease, in some situations AMD may become more acidic. Therefore, treatment systems must be designed to cope with both seasonal and long-term variations in AMD chemistry.

Vestigial and juvenile acidity

Acidity in mine drainage can be characterized as either "vestigial" or "juvenile". Vestigial acidity is generated as water fills an abandoned mine and secondary mineral precipitates of metals and sulphate released during sulphide weathering are dissolved (Younger, 1993; Younger and Sherwood, 1993). In contrast, juvenile acidity is generated by ongoing sulphide weathering resulting from fluctuations of the water table (Younger, 1997). Vestigial acidity has been observed to decay exponentially, often dissipating within 40 years of mine closure, whereas juvenile acidity may be generated for centuries, until the pyrite is consumed (Younger, 1997). In the UK, it has been observed that discharges from recently flooded mines may be net-acidic, but after the "first flush", drainage acidity and metal concentrations may decrease (Younger, 1997).

Longevity of AMD formation depends on the dissipation of vestigial acidity, which is controlled by mine volume, hydraulic connectivity, and the rate of recharge, and on the generation of juvenile acidity (Younger, 1997). When mine workings flood and remain filled with water, dissolved oxygen eventually becomes limiting, and pyrite oxidation ceases (Evangelou, 1998). However, where water levels fluctuate, a self-renewing cycle is observed. Pyrite oxidation leads to the formation of iron hydroxysulphate solids in the unsaturated zone, which subsequently dissolve when the water level rises. When the water level drops, oxidation occurs on the newly exposed pyrite surfaces (Younger, 1997). Juvenile acidity is often attributed more to mineral acidity from dissolved iron, manganese, or aluminum (Hedin et al., 1994) than to hydrogen ion acidity. Iron concentrations may remain high owing to juvenile acidity production if the host rock is rich in pyrite or the water table fluctuation is "vigorous" (Younger, 1997).

The former British Coal Corporation devised a practical rule of thumb for predicting the concentration of iron in mine drainage from flooded mines, which stated that the iron concentration would be decreased by 50% in the time required for the mine workings to fill with water, and again in each subsequent pore volume flushing (Glover, 1983). Younger (1997) observed that in coal workings located in the Midland Valley of Scotland, an approximately exponential improvement in water quality took place in all mines, and drainage reached an asymptotic iron concentration of approximately 10 mg/L after around 40 years. Wood et al. (1999) found similar results in Scottish coal mine drainages. At the Wheal Jane mine in Cornwall, UK, vestigial acidity also decreased exponentially and iron concentrations in the pumped water decreased from 1600 mg/L to 400 mg/L over the period 1992 to 1995, indicating that the system was probably approaching a dynamic equilibrium, as the initial products of pyrite oxidation were flushed out of the groundwater (Banks et al., 1997). Iron concentrations were predicted to reach an asymptotic level as high as 200 mg/L owing to the high content of pyrite in the host rock (data from UK Environment Agency, presented in Younger (1997)). This means that after an initial rapid decrease, iron concentrations in Wheal Jane drainage will stabilize at approximately 200 mg/L, and decrease very slowly after reaching that concentration. In water quality records from 1965 to 1995 of the Garth Tonmawr discharge in the South Wales coalfields, an exponential decrease in vestigial acidity is evident, and an asymptotic iron concentration of approximately 30 mg/L and pH between 4 and 6 was reached after 20 years (data from UK Environment Agency, presented in Younger (1997)).

Acid-base accounting (ABA) (Smith et al., 1976) is an accurate method of predicting post-mining drainage quality from surface mines and has been successfully employed (Skousen et al., 2002; Akabzaa, 2007; Liao et al., 2007; Pinetown et al., 2007). A net acid generation method was used to predict the acid generating potential of mines in China, where Liao et al. (2007) concluded that mine sites with high sulphur content and/or low acid neutralization capacity were those with the highest acid forming potential. However, Banks et al. (1997) argued that mine water quality depends on, among other factors, the relative rates of reaction and consumption of carbonates, silicates, and sulphides, and therefore cannot be predicted solely by using ABA. Kinetic or leaching tests may be used to supplement ABA data, and are useful in situations where the acid-producing and acid-neutralizing materials present are nearly balanced (Skousen et al., 2002).

2.3. Factors Affecting Metal Concentrations

The concentrations of metals in AMD vary with the host rock composition and the quantity and type of sulphide minerals present (Younger, 2000a). Static and kinetic tests are available to predict AMD formation based on the theory that formation depends primarily upon geochemical characteristics of the site (Salomons, 1995). However, the interplay of geochemical reactions, biotic reactions, and hydrological factors is complex and makes accurate modelling and prediction difficult.

The pH conditions have a significant impact on metal concentrations. An extreme example of this is found at Richmond Mine, Iron Mountain, in Northern California, where extremely acidic conditions result in high metal and sulphate concentrations in mine drainage. Negative pH has been measured at this site (Nordstrom and Alpers, 1999; Nordstrom et al., 2000), resulting in iron concentrations as high as 141 g/L, copper, zinc, arsenic, and cadmium in concentrations from tenths to tens of grams per liter, and sulphate concentrations as high as 760 g/L (Nordstrom et al., 2000). Seasonal factors such as precipitation also affect AMD chemistry. For example, in some regions evaporitic salts may precipitate during the summer and then dissolve in the autumn, creating significant fluctuations in contaminant concentrations (Canovas et al., 2007).

In anoxic AMD at pH >6, the reduced species Fe^{2+} and Mn^{2+} are present in higher concentrations than oxidized Fe^{3+} and Mn^{4+}, because the reduced species are more stable at higher pH (Johnson and Hallberg, 2005; Mariner et al., 2008). Once AMD is introduced into oxygenated surface waters, several geochemical processes may occur that affect metal concentrations. In streams containing iron and aluminum, neutralization by dilution or convergence with more alkaline streams leads to precipitation of iron and aluminum oxyhydroxides and hydroxysulphates. Depending on the stream pH, other dissolved metals may then be adsorbed onto the precipitates (Canovas et al., 2007; Desbarats and Dirom, 2007; Lee and Faure, 2007). Co-precipitation is an important natural attenuation mechanism (Wolkersdorfer 2008). Oxides, hydroxides, and hydroxysulphates of iron, aluminum, and manganese are important carriers of toxic trace elements in AMD (Lee et al., 2002; Dolenec et al., 2007). Precipitates in AMD may be deposited in lakes, reservoirs, and depositional areas of rivers (Munk et al., 2002) and soils (Zhou et al., 2007). Munk et al. (2002) investigated the changes in water chemistry of a stream receiving AMD containing iron, aluminum, zinc, copper, lead, nickel, sulphate, and other metals. Removal of iron occurred by oxidation of ferrous iron to the ferric form and subsequent precipitation of iron oxyhydroxides, which form ochreous deposits on the streambed. As the stream (pH ~3.0) mixed with another stream (pH ~7.0), the pH was buffered to 6.3 in the mixing zone, and aluminum oxyhydroxide precipitates formed that were rich in sulphate. These precipitates, and associated organic carbon complexes, adsorbed trace metals from the stream, and had high concentrations of zinc, copper, lead, and nickel. Highest concentrations of trace metals were observed in the mixing zone of the streams, and a flocculent tan-to-white precipitate was observed on the streambed of the mixing zone as well as further downstream, where pH dropped to 5.2-5.3. The adsorption of zinc, copper, lead, and nickel at pH 6.3 was facilitated by the presence of sulphate, which may form ternary bridging complexes between the sorbent and the sorbate cation (Munk et al., 2002). It has previously been reported (Webster et al., 1998) that sulphate enhances sorption of trace metals onto natural mine drainage precipitates as well as onto schwertmannite and ferrihydrite. Metal sorption was also observed to be pH-dependent, with preferential sorption occurring with increasing pH in the order lead, copper, zinc, and nickel. Sulphate removal was mainly by precipitation as aluminum hydroxysulphate between pH 4 and 5, and some sulphate was removed by sorption to microcrystalline gibbsite.

Metal transport in groundwater is difficult to model or predict (Malmström et al., 2008), although various novel approaches have been proposed (Collon et al., 2006). A contaminant may be precipitated and retained over variable timescales, only to be released in unpredictable quantities and concentrations. Malmström et al. (2008) modelled reactive transport of zinc in heterogeneous groundwater and reported that zinc was likely to be precipitated as smithsonite ($ZnCO_3$), which may delay zinc arrival downstream and cause peaks in concentration once it is released. Analysis of contaminant transport may be made more difficult if the amount of precipitates has high spatial variability.

Metal removal from AMD depends on the concentrations of different metal ions, ionic strength, temperature, reduction potential, and the concentration of complexing agents such as humic acids (Kalin et al., 2006). Precipitation of metals, a key mechanism in AMD treatment, is pH-dependent. In AMD treatment systems, two approaches may be used for removing metals; a multi-stage treatment may be designed to precipitate each metal at its optimum pH, or a single pH may be used that allows the removal of the majority of metals by co-precipitation (Coulton et al., 2003a). Multi-stage treatment systems must be carefully designed to achieve effective metal removal when treating AMD with a range of contaminants.

2.4. Sulphate Chemistry in AMD Treatment

Sulphate may be present in AMD in moderate (10^3 mg/L) to high (10^6 mg/L) concentrations, and its analysis must be included as part of site characterization (Liang and Thomson, 2008). Analysis is normally by either ion chromatography or inductively-coupled plasma atomic emission spectrometry (ICP-AES) (Reisman et al., 2007), with sulphur isotopes used to evaluate sulphate loading from different sources (Trettin et al., 2007). Sulphate removal from AMD by active treatment may be achieved through neutralization and precipitation using calcium added as lime, or other reagents, or by the use of sulphate-reducing bacteria (SRB) to encourage sulphide precipitation (Dvorak et al., 1992).

Sulphate reduction is the key aspect of AMD remediation in many treatment systems (Sections 3.5, 4.4). Sulphate-reducing bacteria catalyze the reduction of sulphate (Eq. 10), a process that consumes hydrogen ions, increases the pH and alkalinity of the AMD, and generates hydrogen sulphide (H_2S). The electron donor may be supplied by an organic material (denoted as CH_2O) or by hydrogen gas (yielding hydroxide ions) (Eq. 11). Insoluble metal sulphides may form with divalent metal ions such as zinc (Eq. 12).

$$SO_4^{2-} + 2\ CH_2O + 2\ H^+ \rightarrow H_2S + 2\ H_2CO_3 \qquad (10)$$

$$2\ SO_4^{2-} + 8\ H_2 \rightarrow H_2S + HS^- + 5\ H_2O + 3\ OH^- \qquad (11)$$

$$Zn^{2+} + H_2S \rightarrow ZnS \downarrow + 2\ H^+ \qquad (12)$$

Removal of metals as sulphides depends on pH, the solubility of the metal sulphide, which is typically very low in water, and the concentration of the dissolved metal (Utgikar et al., 2000). If reducing conditions are maintained, metal sulphides will remain insoluble, and this process becomes a more effective sink for heavy metals than precipitation as hydroxides or other processes (Woulds and Ngwenya, 2004). Although the formation of metal monosulphides releases hydrogen ions as hydrogen sulphide dissociates, the metal acidity of the AMD is decreased and the alkalinity from the sulphate reduction neutralizes the acidity. In some cases, however, excess sulphate may need to be reduced to prevent a decrease in pH (Kaksonen and Puhakka, 2007).

3. ACTIVE TREATMENT SYSTEMS

Active treatment systems require ongoing inputs of energy and/or reagents (Younger et al., 2002). Although these systems need a higher level of maintenance for successful operation, they have a smaller footprint than passive systems and are often used where land availability is limited. Active systems may be the only feasible option for treating certain AMD wastewaters that would be difficult to treat with passive systems (e.g. owing to pH extremes or very high metal concentrations). The most widely used active treatment systems include oxidation and chemical neutralization (Section 3.1), flotation (Section 3.2), membrane filtration (Section 3.3), ion exchange (Section 3.4), sulphate reducing bacteria-based systems (Section 3.5) and adsorption (Section 3.6).

3.1 Oxidation and Chemical Neutralization

The most common active treatment system is reactor-based chemical neutralization followed by precipitation of metals (e.g. iron, manganese, and aluminum) as hydroxides or carbonates. In the typical two-stage process, AMD is neutralized and aerated to encourage metal precipitation, with subsequent solids/liquid separation in a clarifier (Figure 1) (Coulton et al., 2003a).

In the first reaction vessel, AMD is dosed with an alkaline material to raise the pH (e.g. calcium hydroxide (hydrated lime, $Ca(OH)_2$); calcium oxide (quicklime, CaO); calcium carbonate (limestone, $CaCO_3$); sodium carbonate (soda ash, Na_2CO_3); sodium hydroxide (caustic soda, $NaOH$); magnesium oxide (MgO); or magnesium hydroxide ($Mg(OH)_2$)). Hydrated lime and quicklime are commonly used reagents, owing to their relatively low cost, high reactivity, and effectiveness in treating metal-laden AMD. Quicklime becomes hydrated and forms hydrated lime in the presence of water, and then dissolves, releasing hydroxide ions and raising the pH of the AMD to ~ pH 10.5 (Eqs. 13-14) (Cohen, 2006). Quicklime may be used to treat inorganic effluent with metal concentrations higher than 1000 mg/L (Kurniawan et al., 2006).

$$CaO + H_2O \rightarrow Ca(OH)_2 \quad (13)$$

$$Ca(OH)_2 \rightarrow Ca^{2+} + 2OH^- \quad (14)$$

The reaction that occurs when limestone ($CaCO_3$) mixes with AMD adds bicarbonate alkalinity to AMD (Eqs. 15-17) (Nairn et al., 1992).

$$CaCO_3 + 2 H^+ \rightarrow Ca^{2+} + H_2CO_3 \quad (15)$$

Dissolved carbon dioxide (H_2CO_3) (Eq. 16) acts as a weak acid and reacts with limestone to form bicarbonate alkalinity (Eq. 17).

$$CO_2 + H_2O \leftrightarrow H_2CO_3 \leftrightarrow H^+ + HCO_3^- \quad (16)$$

$$CaCO_3 + H_2CO_3 \rightarrow Ca^{2+} + 2 HCO_3^- \quad (17)$$

Limestone dissolution may also increase pH through the release of hydroxide ions (Eq. 18).

$$CaCO_3 + H_2O \rightarrow Ca^{2+} + HCO_3^- + OH^- \quad (18)$$

For treatment of AMD containing ferrous iron, aeration and/or dosing with a chemical oxidizing agent (e.g. hydrogen peroxide, calcium peroxide, or ozone) is required to encourage oxidation of ferrous iron to the ferric form (Rao et al., 1995; Coulton et al., 2003a). Chemical oxidizing reagents may be used to speed oxidation reactions when metal concentrations are low. In the second stage of treatment, the addition of flocculants or coagulants causes ferric iron and other metals to precipitate out of solution, typically as hydroxides (as shown for copper in Eq. 19), either in clarifiers or settlement ponds.

$$Cu^{2+} + 2OH^- \rightarrow Cu(OH)_2 \quad (19)$$

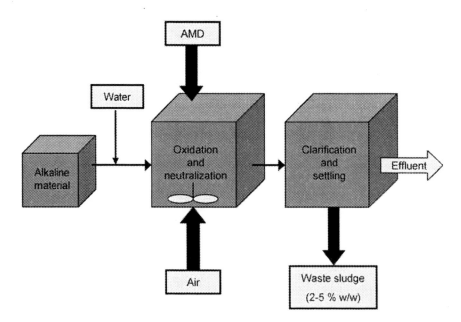

Figure 1. Schematic diagram of a conventional oxidation and chemical neutralization system

Precipitation and pH

Precipitation of iron from net-alkaline mine drainage may be achieved by oxidation and subsequent precipitation of ferric hydroxides, with little or no neutralizing reagent required (Coulton et al., 2003a). In net-acidic AMD, ferric and aluminum hydroxides precipitate at pH ~3 and become highly stable above pH ~5 (Kalin et al., 2006). In contrast, divalent metals precipitate at strongly alkaline pH values (e.g. nickel at pH ~8, ferrous iron at pH 8-9, and zinc at pH >9) (Kalin et al., 2006). However, the range of pH values for optimum metal precipitation varies. Using ion exchange and precipitation, Papadopoulos et al. (2004) reported 71% removal of nickel at pH 7.5 and 85% removal at pH 10.5, from an initial concentration of 51.6 mg/L. Tünay and Kabdasli (1994) observed 80% copper removal (initial concentration 16 mg/L) using magnesium hydroxide at pH 9.5, and 96% cadmium removal (initial concentration 37 mg/L) using ferric hydroxide at pH 11. Using hydrated lime precipitation, Charerntanyarak (1999) achieved >99% removal efficiency of zinc and cadmium at initial concentrations of 450 mg/L and 150 mg/L, respectively, and recommended optimum pH 11. Alkali dosing can be regulated to keep system pH within an optimum range (Yong Gan et al., 2005). Alternatively, metals may be removed by co-precipitation with iron, avoiding the requirement for high pH (Coulton et al., 2003a).

Manganese removal by hydroxide precipitation is difficult because manganese must be oxidized from the divalent to the trivalent or tetravalent form (Hedin et al., 1994). Oxidation of manganese is abiotic at pH >8, and the uncatalyzed oxidation is very slow at pH <9 (Stumm and Morgan, 1981; Brezonik, 1994). Thus, a relatively high pH is required to precipitate manganese hydroxides or oxyhydroxides (Eq. 20) (Hedin et al., 1994), which tend to oxidize to the more stable manganese oxide (Gazea et al., 1996). Charerntanyarak (1999) observed 99.3% removal of manganese at an initial concentration 1085 mg/L using hydrated lime precipitation at pH 11 and subsequent coagulation. Alternatively, reagents such as carbonates can be used to precipitate manganese at lower pH (Coulton et al., 2003a). In an alkaline environment, manganese may precipitate as a carbonate and then further oxidize to manganese oxide (Eqs. 21-22) (Diehl and Stumm, 1984).

$$4\ Mn^{2+} + O_2 + 6\ H_2O \rightarrow 4\ MnOOH + 8\ H^+ \tag{20}$$

$$Mn^{2+} + HCO_3^- \rightarrow MnCO_3\ (s) + H^+ \tag{21}$$

$$2\ MnCO_3 + O_2 \rightarrow 2\ MnO_2\ (s) + 2\ CO_2 \tag{22}$$

Sibrell et al. (2007) reported zinc and manganese removal via a carbonate co-precipitation process. The process was designed to increase chemical stability of sludges, as well as to remove metals more efficiently.

Problems with limestone and alternative techniques

Although limestone is a relatively inexpensive and effective neutralizing agent, limestone-based systems are not recommended for treating highly acidic, iron-rich (>5mg/L) AMD (Skousen et al., 1995), owing to associated armoring with ferric oxyhydroxide coatings. Armoring is caused by metal oxyhydroxides, hydroxysulphates, or gypsum (calcium sulphate, $CaSO_4 \cdot 2H_2O$). For example, aluminum hydroxysulphate precipitates can form at pH >5 (Nordstrom and Alpers, 1999). Therefore, treatment of aluminum-rich AMD in limestone-based systems may become ineffective, owing to aluminum armoring (Watzlaf et al., 2000; Hammarstrom et al., 2003).

Carbonate reagents only raise the pH to ~8.5, compared with hydroxides, which typically raise the pH to ~10; therefore, carbonate reagents may not be effective at precipitating metals that require a high pH. Coulton et al. (2003a) reviewed the theoretical costs, doses, and effectiveness of commonly used precipitation reagents. The precipitation process is usually rapid, but coagulation and settlement may be less efficient in cases where the iron concentration of the influent is low. Ion exchange resins, filtration columns, flotation or adsorption techniques may be incorporated into the system to increase metal removal.

Sodium hydroxide has been used to neutralize AMD and selectively recover copper, iron, and manganese, with removal rates as high as 90%, as well as zinc and nickel, with removal rates less than 80% (Pavlović et al., 2007). Sodium hydroxide may be used to treat AMD with high sulphate concentrations (>2000 mg/L) because it neutralizes the effluent without precipitating gypsum. However, because sodium hydroxide reacts quickly with AMD, the precipitate is fine and settles slowly (Coulton et al., 2003a). Magnesium hydroxide can also be used to neutralize AMD, and although it is more expensive than lime, it has the advantages of producing a denser sludge and not precipitating sulphate (e.g. gypsum). However, because the reaction kinetics are slower, a larger reaction vessel is required (Coulton et al., 2003a).

Innovative processes have been developed to increase the efficiency of conventional neutralization treatment systems. The Savmin process (Ramsay, 1998) uses five stages: (i) heavy metal and magnesium precipitation at pH ~12 (attained with addition of calcium hydroxide), (ii) gypsum desupersaturation, (iii) ettringite precipitation for removal of calcium and sulphate at pH 11.4-12.4 (attained with addition of aluminum hydroxide), (iv) carbonation (i.e. addition of carbon dioxide), and (v) recycling of aluminum hydroxide. This process is followed by a multistage solid-liquid separation process.

Herrera et al. (2007) evaluated a two-step neutralization process designed to improve the sludge volume index (SVI), initial settling rate, and solids content of hydroxide sludge (Igarashi et al., 2006). In the first neutralization step, either magnesium oxide (MgO) or calcium carbonate ($CaCO_3$) is used to raise the pH to ~4.8 and precipitate metal hydroxides of low solubility (e.g. ferric iron, copper, aluminum, and arsenic), which are then removed in a settling tank. In the second neutralization step, sodium hydroxide is added to increase the pH to 8.5 and precipitate ferrite sludge, which contains ferrous and ferric hydroxides and the remaining heavy metals precipitated as hydroxides (e.g. ferrous iron, silicon, and zinc). The sludge is then removed in a clarifier, and some sludge is recycled through the ferrite formation tank. The system was used to treat AMD from an abandoned pyrite mine, which is currently treated in a conventional neutralization system of four neutralization tanks with slaked lime, two settling tanks, sludge recycling and a polishing pond. Using the two-step neutralization ferrite-formation process, Herrera et al. (2007) reported effective metal removal from AMD (Table 2) and improvement of SVI, initial settling rate, and solids content. The use of magnesium oxide as the first neutralizer enhanced ferrite formation, which is a relevant parameter for the industrial resale value of the sludge.

Table 2. AMD chemistry before and after treatment in a two-step neutralization process using magnesium oxide (MgO) and calcium carbonate (CaCO$_3$) (reproduced from Herrera et al. (2007) with permission of Elsevier Ltd.)

First neutralizing agent		pH	Fe^{2+} mg/L	Fe^{3+} mg/L	Al mg/L	Si mg/L	Ca Mg/L	Mg mg/L	SO_4 mg/L	Cu mg/L	Zn mg/L	As µg/L
MgO	Inflow	2.1	588	715	45.1	22.9	96.6	27.8	2818	13.1	15.0	862
	Outflow	8.2	1.04	0.9	0.1	1.6	85.2	325	2260	0.0	0.2	1.8
CaCO$_3$	Inflow	2.1	579	739	46.1	23.4	135	41.9	2800	12.0	15.7	823
	Outflow	8.3	0.79	1.0	0.4	2.3	347	32.7	2385	0.0	0.2	1.4

Hydroxide sludge

The oxidation and chemical neutralization process increases the pH of AMD by adding alkalinity and reducing metal acidity by removing metals from solution. Large volumes of alkaline materials are usually required to maintain optimum conditions for metal precipitation. After solids are separated from the treated effluent in clarifier/thickener tanks or lagoons, the residual sludge is very thin (2-5% w/w) (Coulton et al., 2003a) and difficult to dewater owing to the floc-like nature of iron oxide precipitates (Brown et al., 1995). Sludge typically has high concentrations of aluminum and iron hydroxides and oxyhydroxides and requires careful handling and disposal. In addition, the sludge typically contains calcium carbonate, and may contain magnesium sulphate or gypsum (Sibrell and Watten, 2003; Kalin et al., 2006). Research is needed to examine the stability and potential disposal hazards of the secondary wastes generated by this widely-used conventional treatment process.

The use of limestone in the neutralization process offers the advantages of lower sludge volumes and improved settleability and filtering compared to sodium hydroxide, lime, or ammonium hydroxide (Dempsey and Jeon, 2001; Sibrell and Watten, 2003). Vacuum filtration, continuous pressure dewatering, frame- and plate-dewatering, and centrifuge separation may all be used for dewatering to reduce sludge volumes. Sibrell and Watten (2003) recommended the use of a settling/percolation pond which can achieve a 25% solids concentration. It is ideal for use where land availability is not a constraining factor, and has low operating costs compared to alternatives such as vacuum or pressure filtration.

The high density sludge (HDS) process has been developed to reduce the water content and volume of hydroxide sludge for final disposal (Kostenbader and Haines, 1970; Murdock et al., 1994). Although typically used for AMD with high concentrations of iron, the HDS process may also be used for AMD with low iron concentrations and high concentrations of other metals such as zinc, copper, or aluminum (Coulton et al., 2003a). The HDS process employs a multistage neutralization process with a fraction of the sludge recirculated, which increases the density of the sludge within the reactors by slowing down neutralization and allowing precipitates to form, with existing solids serving as seed particles (Yong Gan et al., 2005). In the HDS process, a much higher solids content (10-30%) is achieved compared to conventional neutralization processes (Kuyucak, 1998). Higher solids contents of approximately 30% (w/w) are obtained by recycling sludge through the treatment plant and with further treatment in a drying bed this can be increased to 50%. Processing in a centrifuge or filter press can increase solids even further to 50-80% (Coulton et al., 2003a). High density sludge has been observed to have improved settleability, higher metal content and improved crystallinity, with an associated reduced leaching potential (Zinck, 2004). High density sludge tends to be high in gypsum (Aubé and Zinck, 2003). Zinck (2004) found that over time, the alkalinity of HDS sludges decreased, gypsum and calcite content increased, and natural dewatering further increased the density.

Different processes may be responsible for increasing the density of the sludge. Free water molecules may be removed owing to changes in the particle shape, or precipitates with fewer water molecules per solid particle may be formed (Yong Gan et al., 2005). Various system parameters have been tested for their relative importance in the HDS process. Yong Gan et al. (2005) examined the effect of operating pH, alkali dosing sequence, and flocculant addition on the densification of ferric hydroxide sludge. Iron chloride was neutralized for the experimental procedure to test ferric hydroxide precipitation. They suggested that the key to better performance of an HDS process is the alkali dosing regime, with better performance observed when the pH was maintained within an optimum range (pH 3.5-4.5) by variable rather than single alkali dosing. Single dosing caused a rapid change in pH that was detrimental to performance. For enhanced system performance, Yong Gan et al. (2005) recommended addition of alkaline material (fresh lime) in at least two doses and prolonged mixing of the sludges (fresh and recycled) at pH 3.5-4.5. They reported that calcium hydroxide was more effective than sodium hydroxide in producing sludge with higher solids contents. High density sludge of final solids content 15% (w/w) was obtained using uneven dosing.

Case study: High density sludge process, Wheal Jane, UK

A HDS system has been installed at the Wheal Jane tin mine in Cornwall, UK. The mine was abandoned in 1991, and an emergency treatment system was installed after water in the mine had rebounded to surface level (Coulton et al., 2003b). In 1992, the failure of an adit plug caused an accidental release of an estimated 50×10^6 L AMD containing in excess of 3500 mg/L of dissolved metals (e.g. iron, zinc, cadmium, arsenic, and aluminum) into the Fal Estuary. This incident led to a review to determine the best long-term treatment system, and a lime-based HDS system was selected as the most cost-effective active system for long-term treatment.

Mine water is continuously pumped out of the mine to prevent another uncontrolled release. Some of this pumped water is dosed with lime and discharged into a tailings dam, where metal precipitates settle out of solution and are stored as sludge. Supernatant water from the tailings dam, as well as the remaining pumped mine drainage are treated together in the HDS treatment plant. Drainage volumes vary seasonally from 110 L/s (August to September) to over 330 L/s (January to March), and have typical AMD characteristics of pH ~3.5, and average metal concentrations of 206 mg Fe/L, 6.0 mg Mn/L, 51 mg Zn/L, 23 mg Al/L, 3.0 mg As/L, 0.80 mg Cu/L, 0.55 mg Ni/L, 0.15 mg Pb/L, and 0.056 mg Cd/L.

Because of the variation in drainage volume, two identical treatment streams were installed, each with the capacity to treat 230 L/s. The HDS systems use iron, which is present at high concentrations in the AMD, as the basis for forming HDS (Figure 2). In the Stage 1 reactors, the pH of raw AMD is raised by recirculated sludge from pH 3.5 to 6-8, and the majority of dissolved metals are precipitated.

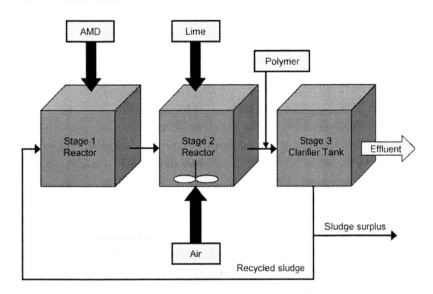

Figure 2. Schematic diagram of the high density sludge (HDS) treatment system used to treat AMD from Wheal Jane mine, UK (adapted from Coulton et al. (2003b))

In the Stage 2 reactors, lime slurry is added to raise the pH to 9.15-9.35, to ensure maximum Mn removal (Coulton et al., 2003b). Air is blown into the reactors, and a turbine and propeller-type impeller are operated to enhance mixing and oxidation of ferrous iron to ferric iron. Next, a polymer flocculant is added via in-line mixers to encourage separation of the solid and liquid phases. The treated water flows into the Stage 3 lamella clarifier tanks, and from this stage, thickened solids (sludge) may be recirculated back to the Stage 1 reactors or discharged as surplus with a solids concentration between 15 and 25% (w/w). Surplus sludge is pumped to a holding tank and then ultimately pumped to the tailings dam, where settling occurs, resulting in a final solids concentration of around 50% (w/w). After almost two years of operation and some process optimization, the two-stage treatment system has proven to successfully treat AMD, producing a final effluent with metal and solids concentrations that are compliant with discharge consents. During the first 22 months of operation, average metals removal efficiency was >99%. Average concentrations of iron decreased from 206 to 0.81 mg/L, manganese decreased from 6.0 to 0.31 mg/L, and zinc decreased from 51 to 0.19 mg/L (Coulton et al., 2003b).

Pulsed limestone bed

The pulsed limestone bed (PLB) process was specifically designed to reduce armoring and improve limestone dissolution rates (Watten, 1999; Sibrell et al., 2000; Sibrell and Watten, 2003). Pulsing of limestone in a fluidized bed creates a high attrition rate at the limestone surface, abrading any armoring that forms, and carbon dioxide is added to the process to further improve limestone reactivity. If the acidity of the influent AMD is high, sufficient carbon dioxide may be generated during treatment and recycled through the system to eliminate the need for additional carbon dioxide supply. Pulsed-bed reactors treating AMD with moderate acidity (300 mg/L as $CaCO_3$) and iron concentrations (30 mg/L) have performed successfully in field tests (Sibrell et al., 2000). Santomartino and Webb (2007) also reported increased effectiveness of limestone neutralization systems when the limestone was agitated to remove ferric hydroxide precipitate.

Highly acidic, iron-rich coal mine AMD (pH ~2.9) that is unsuitable for treatment in a limestone compost wetland (Hedin et al., 1994) was used to test and compare constant and pulsed flow processes (Hammarstrom et al., 2003). In the system, carbon dioxide generated by the reaction of limestone with AMD is recovered and recycled to the incoming AMD by perforated tray scrubber towers to increase limestone dissolution, and no additional carbon dioxide supply is required. The flow regime through the fluidized bed is pulsed rather than the more commonly used constant flow. Pulsed flow encourages particle-particle abrasion, which scours the limestone and abrades coatings forming on the limestone surfaces. Acid mine drainage flows faster through a fluidized bed configuration than a normal packed bed configuration, causing precipitates to be continuously flushed out of the system and minimizing the plugging problems normally associated with packed bed reactors. After 48 h, limestone in the constant flow reactor was completely armored with a gypsum coating overlain by an iron and aluminum hydroxysulphate coating. The pH was increased temporarily to 7.5, but dropped to pH 3.3 after 48 h. In the pulsed bed reactor, pH was increased to pH >6, and armor formation on the limestone was minimized. No gypsum coating was observed on the limestone from the pulsed bed reactor, and the residual grains of refractory minerals that were armored had thicker coatings with higher aluminum content. Hammarstrom et al. (2003) concluded that no inhibition of limestone dissolution by armoring by gypsum and/or metal oxyhydroxide and hydroxysulphate precipitates occurred in the pulsed bed reactor system. After treatment in the pulsed bed system, metal hydroxide precipitates were removed in a settling tank. Final effluent quality was improved during three months of plant operation. The effluent was net-alkaline with pH ~6.5, iron removal was 75% (from 167 to 42 mg/L) and aluminum removal was 98% (from 56 to 1.0 mg/L). Iron removal was dependent on the seasonal variations in ferrous iron concentration and the kinetics of ferrous iron oxidation (Hammarstrom et al., 2003) (Section 1.2). Removal of manganese was poor (<5%, from 9.8 to 9.6 mg/L), because the system pH was too low for manganese oxidation and precipitation.

Chelating reagents

Commercial chelating reagents have been developed as an alternative means of metal removal by precipitation. Problems with proprietary chelating reagents include high costs and sole-source availability (Coulton et al., 2003a). Also, although they may be relatively easy to obtain and use, not all chelating reagents are specifically designed to bind the targeted heavy metals, and many form unstable metal-ligand precipitates that pose the risk of re-releasing metals once released to the environment (Matlock et al., 2002).

The effectiveness of three commonly used chelating reagents, 2,4,6-trimercaptotriazine (TMT, $Na_3C_3N_3S_3 \cdot 9H_2O$), Thio-Red® potassium/sodium thiocarbonate (STC, $[Na,K]_2CS_3 \cdot nH_2O$), and HMP-2000 sodium dimethyldithiocarbamate (SDTC, $C_3H_6NS_2Na \cdot 2H_2O$) has been examined by Matlock et al. (2002). TMT is used for precipitating divalent and univalent heavy metals (Henke et al., 2000; Matlock et al., 2001), and STC precipitates divalent heavy metals through the formation of metal sulphides (e.g. CuS and HgS) (Henke, 1998). Performance was evaluated at stoichiometric doses and at 10% molar excess doses. The three reagents were reported to be ineffective at decreasing concentrations of cadmium, lead, mercury or ferrous iron from 50.00 ppm to the relevant US

EPA discharge limits (e.g. 1.0ppm Cd, 5.0ppm Pb, 0.2ppm Hg, and 2.0 ppm Fe^{2+}) (CFR, 1994). In addition, Henke (1998) reported that a byproduct of the metal sulphide precipitation with STC is the toxic carbon disulphide, and SDTC decomposes into toxic secondary compounds, including tetramethylthiuram and thiram (Matlock et al., 2002). Extreme care is needed in the use and disposal of SDTC (CAS No. 128-04-1), which is a recognized developmental toxicant and suspected of being both a carcinogen and neurotoxicant. Improved ligands, designed using multiple binding sites and mimicking biological binding mechanisms, have been proposed as alternatives to commercial chelating reagents (Matlock et al., 2002; Blue et al., 2008). These improved ligands are highly selective, more effective at lower dosages, and effective at decreasing metal concentrations to <1 ppb.

3.2. Flotation

Flotation, particularly dissolved-air flotation, is widely used to remove metals from contaminated waters by bubble separation (Matis et al., 2003). Other flotation methods include dispersed-air flotation, vacuum air flotation, electroflotation, and biological flotation (Kurniawan et al., 2006). The main advantages of flotation are the relatively short hydraulic retention times and high removal efficiency of small particles, although the effluent may require further treatment to remove residual metal contamination (Kurniawan et al., 2006).

Removal efficiency of low concentrations (<5 mg/L) of copper, nickel, and zinc from synthetic wastewater at pH 5.5 was >98% in laboratory scale flotation systems that used ferric hydroxide as a precipitating agent and chabazite as a collector (adsorptive particulate) (Table 3) (Rubio and Tessele, 1997). Flotation of zinc and copper using sodium dodecyl sulphate (SDS) as a precipitating agent and zeolite as a collector in sorptive flotation systems has also been used to achieve 99% removal of zinc (initial concentration 50 mg/L) at pH 6.0 and 97% removal of copper (initial concentration 500 mg/L) at pH 5.0 (Table 3) (Zamboulis et al., 2004). Hybrid systems have been developed that combine flotation with other processes such as filtration. For example, Mavrov et al. (2003) tested a system that combined adsorption, membrane separation, and flotation to remove copper, zinc, and nickel at initial concentrations of ~60 mg/L. Removal efficiencies were 97% for both copper and zinc (Table 3) and 84% for nickel.

Solvent extraction and electrowinning processes have been developed to recover copper from mine drainage, but are only economical at very high metal concentrations (>1 g/L), with metal extraction from dilute AMD very problematic. Solvent extraction is enhanced by flotation methods. Air bubbles coated with solvent provide high specific surface area and increase the aqueous/organic ratio, thus improving treatment efficiency (Tarkan and Finch, 2005). This combined technology is known as air-assisted solvent extraction (AASX) (Chen et al., 2003). Tarkan and Finch (2005) examined AASX using bubbles coated with a kerosene-based solvent. Findings indicated that AASX enhances phase separation after treatment and may provide an alternative method of solvent extraction to achieve metal recovery from dilute (<100 mg/L) metal-contaminated solutions.

Table 3. Comparison of metal removal efficiencies using flotation (reproduced from Kurniawan et al. (2006) with permission of Elsevier Ltd.)

Species	Collector	Optimum dose of collector (g/L)	Precipitant	Optimum dose of precipitant (mg/L)	Initial metal concentration (mg/L)	Optimum pH	Removal efficiency (%)	References
Cu^{2+}	Chabazite	0.500	$Fe(OH)_3$	30.9	3.5	5.5	98.26	Rubio and Tessele, 1997
Ni^{2+}			$Fe(OH)_3$	20.0	2.0	5.5	98.6	Rubio and Tessele, 1997
Zn^{2+}			$Fe(OH)_3$	20.0	2.0	5.5	98.6	Rubio and Tessele, 1997
Cu^{2+}	Trien	0.015	SDS	NA[a]	12.7	6.7	85	Doyle and Liu, 2003
Ni^{2+}	Trien	0.015	SDS	NA	5.87	7.0	70	Doyle and Liu, 2003
Cr^{6+}	Hydrotalcite		Magnafloc	3.0	58.8	NA	95	Lazaridis et al., 2001
Cu^{2+}	CTABr	0.020	NA	NA	474.0	8-10	99.99	Blöcher et al., 2003
Ni^{2+}			NA	NA	3.3	8-10	98.50	Blöcher et al., 2003
Zn^{2+}			NA	NA	167	8-10	99.97	Blöcher et al., 2003
Cu^{2+}	Zeolite	0.8	NA	NA	60	8-10	97	Mavrov et al., 2003
Zn^{2+}	Zeolite	0.8	NA	NA	60	8-10	97	Mavrov et al., 2003
Zn^{2+}	Zeolite	2	SDS	40.0	50	6.0	99	Zamboulis et al., 2004
		2	HDTMA	20-40	50	9.0	96	Zamboulis et al., 2004
Cu^{2+}		4	SDS	50.0	500	5.0	97	Zamboulis et al., 2004
Cr^{6+}	Surfactin-105	0.04	$Fe(OH)_3$	600	50	4.0	98	Zouboulis et al., 2003
Zn^{2+}	Lycheny-sin-A	0.04	$Fe(OH)_3$	600	50	4.0	100	Zouboulis et al., 2003
Zn^{2+}	SDS	0.05	NA	20.0	50	7-9	100	Matis et al., 2004

[a] NA: Data not available

3.3. Membrane Filtration

Membrane filtration systems may utilize ultrafiltration, nanofiltration, or reverse osmosis to remove metals, organic compounds, and suspended solids from wastewater (Kurniawan et al., 2006). Ultrafiltration (UF) is the use of a permeable membrane (pore size 5-20 nm) to separate heavy metals, macromolecules, and suspended solids from solution. Nanofiltration (NF) uses steric (sieving) processes as well as electrical potential effects to achieve metal removal. Reverse osmosis (RO) is a pressure-driven membrane process. Each process is effective for removing copper, zinc, cobalt, chromium, cadmium, and nickel from wastewater, with possible applications in AMD treatment.

Ultrafiltration

Removal efficiency of UF depends on membrane characteristics such as charge capacity, pH, and ion concentrations in the effluent (Karate and Marathe, 2008). A membrane filter enhanced with chitosan (a biopolymer) was tested for copper and zinc removal efficiency (Juang and Shiau, 2000). Filtration at 2 bar of pressure was shown to be effective at treating synthetic wastewater (Table 4) with the chitosan increasing removal rates by a factor of 6-10 compared to filtration by unenhanced membranes. The removal of nickel has been investigated in micellar-enhanced ultrafiltration. Yurlova et al. (2002) reported almost complete removal of nickel from synthetic solution at 4 bar of pressure with 1g/L SDS concentration to form the micelles (Table 4). Akita et al. (1999) achieved almost 60% nickel removal from synthetic wastewater using micellar-enhanced ultrafiltration (Table 4). Membranes made of $ZnAl_2O_4 - TiO_2$, a low cost material, have also been used successfully to remove cadmium and chromium ions from solution (Table 4) (Saffaj et al., 2004).

Nanofiltration

In nanofiltration systems, anions in the membrane surface reject cations in the effluent, and larger neutral solutes and salts are filtered by the small membrane pores. Nanofiltration has been used to remove metals from AMD (Zhong et al., 2007), and is effective in treating effluents contaminated with high metal concentrations. Ahn et al. (1999) reported 94% nickel removal from electroplating wastewater using an optimum pressure of 2.9 bar (Table 4).

Reverse osmosis

Ozaki et al. (2002) reported almost complete removal of copper and nickel from real and synthetic electroplating wastewater using ultra-low-pressure reverse osmosis with a polyamide membrane skin at 5 bar of pressure (Table 4). Qin et al. (2002) also used polyamide as the skin material, and achieved 97% removal of nickel from a metal plating wastewater at 15 bar pressure (Table 4). Qdais and Moussa (2004) reported 98% copper removal from synthetic wastewater by reverse osmosis at 7 bar of pressure, compared to 90% Cu^{2+} removal by nanofiltration.

Membrane filtration systems are widely used to treat industrial effluents, and achieve excellent removal efficiencies. However, the systems may not be easily adapted to treat AMD, which contain many different metals at widely varying concentrations as well potentially high concentrations of suspended solids. Ultrafiltration has the advantages of a relatively low space requirement and low driving force (2-5 bar), but the system performance may be diminished by membrane fouling (Choi et al., 2005; Kurniawan et al., 2006). Nanofiltration achieves effective metal removal at relatively low pressures (3-5 bar) compared to ultrafiltration and reverse osmosis, but further research into metal removal by this method is needed. Reverse osmosis offers excellent metal removal rates, but high energy inputs are needed to achieve the high pressures required (up to 20-100 bar). Ujang and Anderson (1996) described some advantages of reverse osmosis, which include resistance to biological attack, the ability to withstand high temperatures, chemical stability, and high water flux rate. However, membrane fouling, scaling by calcium carbonate and gypsum, and decreased flow rates associated with system age all reduce performance capabilities.

Table 4. Comparison of metal rejection rates using ultrafiltration (UF), nanofiltration (NF) and reverse osmosis (RO) (reproduced from Kurniawan et al. (2006) with permission of Elsevier Ltd.)

Kind of Process	Type of Membrane	Species	Pressure (bar)	Initial metal concentration (mg/L)	Optimum pH	Rejection rates (%)	References
UF	YM1	Co^{2+}	NA[a]	29.47	5-7	100	Akita et al., 1999
		Ni^{2+}	NA	29.35	5-7	60	Akita et al., 1999
UF	YM10	Cu^{2+}	2	78.74	8.5-9.5	100	Juang and Shiau, 2000
		Zn^{2+}	2	81.10	8.5-9.5	95	Juang and Shiau, 2000
UF	Carbosep M2	Cr^{3+}	3	20.00	6.0	95	Aliane et al., 2001
UF	UPM-20	Ni^{2+}	4	10.00	NA	100	Yurlova et al., 2002
UF	HL	Ni^{2+}	2-5	25.00	NA	99.9	Kryvoruchko et al., 2002
		Co^{2+}	2-5	25.00	NA	95.0	Kryvoruchko et al., 2002
UF	$ZnAl_2O_4$-TiO_2	Cd^{2+}	10	112.00	5.1	93	Saffaj et al., 2004
		Cr^{3+}	10	52	3.6	86	Saffaj et al., 2004
NF	HL	Ni^{2+}	4	NA	4-8	96.49	Mohammad et al., 2004
NF	NTR-7250	Ni^{2+}	2.9	2000	3-7	94	Ahn et al., 1999
RO	Polyamide	Cu^{2+}	7	200	4-11	98	Qdais and Moussa, 2004
		Cd^{2+}	7	200	4-11	99	Qdais and Moussa, 2004
RO	Sulfonated polysulfone	Cu^{2+}	4.5	NA	3-5	98	Ujang and Anderson, 1996
		Zn^{2+}	4.5	NA	3-5	99	Ujang and Anderson, 1996
RO	ULPROM	Cu^{2+}	5	50	7-9	100	Ozaki et al., 2002
		Ni^{2+}	5	50	7-9	100	Ozaki et al., 2002
RO	CPA2	Ni^{2+}	15	21	7.0	97	Qin et al., 2002

[a]NA: Data not available

3.4. Ion Exchange

Exchange of ions from solid and liquid phases (i.e. from an exchange resin and an electrolytic solution), can be used to remove and recover heavy metals from industrial effluents and AMD with concentrations of copper, nickel, zinc, cadmium, and chromium ranging from 10-100 mg/L over a pH range of 2-6. (Álvarez-Ayuso et al., 2003; Dąbrowski et al., 2004; Kurniawan et al., 2006). Complete removal of copper, nickel, chromium, cadmium, and zinc, at initial concentrations of 100 mg/L respectively, has been reported (Álvarez-Ayuso et al., 2003). After exchange, the resin may be treated with reagents to recover valuable metals (Kurniawan et al., 2006). Ion exchange is also an effective method of removing dissolved salts (e.g. magnesium sulphate and gypsum), from AMD in the form of solid sulphates or hydroxides (Feng et al., 2000). An advantage of ion exchange is that no sludge is produced. In addition, the system is relatively portable and rapid. Some pretreatment may be necessary, such as filtration or other methods to remove suspended solids.

Feng et al. (2000) investigated the efficiency of heavy metal precipitation with lime and sulphide-carrier magnetic separation followed by ion exchange for treating AMD from a gold mine. The AMD was first oxidized with hydrogen peroxide, and then lime and iron oxide were added to increase the pH from 1.65 to ~5. The pH was subsequently increased to ~8 by addition of sodium sulphide and additional lime. Precipitates were filtered magnetically. After this treatment, chromium, copper, iron, manganese, nickel, lead, cadmium, zinc, titanium, zirconium, and cobalt were all removed to target concentrations. Concentrations of calcium, magnesium, sodium, potassium, sulphate, fluorite, chloride and bromide remained high. The resin Amberlite IR 120H was used for cation removal, and adsorption selectivity was observed to be $Ca^{2+}>Mg^{2+}>K^+>Na^+$. Three anion exchange resins, Duolite A375, Duolite 161, and Amberlite IRA67, were used for anion removal, and adsorption selectivity was observed to be $SO_4^{2-}>Br^->Cl^->F^-$. Using sulphuric acid as the cation resin regenerator, and sodium hydroxide and lime solution as the anion resin regenerator, high quality gypsum was obtained as a byproduct that could be sold as a raw material.

3.5. Sulphate Reducing Bacteria-Based Systems

Metals can be precipitated from solution by the addition of reactive sulphides, but the cost and safety issues associated with supplying and storing these chemicals (e.g. hydrogen sulphide, sodium hydrosulphide, or sodium sulphide) are significant drawbacks to the adoption of this treatment technology (Cohen, 2006; Huisman et al., 2006). In contrast, systems in which biogenic hydrogen sulphide is formed on-site by sulphate reducing bacteria (SRB) may be more economical and pose less of an environmental hazard than systems in which sulphide is supplied from chemical sources. Treatment based on microbial sulphate reduction and biogenic hydrogen sulphide production is becoming more common as the understanding of microbial reductive processes improves (Postgate, 1984; Odom and Singleton, 1993; Kaksonen and Puhakka, 2007). Sulphate reducing bacteria oxidize organic matter under anaerobic conditions, using sulphate as a terminal electron acceptor and reducing it to sulphide and bicarbonate ions (Section 2.4). Acidity decreases as hydrogen ions are consumed during sulphate reduction, AMD is further neutralized by biogenic bicarbonate alkalinity, and dissolved metals are selectively removed as insoluble sulphide precipitates by reacting with hydrogen sulphide. Biogenic hydrogen sulphide reacts readily with heavy metals at pH ≥ 3 to form insoluble metal sulphides (Gazea et al., 1996), whereas oxidation and hydrolysis may be ineffective for precipitating some metals (e.g. zinc) at a pH <8. However, limited metal removal as hydroxides does occur in SRB-based systems. For example, aluminum hydroxides are stable in the pH range 5-8 (the typical operational range of SRB-based systems), so it is expected that some aluminum will be precipitated as hydroxides (Nordwick and Bless, 2008).

Huisman et al. (2006) described the advantages of metal precipitation in the form of metal sulphides. For example, sulphide sludge volumes may be 6-10 times less than the sludge volume from conventional neutralization and precipitation systems. In addition, sulphide sludges are more dense and stable than hydroxide sludges, and have better thickening and dewatering characteristics. Metal sulphides have low solubilities over a broad pH range, which results in lower effluent concentrations of toxic metals (0.01-1 ppm) with the possibility of selective metal precipitation. Sulphide precipitation is less sensitive to the presence of chelating agents and other complexes than hydroxide precipitation. Finally, chromates and dichromates may be removed without reduction of chromium to the trivalent state.

Active systems that employ SRB to reduce sulphate and generate alkalinity are known as SRB bioreactors. Elemental sulphur may be supplied to a bioreactor to produce hydrogen sulphide gas if no sulphate-containing stream is available. Kaksonen and Puhakka (2007) have reviewed the benefits and drawbacks of the various SRB bioreactor types, including membrane bioreactors, upflow anaerobic sludge blanket reactors, fluidized-bed reactors, and continuously stirred tank reactors. Systems must be designed to prevent toxic concentrations of heavy metals from making direct contact with the SRB and inhibiting microbial activity and affecting species diversity (Hao et al., 1994; Utgikar et al., 2000). If the influent metal loading is greater than the sulphate concentration within the bioreactor, SRB activity may be inhibited. If metal ion concentrations become too high, microbial diversity and abundance will be reduced (Temple and LeRoux, 1964; Babich and Stotsky, 1985; White et al., 1997). In cases where acidity or metal concentrations are high enough to inhibit bioreactor performance, pretreatment by neutralization, dilution, or biosorption may be necessary (Section 3.6) (Utgikar et al., 2000; Kaksonen and Puhakka, 2007). Anaerobic, reducing conditions must be maintained in the system for SRB to thrive.

The pH of influent AMD is a critical parameter affecting system performance (Cohen, 2006). Although SRB can survive in environments with pH ranging from <5 to 9.5 (Postgate, 1984), the optimum pH for SRB growth and activity is generally believed to be pH 5-8 (Cohen and Staub, 1992; Dvorak et al., 1992; Hao et al., 1996; Hedin, 1996; Willow and Cohen, 2003; Johnson et al., 2006; Neculita et al., 2007). In contrast, Elliott et al. (1998) reported SRB activity in an anaerobic upflow bioreactor at pH 3.0. Highly acidic AMD may cause system pH to drop, limiting sulphate reduction and treatment capacity.

SRB bioreactors: System configurations

Many different biological sulphate reduction treatment plant designs have been developed. A key advantage of active bioreactors over passive systems is that performance can be optimized by varying both hydraulic retention time and sludge retention time. This results in more compact treatment units with smaller reactor volumes and lower capital costs. Kaksonen and Puhakka (2007) have reviewed different reactor designs, including batch reactors, sequencing batch reactors, continuously stirred tank reactors, anaerobic contact processes, anaerobic baffled reactors, anaerobic filters, fluidized-bed reactors, gas lift reactors, upflow anaerobic sludge blanket reactors, anaerobic hybrid reactors, and membrane bioreactors. Both single-stage and multiple-stage process configurations have been used in SRB bioreactors. Single-stage configurations are generally not as effective for treatment of AMD with very high acidity or metal concentrations, and the addition of alkaline material may be required. Multiple-stage processes may recycle treated effluent to dilute the influent AMD or use separate reactors for metal precipitation and sulphate reduction, thus protecting the SRB from high acidity or metal loading. The use of separate reactors for biogenic sulphide production and metal sulphide precipitation has been reported to be more effective than a single stage method for copper removal (Luptakova and Kusnierova, 2005). The sulphide sludge may be recovered after treatment by backwashing, removal from the bottom of the bioreactor, or in a downstream clarifier, depending on the system configuration (Kaksonen and Puhakka, 2007).

Sierra-Alvarez et al. (2007) designed a treatment system integrating a SRB-bioreactor with a fluidized-bed crystallization reactor containing fine sand. The two-stage system was designed to minimize metal concentrations in the bioreactor influent, thus avoiding mineral toxicity and metal overloading. The use of a crystallization reactor instead of a coagulation or flocculation chamber was intended to reduce maintenance requirements and eliminate the need for sludge dewatering. The sand provided a large mineral surface for copper nucleation and crystal growth and retention. To maintain an environment more favorable to SRB than to methanogenic bacteria, the bioreactor was operated at pH 7.4-8.0 (O'Flaherty et al., 1998; Visser et al., 1996), and sulphate concentrations were kept in excess at 0.60 ± 0.09 g COD/ g SO_4 (Colleran et al., 1995). Soluble copper was added as copper (II) chloride, in concentrations ranging from approximately 4.2 to 66.4 mg/L copper, was recovered as a purified copper-sulfide mineral (covellite, CuS), with copper removal occurring primarily in the crystallization reactor. Soluble copper removal efficiencies exceeded 99% (Sierra-Alvarez et al., 2007).

Metal recovery using SRB systems

Selective metal precipitation is possible with some bioreactor configurations. Treated effluent containing hydrogen sulphide is recycled to a metal precipitation reaction vessel. Valuable metals, such as copper, zinc, nickel and cobalt, may be recovered in a purified form by controlling the pH of the system. To increase the pH for selective metal precipitation, alkalinity may be added by a chemical neutralization step or water recycling (Kaksonen and Puhakka, 2007). Tabak et al. (2003) designed a 6-stage process used for selective precipitation of copper, zinc, ferrous iron, and manganese as sulphides using biogenic hydrogen sulphide, and the precipitation of aluminum and ferric iron as hydroxides using sodium hydroxide.

The Paques BV THIOTEQ® system is an upflow anaerobic sludge blanket reactor that uses a two-stage process for metal recovery (Figure 3) (Huisman et al., 2006).

The system is off-line, which differs from in-line systems in that the AMD passes through the chemical stage only and does not enter the biological stage. In the biological stage, elemental sulphur or concentrated waste sulphuric acid is used for biogenic hydrogen sulphide production, and the electron donor is acetic acid, ethanol, hydrogen, or another organic compound. Hydrogen sulphide is transported to the chemical stage in a carrier gas comprising a mixture of carbon dioxide and nitrogen gas. In the chemical stage, which is essentially a gas/liquid contactor, metals in the AMD react with sulphide to form insoluble metal sulphide precipitates. Copper usually precipitates without any pH adjustment or precipitation of other heavy metals, and subsequently copper sulphide content is usually high (>90%). Zinc and nickel are recovered with the addition of process stages, including addition of alkaline material to achieve the optimum pH for metal precipitation. After precipitation, metal sulphide sludge is passed through a clarifier and then dewatered in a filter press, after which it may be sent to smelters. Treated effluent may be discharged to the environment or recycled for process use.

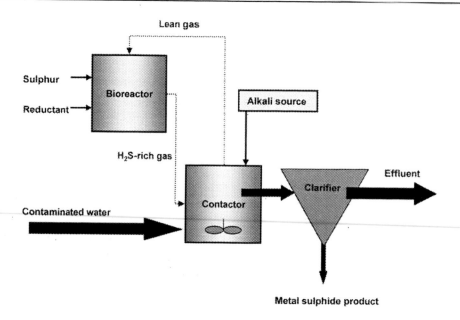

Figure 3. Schematic diagram of the THIOTEQ® process for water treatment and recovery of one metal (reproduced from Huisman et al. (2006) with permission of Elsevier, Ltd.)

The THIOTEQ® system was installed upstream of an existing lime neutralization plant treating AMD from an abandoned zinc mine (Huisman et al., 2006). Influent pH was ~3.7, and the AMD contained ~450 mg Zn/L, ~30 mg Cu/L, ~150 mg Fe/L, and ~3800 mg SO_4/L. The plant treated 700 m^3 per day. Selective metal precipitation and recovery were used to achieve copper removal to <0.01 ppm, and a saleable zinc concentrate. In one 10-month period of operation, almost 35 tonnes of saleable zinc concentrate containing copper, cadmium, and lead were recovered (Huisman et al., 2006). The effluent from the THIOTEQ® then entered the lime neutralization plant for iron and aluminum removal. With this type of integrated system, lime requirements and the volume of hydroxide-laden sludge are reduced. Reagent savings can be particularly significant if zinc and/or nickel are present in AMD, because both metals require a high pH for removal in a lime plant but can be recovered at neutral or slightly acidic pH as sulphides (Huisman et al., 2006).

The BioSulphide® process is another commercially available SRB system that is used to selectively recover metal sulphides. Raw AMD flows into the chemical circuit, where it mixes with effluent recycled from the biological circuit (i.e. the SRB bioreactor) that contains excess hydrogen sulphide and alkalinity. Metal sulphide precipitates are recovered by clarification and filtration, resulting in a filter cake (Adams et al., 2008). Sulphate is supplied to the SRB by the effluent from the chemical circuit (Rowley et al., 1997) or as elemental sulphur (Adams et al., 2008).

Groundwater remediation at the Budelco zinc refinery in the Netherlands is being carried out using an SRB system (Scheeren et al., 1993). The system, known commercially as Thiopaq®, uses two biological steps. The first step uses SRB in an ethanol-fed reactor to generate alkalinity and hydrogen sulphide, which reacts to form metal sulphide precipitates in the reactor. The next step is aerobic filtration where excess sulphide is oxidized by SRB to elemental sulphur, which is then used for other industrial or agricultural applications. Zinc sulphide is recovered and recycled to the roaster feed at the refinery (Johnson and Hallberg, 2005).

3.6. Adsorption

Metal removal may be achieved by a variety of adsorptive techniques. Sorptive flotation, adsorption onto inorganic materials, and biosorption may all be used for metal removal with the possibility of metal recovery.

Sorptive flotation
Metal removal by sorptive flotation is achieved by scavenging metal ions with adsorbent materials and subsequently separating the metal-laden particles from solution by a flotation stage (Zouboulis and Matis, 1997; Lazaridis et al., 2001). Adsorbent materials are typically fine- or ultra-fine particles and can be inorganic, biological, industrial solid byproducts, or conventional adsorbents (e.g. powdered activated carbon or zeolites) (Zouboulis and Matis, 1997; Wingenfelder et al., 2005). Dissolved-air, electrolytic, or dispersed flotation methods can all be used for their removal from the treated liquid phase. Metals can either be recovered from the foam concentrate or be released from adsorbent materials using a metal eluant (Zouboulis and Matis, 1997).

Adsorption onto inorganic materials
Materials such as fly ash, natural clinker, synthetic zeolites (Rios et al., 2008), loess (Tang et al., 2008), lignite (Mohan and Chander, 2006), and activated carbon (Kurniawan et al., 2006) are all widely employed as adsorbents. Geremias et al. (2008) described the use of calcinated coal mine waste comprising 66.2% SiO_2, 5.8% Fe_2O_3, and 23% Al_2O_3 to treat AMD with initial average metal concentrations of 15.9 mg Fe/L, 30.6 mg Al/L, and 11.4 mg Mn/L. The waste increased the pH from 2.6 to 7.8 and achieved metal removal rates of 100% for ferric iron and aluminum, and 89% for manganese. As H_3O^+ ions were adsorbed by the metal oxides in the waste, the pH increased and dissolved metals precipitated as oxides and hydroxides. A key advantage of a batch flow treatment system using coal mine waste is that the need for initial neutralization stage or final flocculation stage is eliminated.

Biosorption
Adsorption of metals onto biomass, or biosorption, is an alternative method of removing metals from solution. The uptake mechanism may be chemisorption, complexation, surface adsorption, ion exchange, microprecipitation, adsorption-complexation on the surface or pores, diffusion through pores, or heavy metal hydroxide condensation onto the biosurface (Kasan, 1993; Demirbas, 2008; Sud et al., 2008). The main mechanism of biosorption may be ion exchange involving carboxyl and sulphate groups in the biomass (Seki and Suzuki, 1998).
Materials tested as biosorbents include algal biomass (Kratochvil and Volesky, 1998; Yu et al., 1999; Kaewsarn, 2002; Davis et al., 2003), bacteria (Chang and Chen, 1998; Haferburg et al., 2007a, 2007b), fungi (Subudhi and Kar, 1996; Haferburg et al., 2007b), activated sludge (Stoveland and Lester, 1980; Tien and Huang, 1987; Solari et al., 1996; Wang et al., 1999; Utgikar et al., 2000), peat, plant cells and plant organs (Stark et al., 1999).

Haferburg et al. (2007b) investigated the use of fungi and bacteria as biosorbents and identified two strains with good aluminum and uranium retention capacity that could grow in media contaminated with diluted AMD. Metal removal rates of up to 80% for aluminum and copper, and >60% for uranium, have been achieved using bacteria as biosorbents (Haferburg et al., 2007a). Nelson et al. (1981) concluded that biosorption by living activated sludge was primarily by physico-chemical processes rather than active biological transport processes, and was dependent on system pH and biomass age, with increased metal uptake associated with increased biomass age (Stoveland and Lester, 1980). The C/N ratio of the biomass may also affect metal adsorption (Tien and Huang, 1987). Problems affecting the use of living biosorbents include toxicity of metals to microorganisms, and maintaining favorable conditions to keep the microorganisms alive, especially the pH (Solari et al., 1996).

The use of non-living biosorbents avoids these problems. The metal adsorption capacities of non-viable microbial and plant-derived biomass were reviewed by Ahluwalia and Goyal (2007). Davis et al. (2003) concluded that brown algae are the most effective low-cost biosorbent, and provided an overview of its biochemical properties and adsorption behavior. Byproducts of processing or agriculture are also efficient, inexpensive and readily available sources of adsorbent materials. Sud et al. (2008) have explored the use of agricultural waste materials, such as sugar-industry waste and sawdust, as biosorbents. Li et al. (2008) tested chemically-modified orange peel cellulose as a biosorbent and reported adsorption capacities of 1.23, 1.28, 1.21, and 1.13 mol/kg for cobalt, nickel, zinc, and cadmium, respectively. Immobilized biomass beads, made of materials such as dried sphagnum peat moss held in a matrix, can also be used to treat AMD (Spinti et al., 1995).

The effectiveness of non-living, non-viable activated sludge as a biosorbent of zinc and copper in a packed-bed column adsorber using feed solutions at pH 3.0 and 5.5 was compared by Utgikar et al. (2000). Based on a decrease in solution pH from 5.5 to 3.8 as adsorption occurred, they observed that hydrogen ion--metal ion exchange was the primary mechanism for metal adsorption in the feed solution with the higher pH. However, the pH of the acidic feed solution remained unchanged at pH 3.0. Further investigations into changes in sulphate concentrations in both solutions revealed that while sulphate concentration remained constant at the equilibrium pH of 3.8, it decreased significantly at pH 3.0 indicating that sulphate ions were being adsorbed simultaneously with metals. Metal uptake capacity at equilibrium was 3.4 and 5.9 mg/g (dry biomass) of zinc and copper respectively at pH 3.8, compared to 2.5 and 1.9 mg/g (dry biomass) of zinc and copper respectively at pH 3.0. The biomass had a neutralizing effect on the AMD, which is likely to be caused by the exchange of hydrogen ions from the AMD with light metal ions in the biomass. This neutralization capacity decreased with the number of times the biomass was used (Utgikar et al., 2000).

Andres and Francisco (2008) reported that sewage sludge retained iron, manganese, and lead from mine tailings drainage. However, copper and zinc retention was variable, and nickel was mobilized during treatment, possibly owing to high concentrations of leachable nickel in the sewage sludge.

3.7. Emulsion Liquid Membrane Technology

Emulsion liquid membrane (ELM) technology involves the use of a chelating carrier extractant and surfactant to form the membrane and an acidic stripping solution. The surfactant may be dissolved in an organic diluent such as kerosene to minimize evaporative losses during vigorous stirring. Emulsion liquid membrane technology has been tested for the removal of zinc (Valenzuela et al., 2005c, 2007), copper (Valenzuela et al., 2005a), chromium (Bhowal and Dhatta 2001; Chakraborty et al., 2005), gold (Kargari et al., 2004), and gallium (Kumbasar and Tutkun 2004). Valenzuela et al. (2009) evaluated copper removal by ELM technology, using a sample of real AMD from a copper mine in Chile with pH 1-5, ~350 mg Cu/L, ~250 mg Zn/L, ~280 mg Fe/L, and ~4000-5000 mg SO_4/L. The AMD was treated with an ELM contactor in a stirred tank, in contact with an organic solvent solution of carrier extractant (3% w/w LIX-860 N-IC 5-nonylsalicylaldoxime) and surfactant (2% w/w Span-80 sorbitan monooleate) in a kerosene diluent (Valenzuela et al., 2005b), and sulphuric acid was used as the stripping liquor. Above pH 2.0, copper removal was rapid, with removal efficiency >98% at pH 2.14. Zinc and iron were not removed at pH 2.1. The higher the initial pH of the AMD, the more rapid the copper removal process became. Above pH 4.0, iron removal was observed. Valenzuela et al. (2009) suggested that higher copper concentrations would require the use of a higher concentration of extractant in the membranes, but that the process kinetics are faster in ELM systems and so require less extractant for the same extraction than conventional solvent extraction using mixer-settler reactors. The final recovery stage is a copper sulphate crystallization process.

4. PASSIVE TREATMENT SYSTEMS

Systems that do not require regular maintenance or chemical and power inputs are known as passive treatment systems (Hedin et al., 1994; Ziemkiewicz et al., 2003; Johnson and Hallberg, 2005). Passive systems use natural energy sources, such as microbial metabolic energy, photosynthesis, or a topographical gradient (PIRAMID Consortium, 2003). Passive systems are less expensive to maintain than conventional, active systems (Cohen, 2006), and are often the only practical systems for remote or inaccessible mine sites. However, contaminant removal processes in passive systems are slower and require longer retention times and larger land areas, which makes the footprint of passive systems much larger (Hedin et al., 1994; Gazea et al., 1996). Complete site characterization, including hydrological, topographical, chemical, and geological parameters, must be carried out to design an effective passive system to treat AMD (Gray, 1997). In particular, the longevity of AMD formation must be taken into account (Younger, 1997; Ziemkiewicz et al., 2003) (Section 2.2).

4.1. Constructed Wetlands

Constructed wetlands are the most commonly used passive treatment systems for AMD (Hedin et al., 1994; Gazea et al., 1996; Younger et al., 2002; Sheoran and Sheoran, 2006). These can be loosely grouped as aerobic and compost (anoxic) wetlands, with both systems utilizing natural biotic and abiotic processes to neutralize AMD and remove metals and sulphate (Kosolapov et al., 2004). Both types of wetland may be either used alone, as a part of an integrated passive system, or as a final polishing stage after active treatment. Constructed wetlands can be further differentiated according to flow regime and redox conditions within the substrate (Gray, 2004). In aerobic wetlands, AMD moves through the system primarily by low-velocity surface flow over the shallow substrate and through the wetland vegetation. Metal removal in aerobic wetlands is primarily by oxidation and precipitation of metals as oxides, hydroxides, and oxyhydroxides, with some sulphide precipitation in areas of the substrate where the redox potential is low. In compost wetlands, the flow regime is primarily subsurface and horizontal through a deeper, anoxic substrate comprising a mixture of organic materials and possibly limestone. Reducing conditions are maintained in the substrate, and metals are removed primarily by precipitation as sulphides or by adsorption onto the substrate material.

Metal removal in constructed wetlands involves many mechanisms, including filtering of suspended particles and colloidal metal hydroxides, sedimentation, adsorption, precipitation and co-precipitation, cation exchange, photodegradation, plant uptake, formation of iron plaque on roots and rhizomes owing to oxygenation of the rhizosphere, and microbial activity (Gazea et al., 1996; Walton-Day, 1999; Batty and Younger, 2002; Ziemkiewicz et al., 2003; Kalin et al., 2006; Sheoran and Sheoran, 2006). After adsorption onto colloids or particulate matter, metals may settle to the sediment, where microbially-mediated biomineralization occurs (Kalin, 2004; Rai, 2008). Biomineralization is an anaerobic process that mineralizes organically bound metals into stable compounds. During the establishment period of a wetland, adsorption of metals onto organic matter is at a maximum while adsorption sites are in excess (Batty et al., 2008). Dissolved metals in AMD may be exchanged for carboxylic or phenolic acid groups on humic and fulvic acids, thus remaining, at least temporarily, in the substrate (Gazea et al., 1996). The adsorption of metals onto the wetland substrate is pH dependent, with an optimum range of pH 4-6 (Gazea et al., 1996). Kerndorf and Schnitzer (1980) reported the following adsorption order of metals onto humic acids, ranging from 100% to 10%, at pH 4.7: Hg=Fe=Pb=Cu=Al=Cr>Cd>Ni=Zn>Co>Mn. As sorption sites become saturated, competition for sites may lead to desorption of weakly sorbed metals in ion exchange processes (Cohen, 2006). Therefore, this mechanism is effective only for as long as sorption sites are available. As the wetland system ages, other mechanisms, such as sulphide precipitation, become more important (Sheoran and Sheoran, 2006).

Wetlands must be designed to allow optimum flows through the system while maintaining sufficient residence time for metal removal (Cohen, 2006). Both aerobic and compost wetlands require large land areas to allow for the long residence times required. Hedin et al. (1994) recommended sizing of beds based on removal rates of 3.5-7.0 g acidity/m^2/day or 5.0 g Fe/m^2/day for compost wetlands treating net-acidic (pH 4-5.5) waters. A removal rate of 10-20 g Fe/m^2/day has been suggested for sizing aerobic wetlands treating net-alkaline (pH >5.5) drainage (PIRAMID Consortium 2003); however, removal rates have been found to vary spatially within a wetland system (Hall and Puhlmann, 2005), with removal efficiencies primarily dependent on temperature, hydraulic loading, and contaminant loading (Mayes et al., 2008). The area required for iron removal increases as the inflow concentration decreases (Hedin et al., 1994; Batty et al., 2008). Different approaches have been suggested for calculating area requirements for wetlands, including volume-adjusted contaminant removal rates (Manyin et al., 1997) and a first-order kinetic model based on oxidation of ferrous iron (Tarutis et al., 1999). Reactions other than iron oxidation, such as precipitation and sedimentation, have been recommended for use when calculating removal rates for system design, and biotic processes by microorganisms and plants should also be considered (Batty and Younger, 2002; Younger et al., 2002). Most commonly, area requirements are calculated based on area-adjusted contaminant removal rates (Hoover and Rightnour, 2002; PIRAMID Consortium, 2003). Mayes et al. (2008) summarized recently-reported removal rates, based on zero-order kinetics, for various metals and acidity in different wetland systems (Eq. 22, Table 5). These rates do not assume constant flow or steady influent contaminant concentrations.

Table 5. Examples of documented pollutant removal rates in wetlands (reproduced from Mayes et al. (2008) with permission of Elsevier Ltd.)

Pollutant	Type of wetland	Typical removal rates ($g/m^2/day$)	Comments	Example references
Acidity	Compost	3.5-7	Early assessment of range in performance from the USA.	Hedin et al. (1994)
Al	Aerobic	0.1	Removal as hydroxide occurs in both aerobic and anaerobic conditions. Far slower removal rates at low pH.	Hoover and Rightnour (2002)
As	Aerobic	18	Bacterially catalyzed under acid conditions.	PIRAMID Consortium (2003)
Cd	Compost	0.02	Immobilized as reduced sulphide (greenockite: CdS) within anoxic substrate.	Ettner (1999)
Cu	Aerobic	0.05	Value from volunteer site so likely a minimum value.	Brown (1997)
	Compost	10	Cu removal likely to be as carbonate phase formed through reaction with respired microbial CO_2.	PIRAMID Consortium (2003)
Fe	Aerobic	10	Value typically quoted for net-alkaline waters	Hedin et al. (1994)
	Compost	0.8	Acidic (pH <3) colliery spoil drainage	Heal and Salt (1999)
Mn	Aerobic	0.5	Alternative Mn oxidizing bacteria systems achieve higher removal rates	Hedin et al. (1994)
	Compost	0.01	Acidic (pH <3) colliery spoil drainage	Heal and Salt (1999)
Ni	Aerobic	0.04	Preliminary data	Eger et al. (1994)
	Compost	2	Immobilized as millerite (NiS) in anoxic substrate	Ettner (1999)
Se	Aerobic	0.01	Se common to fly ash leachates	Hoover and Rightnour (2002)
U	Aerobic	0.1	Single value from volunteer site at Boršt (Slovenia)	PIRAMID Consortium (2003)
Zn	Aerobic	7	Aerobic reedbed with floating algal mats. Strong seasonal variation in performance. May be a net exporter in winter.	Kalin (1998)

$$R_A = ((Q_d(C_i - C_d))/A \qquad (22)$$

where R_A is the area-adjusted contaminant removal rate (g/m^2/day), Q_d is the average daily flow rate (m^3/day), C_i is the influent concentration (g/m^3), C_e is the effluent concentration (g/m^3), and A is the treatment media area (m^2).

Constructed wetlands are low-maintenance and offer the best treatment option in many situations, for example, in inaccessible areas where the discharge rate of AMD is relatively small and water quality is moderate with relatively low metal concentrations (Weider, 1993). It is critical to consider the chemistry, variability, and longevity of AMD to design an effective wetland. Overloading of metal adsorption sites or depletion of available organic carbon may lead to decreased metal removal, diminishing populations of microbial communities, and eventual failure or destruction of the wetland system (Kalin et al., 2006). Accumulation of metal precipitates may alter flow dynamics, leading to short circuiting and reduced system performance (Mayes et al., 2008). Removal efficiencies may also vary because of seasonal effects, such as increased precipitation, freezing, or evaporation. Finally, removal and disposal of non-viable, metal-laden biomass may be periodically required.

Aerobic wetlands

Aerobic wetlands are basins or channels planted with vegetation in a shallow (<0.3 m) substrate of clay, soil, gravel or mine spoil (Ziemkiewicz et al., 2003) that create an oxidizing environment for treating circum-neutral and net-alkaline AMD (Figure 4) (Hedin et al., 1994). Water depth is typically 0.1-0.5 m, and the bed is designed to ensure slow and uniform surface flow of the AMD over the substrate and through the vegetation. The length to width ratio is typically >10:1, allowing sufficient residence time for iron, aluminum, and manganese hydroxides, oxyhydroxides, and oxides to form and precipitate out of solution (Gazea et al., 1996). They are the primary passive system in use for treating circum-neutral coal mine drainage with elevated iron concentrations (Younger, 2000b; Younger et al., 2002; Batty et al., 2008), and have been identified as the best systems for treating AMD with pH >5 (Hellier et al., 1994). Aerobic wetlands are also widely used to polish effluent from active and passive treatment systems (Laine, 1999; Younger, 2000b).

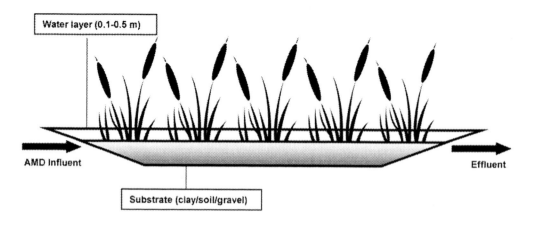

Figure 4. Schematic diagram of an aerobic wetland

Iron is the metal most susceptible to oxidation in aerobic wetland systems (Gazea et al., 1996). Ferrous iron that drains from mine spoil or underground workings is oxidized to ferric iron and hydrolyzed, forming ferric hydroxide, which precipitates as ochre at pH >3 in a wetland system (Johnson and Hallberg, 2005). These reactions generate acidity, so in wetlands treating circum-neutral, iron-laden water, effluent pH may be decreased to pH 2-3 after the ferric iron is hydrolyzed (Gazea et al., 1996). If alkalinity generation is inadequate, the pH will decrease further, iron removal will become less efficient, and the solubility of other metals in the AMD will increase (Whitehead et al., 2005; Mayes et al., 2008). Optimum metal precipitation occurs when the pH is maintained between 5.5-6.5 (Gazea et al., 1996). Hydrolysis of ferric iron to hydroxide precipitates is mainly biological at pH <5, and in net acidic conditions (pH <3), the process slows and iron removal is further diminished (Younger et al., 2002). Also, biological activity and plant performance may be reduced in more acidic conditions (Mayes et al., 2008). To prevent a significant decrease in pH, an alkaline addition step (e.g. an anoxic limestone drain), may be used as part of the treatment system, or limestone may be mixed with the substrate (Younger et al., 2002).

Wetlands are generally ineffective at removing manganese. Manganese forms oxyhydroxides and oxides in aerobic conditions where pH > 8 (Hedin et al., 1994; Hallberg and Johnson, 2005b; Nordwick and Bless, 2008), however, Gouzinis et al. (1998) found that high iron concentrations may inhibit the formation of manganese oxides. Soluble ferrous iron can reductively dissolve manganese oxides, thus preventing precipitation (Villinski et al., 2001) (Eqs. 23-24).

$$MnO_2 + 2\ Fe^{2+} + 2\ H_2O \rightarrow 2\ FeOOH + Mn^{2+} + 2H^+ \quad (23)$$

$$MnOOH + Fe^{2+} \rightarrow FeOOH + Mn^{2+} \quad (24)$$

In alkaline systems, manganese can precipitate as a carbonate, and then further oxidize to manganese oxide (Section 3.1); however, Batty et al. (2008) observed that manganese removal from net-alkaline AMD in constructed wetlands only became significant after iron concentrations had been reduced to <1 mg/L. Hedin et al. (1994) suggested that manganese removal is between 20 and 40 times slower than iron removal in passive treatment systems, and recommended sizing criteria of 0.5-1.0 g Mn/m^2/day for aerobic wetlands.

Wetland vegetation is involved in the removal of metals by various mechanisms such as physical adsorption, chemisorption, microprecipitation, ion exchange, complexation and biological uptake (Kasan, 1993; Schiewer and Volesky, 1995; Kratochvil and Volesky, 1998; Collins et al., 2005; Sheoran and Sheoran, 2006). The overall importance of plants for metal uptake in wetlands, and the traditional assumption that plants were key to metal removal, have been extensively investigated (Cheng et al., 2002; Cohen, 2006). Although accumulation by plants accounts for only a small percentage of metal removal (Mays and Edwards, 2001), plants enhance metal removal rates in aerobic wetlands by regulating flow, maximizing the hydraulic residence time, and releasing oxygen from the root zone (Batty, 2003). Localized oxidizing zones in the root zone may enhance metal oxidation and precipitation (Gazea et al., 1996). Plants also minimize wind-induced erosion of deposited sediments and stabilize the substrate. The effect of wetland plants and biotic processes within the root zone on iron removal are most significant where iron concentrations are low (<2 mg/L) (Batty and Younger, 2002). Processes occurring in the rhizosphere and the associated buffering effects that protect microbial communities from toxic metal concentrations are important in overall wetland performance (Jacob and Otte, 2003; Weber et al., 2008). The rhizosphere also serves as an enriched environment for microbes (Macek et al., 2000). Decayed plant matter is an important source of organic matter on which metals may be adsorbed, and plant cellulosic material may provide nutrients for sulphate-reducing bacteria in anaerobic systems (Gazea et al., 1996).

To optimize performance, plants that have a high metal bioaccumulation potential and are tolerant to prolonged high metal concentrations should be selected (Rai, 2008). *Typha latifolia* (broadleaf cattail) and *Typha angustifolia* (narrowleaf cattail) are two plant species commonly used in aerobic wetlands because they are readily available and tolerant of a wide range of conditions. High metal concentrations and acidic conditions have been observed to limit plant growth in wetlands (Batty and Younger, 2004). However, research has indicated that plants can become acclimatized or modify the pH in the rhizosphere (Otte, 2001; Brix et al., 2002), thus continuing to take up metals (Batty, 2005). Decreased leaf decomposition, limiting metal adsorption onto organic matter, may be a problem in acidic environments. Although Batty and Younger (2007) reported that decomposition of *Phragmites australis* (common reed) was not affected by acidic conditions, the decomposition of *Typha latifolia* was inhibited (Kittle et al., 1995). Mayes et al. (2008) recommended further research into the impacts of acidic environments on wetland plants and carbon cycling in wetland systems.

The treatment of net-alkaline AMD in wetlands may result in the pH being lowered, owing to high partial pressures of carbon dioxide (pCO_2) from aerobic and anaerobic microbial respiration both in the water and the substrate. The high pCO_2 can result in increased precipitation of calcium carbonate, which consumes alkalinity (Mayes et al., 2008). Ross (1996) observed that cation exchange and the production of organic acids could also lower the pH of alkaline waters. An aerobic wetland system was used to successfully remediate net-alkaline AMD from abandoned mine workings containing 45 mg Fe/L, 2 mg Mn/L, and 3000 mg SO_4/L at Whittle Colliery, UK (Batty et al., 2008). The system, consisting of oxidation lagoons and aerobic wetlands in series, was fully commissioned in 2002 (Nuttall, 2003). Iron removal, as measured in 2004, was over 98%, with effluent iron concentrations ≤1 mg/L, which is similar to the performance of other aerobic wetland systems (Tarutis et al., 1999). Manganese removal was around 78%, with effluent manganese concentrations ≤0.4 mg/L, and the pH was raised from pH 6.8 to 7.8.

Low pH, which inhibits iron and manganese oxidation and precipitation, high metal concentrations in influent AMD, and the effects of wetland age (e.g. reduced active surface area and depletion of organic matter supply for SRB) all reduce treatment performance (Woulds and Ngwenya, 2004). Other problems associated with aerobic wetlands as they age include possible remobilization of metals and changes in the residence time and flow dynamics of the system (Mayes et al., 2008). It is critical to harvest and either recycle or dispose of the metal-laden biomass to avoid subsequent leaching of metals. Biomass processing options are reviewed by Hetland et al. (2001).

Co-treatment of AMD and sewage effluent

Co-treatment of sewage effluent and AMD is effective because compounds that are high in one effluent stream tend to be low in the other. For example, (i) sewage effluent with relatively high concentrations of suspended solids may enhance iron oxyhydroxide precipitation by encouraging iron, which is present in high concentrations in AMD, to form flocs (Johnson and Younger, 2006); (ii) phosphate, which is present in high concentrations in sewage effluent, is sorbed onto the iron oxyhydroxide precipitates (Cooke, 1994; Drizo, 1999); and (iii) oxyhydroxides may provide attachment sites for nitrifying and denitrifying bacteria (Demin et al., 2002).

Johnson and Younger (2006) evaluated the effectiveness of a pilot-scale aerobic wetland treating secondary sewage effluent in admixture with net-alkaline mine drainage pumped from an abandoned coal shaft. The size of the wetland was based on iron removal because the volume of mine drainage was three times greater than the volume of sewage effluent. Also, iron oxyhydroxide precipitates were expected to be involved in phosphate and nitrogen removal. Some 30 cm depth of topsoil from the site was used as the wetland substrate, which was planted with *Typha latifolia*, *Phragmites australis*, and *Iris pseudacorus* (yellow flag iris). The free water depth was approximately 15 cm. Average influent concentrations were 3 mg Fe/L, 1 mg Mn/L, 14 mg/L biochemical oxygen demand (BOD), 2 mg/L ammoniacal nitrogen (N-NH_3), 23 mg/L suspended solids (SS), and 7 mg/L phosphate. During the initial four months of operation (August-November), the wetland was effective in removing iron (40-80% removal), phosphate (10-50%), BOD (20-75%), and SS (30-70%). Phosphate removal followed the same trend as iron removal, and data indicated that phosphate removal by sorption onto iron oxyhydroxide flocs was occurring. Ammonia removal was initially good (35-50%), but dropped to 5-20% after two months of operation. Nitrate nitrogen (N-NO_3) removal was also initially good (20-85%), but after two months of operation nitrate removal ceased and the wetland began releasing nitrate. This was probably because of colder temperatures in October reducing the activity of both nitrifying and denitrifying bacteria (Wallace et al., 2001). Johnson and Younger (2006) noted a decrease in manganese concentrations, correlated to a decrease in SS concentrations, indicating that manganese was being removed by sorption onto the solids, which is a significant result because manganese removal in wetlands is not typically effective (Hedin and Nairn, 1993). In addition, significant quantities of calcium, sulphate, magnesium and bicarbonate were removed in the wetland.

Compost wetlands

In compost wetlands, AMD flows through an anoxic substrate comprised of organic compost and other organic materials (Figure 5) (Eger et al., 1994; Hedin et al., 1994).

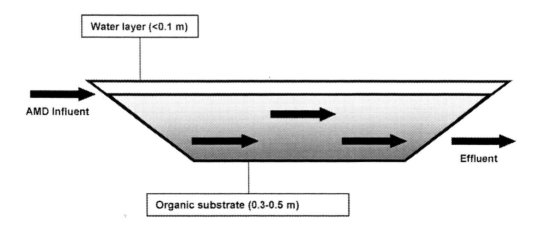

Figure 5. Schematic diagram of a compost wetland

The substrate is deeper than that of aerobic wetlands (~0.3-0.45 m) (Hedin et al., 1994), and may consist of soil, peat, straw, woodchips, spent mushroom compost, manure, composted municipal waste, or other readily available organic material. The substrate may also be mixed with limestone, which generates alkalinity and decreases acidity as it dissolves. In compost wetlands, several of the processes observed in aerobic wetlands (e.g. ion exchange and complexation) also occur, owing to sharp redox gradients that may be present from the oxygenated surface to the anoxic deeper layers of the substrate (Mayes et al., 2008). However, in the reducing environments of the organic substrates, sulphate reducing bacteria (SRB) reduce sulphate, oxidize organic matter, and generate alkalinity as they release bicarbonate and hydrogen sulphide as metabolic waste products. Dissolved metals are primarily removed from solution as metal sulphide precipitates, although as carbonates in the substrate are dissolved, metals can also be removed as hydroxides or carbonates and additional alkalinity is generated (Henrot and Weider, 1990; Walker and Hurl, 2001; Ziemkiewicz et al., 2003; Mayes et al., 2008). Ferric hydroxides can also adsorb dissolved metals. Compost wetlands are therefore ideal for treating net-acidic metal-laden AMD with pH 4-5.5 (Hedin et al., 1994; Kleinmann et al., 1998), with acidity removal rates ranging from 10.4-25.0 $g/m^2/day$ (Younger et al., 1997; Jarvis and Younger, 1999; Ziemkiewicz et al., 2003).

The performance of microbial communities is critical when using compost wetlands to treat net-acidic AMD (Mayes et al., 2008). Systems that incorporate an anaerobic alkalinity-generating step (e.g. compost wetlands, sulphate reducing bioreactors, permeable reactive barriers, and SAPS) require a continuous supply of biodegradable organic matter to provide carbon for microbial metabolic processes. Carbon supply may be a limiting factor in system effectiveness and longevity (Eger and Wagner, 2001, 2003). The degradation of cellulose in the substrate has also been identified as a rate-limiting step for sulphide production in SRB-based systems (Logan et al., 2003, 2005), rather than the population of SRB (Pruden et al., 2005, 2006). The availability of nutrients such as phosphate, nitrogen, and other trace elements can also become limiting (Kalin et al., 2006). Sulphate-reducing bacteria have been reported to perform best at pH >4.0, with adequate sulphate concentrations, and no oxidizing agents (e.g. oxygen or ferric iron) (Gazea et al., 1996). Research into the optimum conditions for SRB is needed to identify the acidophilic genera of SRB that can be used to treat AMD at extremely low pH (Hallberg and Johnson, 2005b). Compost wetlands may or may not be planted with vegetation, and it has been suggested that infiltration of plant roots into the substrate may inhibit performance of SRB by introducing oxygen (Gazea et al., 1996, Johnson and Hallberg, 2005). Some compost wetlands are contained entirely underground and support no vegetation.

Problems associated with compost wetlands include clogging of the substrate interstices with sulphides, the depletion of organic carbon, and changes in residence time and flow dynamics as the system ages (Mayes et al., 2008). Treatment efficiency varies seasonally, and increased contaminant removal has been reported in warmer months when bacterial activity is optimum (Heal and Salt, 1999). Also, removal of zinc may be inefficient where AMD is net-alkaline or neutral, and may require larger wetland areas for treatment (Nuttall and Younger, 2000; Yang et al., 2006). Hedin (1996) described some problems associated with vegetated compost wetlands, including preferential flow paths and subsequent decrease in retention time, owing to accumulation of plant litter. Animals such as rodents and rabbits may be attracted to the vegetated areas and dig tunnels, which may disrupt flow regimes.

4.2. Anoxic Limestone Drains

Anoxic limestone drains (ALD) are used to neutralize net-acidic AMD as it flows by gravity through a limestone gravel bed contained in a cell or trench that is buried under clay and lined and sealed with impermeable materials to prevent water and oxygen from entering (Turner and McCoy, 1990; Kleinmann et al., 1998). As the limestone dissolves on contact with AMD (Section 3.1), alkalinity (CO_3^{2-} + HCO_3^- + OH^-) and effluent pH are both increased. Carbon dioxide cannot escape from the system, and limestone dissolution is enhanced with increased partial pressure of pCO_2 (Cravotta, 2003). Alkalinity is therefore added to AMD within the ALD and depending on influent chemistry and residence time can reach concentrations up to 275 mg/L as $CaCO_3$, compared to open systems where alkalinity may only reach 50-60 mg/L as $CaCO_3$ (Watzlaf and Hedin, 1993; Kleinmann et al., 1998). An average acid removal rate of 56 g/day/t limestone has been reported (Ziemkiewicz et al., 2003). The added alkalinity makes subsequent treatment either in a wetland or settling pond more effective, although metal hydrolysis in later treatment stages may increase acidity. Alkalinity added in the ALD acts as a buffer to prevent the pH of the final effluent from decreasing. The ALD is not intended for metal removal, indeed, metal precipitation within an ALD would hinder system performance by reducing drain permeability (Demchak et al., 2001).

Similar to active chemical neutralization and precipitation systems (Section 3.1), the limestone in an ALD is susceptible to armoring by metal hydroxides (Nairn et al., 1991; Watzlaf et al., 2000) or by various compounds of calcium and sulphate (Cravotta, 2003). If armoring occurs, the ALD will eventually fail as limestone dissolution is reduced, permeability is decreased, and alkalinity generation ceases. Iron and aluminum hydroxides may form if influent concentrations are higher than 2mg Fe^{3+}/L or 2mg Al^{3+}/L (Gazea et al., 1996), and even lower permissible concentrations of <1 mg/L Fe^{3+} or Al^{3+} have been recommended (Hedin and Watzlaf, 1994; Hedin et al., 1994). Performance may be poor if the influent AMD is aerated with dissolved oxygen (DO) concentrations >1 mg/L which causes oxidation of ferrous iron (Ziemkiewicz et al., 2003). In anoxic conditions, however, dissolved iron remains in the ferrous state and no iron hydroxide precipitation (as $Fe(OH)_2$) occurs at pH <8.0 (Evangelou, 1995; Demchak et al., 2001). Prior to exposure to the atmosphere, AMD is often anoxic with high concentrations of ferrous iron, so treatment in an ALD is very effective for adding alkalinity (Gazea et al., 1996), although performance is affected by water chemistry, residence time, and pCO_2 (Watzlaf and Hedin, 1993; Evangelou, 1995). Limestone of high purity (>82%) has been reported as most effective (Watzlaf and Hedin, 1993), with a residence time of 15 hours recommended for maximum alkalinity generation. In a sense, ALDs are not truly passive, because they eventually require replacement and maintenance of the reactive limestone surfaces. However, pretreatment in an anoxic pond to reduce DO concentrations will improve both ALD performance and longevity (Johnson and Hallberg, 2005). The performance of both aerobic and compost wetlands are improved by the addition of an anoxic limestone drain (ALD) (Turner and McCoy, 1990; Brodie et al., 1990; Nairn et al., 1991; Watzlaf and Hedin, 1993; Hedin et al., 1994; Kleinmann et al., 1998). In a review of 83 passive systems (ALDs, limestone leach beds, vertical flow wetlands, and anaerobic wetlands) in operation in the US, Ziemkiewicz et al. (2003) concluded that the ALD was the "most consistently efficient passive treatment system in terms of cost per tonne of acid removed".

Cravotta (2003) reviewed the long-term (5-11 years) performance of three ALDs and compared alkalinity production and limestone dissolution within each. The performance data were used to validate a method for predicting alkalinity production in an ALD and calculating the mass of limestone required, based on the initial concentration of alkalinity of the influent AMD, influent flow rate, and density and porosity of the packed bed.

In net-acidic AMD, zinc is mainly present as Zn^{2+} and will readily form hydroxide or sulphide precipitates (Nuttall and Younger, 2000). However, in hard, net-alkaline AMD, zinc does not readily precipitate as a non-carbonate solid. Nuttall and Younger (2000) designed an ALD with an aeration step to treat net-alkaline AMD with high zinc (<10 mg/L) and sulphate concentrations (>100 mg/L) by increasing the pH from 7.5 to 8.2 The combined effects of the closed system (e.g. increased pCO_2 and high calcium and sulphate concentrations) raise the pH and enhance calcite dissolution, forming bicarbonate, hydroxide, and zinc ions, creating ideal conditions for zinc precipitation as smithsonite.

4.3. Open Limestone Channels and Limestone Leach Beds

Open limestone channels, lined with large limestone rocks, operate on the same principles as ALDs, i.e. the AMD flows through a substrate of limestone which dissolves, increasing the pH of the AMD (Cravotta and Trahan, 1999). Similar to ALDs, these systems are prone to armoring by metal hydroxides, and may cease to function after just a few years of operation (Ziemkiewicz et al., 2003). Open limestone channels, with an average acidity removal rate of 22 g/day/t limestone reported by Ziemkiewicz et al. (2003), are effective in treating AMD at sites with slopes >12% (Ziemkiewicz et al., 1997) The slope is important for performance, because greater slopes increase flow velocities and help to minimize armoring. Even in situations where limestone is armored, it continues to dissolve, albeit at only 20 to 50% of the dissolution rate of unarmored limestone (Pearson and McDonnell, 1975; Ziemkiewicz et al., 1994), and AMD acidity may still be reduced by up to 60% within the channel (Ziemkiewicz et al., 1997).

Limestone leach beds, cells containing limestone chips and rocks of various sizes, may be used to treat net-acidic water with low dissolved metal concentration (Black et al., 1999). As the limestone dissolves on contact with AMD, alkalinity is generated, reaching up to 75 mg/L as $CaCO_3$. Average acid removal rates of 28 g/day/t limestone have been reported (Ziemkiewicz et al., 2003). A variation of this system is the slag leach bed which is filled with steel slag fines. This system can be used to treat acidic water that contains no iron, manganese, or aluminum (Simmons et al., 2002). Alkalinity may reach as high as 2,000 mg/L as $CaCO_3$ (Ziemkiewicz et al., 2003).

4.4. SRB-Based Passive Systems

The reductive processes catalyzed by sulphate reducing bacteria (SRB) are key to several passive AMD treatment systems designed to treat both surface and ground waters. As is the case for compost wetlands (Section 4.1), substrate design is key to alkalinity generation and metal removal. Decayed plant matter in the substrate is an important source of organic matter on which metals may be adsorbed, and plant cellulosic material may provide nutrients for SRB (Gazea et al., 1996; Logan et al., 2003, 2005) with additional alkalinity generated by limestone dissolution where required. Systems in which biogenic hydrogen sulphide is formed on-site are more economical and pose less of an environmental hazard than systems in which the sulphide is supplied from chemical addition as this must be transported to and stored on site (Section 3.5).

SRB bioreactor (passive)

Net-acidic, metal-rich, and sulphate-laden AMD can be effectively treated using a passive SRB bioreactor (Gusek, 2002, 2004; Gusek and Wildeman, 2002; Gusek et al., 2008). These systems typically consist of a reaction chamber filled with a substrate comprised of organic material such as compost, manure, sawdust, woodchips, and hay, as well as an alkaline agent. The addition of an organic source that naturally contains SRB, such as compost or manure, may be advantageous. Metals are removed as hydroxides, sulphides, carbonates, silicates, or sulphates, and by sorption onto hydroxides, carbonates, organic matter, other metal precipitates or sediment (McCauley et al., 2009). Recaptured metals can be sold as commercial grade or used to enrich existing ores prior to processing. As with active SRB bioreactors, the performance of a passive SRB bioreactor is affected by various design factors, including substrate type and reactor configuration (Neculita et al., 2007). Willow and Cohen (2003) reported that the metal loading rate capacity of a wet-substrate bioreactor could be enhanced by bringing the pH of the influent AMD to circum- neutral prior to entering the bioreactor. This neutralization is normally achieved by pretreatment using an ALD. Microbial communities may be enriched and performance may be improved by periodic system shutdowns (Whitehead et al., 2005; Mayes et al., 2008).

Sulphate removal rates ranging from 0.06-0.44 mol/m^3/day have been reported for reactors using various mixtures of organic material and crushed limestone (Gusek, 2002; Gibert et al., 2004; McCauley et al., 2009), with metal removal rates of 0.3-0.8 mol/m^3/day reported for bioreactors containing similar substrates (Wildeman et al., 2006; McCauley et al., 2009).

Infiltration beds and anoxic ponds

Infiltration beds typically consist of a covered and lined trench containing organic material to encourage SRB growth and activity. The bed may be covered and lined to create anaerobic conditions ideal for SRB activity (Kaksonen and Puhakka 2007). The organic material supplies nutrients for the SRB and determines hydraulic conductivity through the bed. Riekkola-Vanhanen (1999) reported that an infiltration bed with peat and limestone was effective in removing 94-99% copper, 76-97% zinc, 85-96% iron, 76-96% manganese, and 72-94% sulphate from AMD at the Pyhäsalmi mine in Finland.

Anoxic ponds are either constructed anaerobic cells with organic substrate overlying a gravel base, or open pits with added organic material. They are used upstream of anoxic limestone drains to decrease the dissolved oxygen concentration, reduce ferric iron to the ferrous form, and precipitate aluminum (Gazea et al., 1996; Kaksonen and Puhakka, 2007).

SRB-based systems for groundwater remediation

Acid mine drainage may enter groundwater from mine workings that extend below the water table, or from surface material and tailings by infiltration. Acidity may be buffered in the aquifer by mineral phases and aquifer sediment, but sulphate, ferrous iron, and other contaminants may remain in high concentrations. When the groundwater reaches the surface, ferrous iron becomes oxidized, creating acidic conditions in the receiving waters and increasing metal mobility and bioavailability. Removal of Fe^{2+} from the contaminated plume is a major factor in limiting acid production when the groundwater reaches the surface.

To determine the acid-producing potential of groundwater, the concentration of ions as well as the buffering capacity (e.g. alkalinity) must be considered. Ferrous iron, which forms insoluble oxyhydroxides on oxidation, is usually the most acid-producing ion in mine drainage (Benner et al., 1997). The potential of groundwater to generate mineral acidity may be calculated by subtracting the acid buffering capacity (as molar equivalent of dissolved carbonate alkalinity) from the acid-producing potential (as molar equivalent of acid produced by Fe^{2+} oxidation) (Eq. 25) (Snoeyink and Jenkins, 1980):

$$\text{Potential acidity} = \text{moles/L Alkalinity (as } CaCO_3) - 2(\text{moles/L } Fe^{2+}) \qquad (25)$$

A positive value indicates net acid-producing potential, whereas a negative value indicates a net acid-consuming potential.

Different methods of treating AMD-contaminated groundwater in situ have been developed, including substrate injection to enhance microbial activity in aquifers, and permeable reactive barriers (Ludwig et al., 2002; Lee and Saunders, 2003; Saunders et al., 2008). Substrate injection involves placing material into boreholes to encourage SRB activity. Groudev et al. (1998) reported enhanced SRB activity after an acetate-bearing material and ammonium phosphate were injected into the subsurface near Burgas Copper Mines in Bulgaria. Heavy metals, including copper, lead, and manganese, were precipitated as sulphides. This approach may be ineffective in situations where flows are highly variable, owing to seasonal changes, or if oxygenated water infiltrates the substrate.

Permeable reactive barriers (PRBs) are used for in situ treatment of acidic, metalliferous groundwater (Benner et al., 1997, 1999; Blowes et al., 2000; Amos and Younger, 2003). There are two main PRB design configurations in use, (i) continuous barriers (Amos and Younger, 2003), and (ii) funnel and gate systems, which channel contaminated groundwater through treatment zones (Starr and Cherry, 1994). The continuous barrier system typically consists of a subsurface trench filled with organic matter, which may be mixed with limestone gravel, placed in the flow path of a contaminated plume of groundwater (Richardson and Nicklow, 2002). The barrier extends across the width of the contaminated plume. Metals in the groundwater are adsorbed onto material in the PRB and/or precipitate out of solution as sulphides, hydroxides, and carbonates. To remove metals as sulphides, anaerobic conditions must be maintained, so an impermeable cap must be used. Sulphides formed will be stable as long as the reactive barrier remains below the water table and no oxidizing agents (e.g. oxygen, ferric iron, or nitrates) are introduced into the system.

The barrier substrate, which may contain reactive media such as sewage sludge, green waste compost, woodchips, manure, slurry, or sawdust, must be carefully selected to provide adequate permeability and reactivity while also immobilizing precipitated metals (Waybrant et al., 1998; Benner et al., 1999; Amos and Younger, 2003). Gravel and/or limestone may be added to the substrate to enhance permeability and provide additional alkalinity (Gusek and Wildeman, 2002; Amos and Younger, 2003; Watzlaf et al., 2004; Kaksonen and Puhakka, 2007). Material that may be easily compacted (e.g. manure and straw mixtures) is unsuitable for use in a PRB (Amos and Younger, 2003). Long-term effectiveness of the barriers can be seriously affected either by depletion of organic material in the substrate or by clogging by metal precipitates (Richardson and Nicklow, 2002). The interstices of the substrate can easily become blocked resulting in channels being formed which reduce the hydraulic retention time of the AMD within the system. This is avoided by using homogenous mixtures of reactive media (Amos and Younger, 2003). Reactive mixtures containing limestone chips, compost and slurry screenings have been reported to remove significant concentrations of sulphate (~200-300 mg/L per kg media) (Amos and Younger, 2003).

Case study: Permeable reactive barrier, nickel rim, ontario, canada

A permeable reactive barrier has been used in the sand aquifer underlying the Nickel Rim mine tailings impoundment near Sudbury, Ontario, Canada (Benner et al., 2002). The groundwater was significantly contaminated by a tailings impoundment with high concentrations of ferrous iron, sulphate, and nickel, with a pH between 5 and 6 (Bain et al., 1999). As the groundwater surfaced, the ferrous iron oxidized, forming iron hydroxide precipitates and causing the pH to drop to 2. Selection of a suitable material for the barrier was based on the need for it to be reactive enough to reduce the sulphate concentrations, sufficient permeability to accommodate the groundwater flow, ability to remain permeable and reactive for an extended period (i.e. decades), and be readily available and economical (Benner et al., 1997). A mixture of municipal compost, leaf compost, and wood chips was selected, with pea gravel added to ensure sufficient hydraulic conductivity. The PRB was 15 m wide (perpendicular to groundwater flow), 4 m thick, and 3.6 m deep (Figure 6). It was built into the bedrock below and on both sides of the aquifer to ensure that contaminated water passed through, rather than around, the barrier.

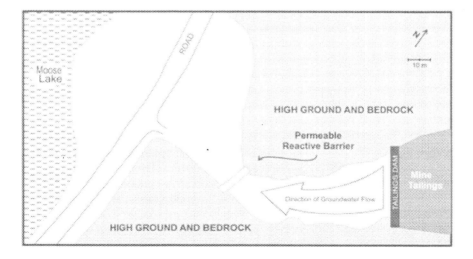

Figure 6. Map view of the permeable reactive barrier (PRB) at Nickel Rim mine, Canada (reproduced from Benner et al. (2002) with permission of Elsevier, Ltd.)

Nine months after installation, groundwater quality showed a significant improvement, with the pH increased from 5.8 to 7.0, reduced concentrations of iron (from 250-1300 to 1-40 mg/L) and sulphate (from 2400-4600 to 200-3600 mg/L), but increased alkalinity (from 0-50 to 600-200 mg/L as $CaCO_3$) (Benner et al., 1997). The iron and sulphate were removed primarily by the precipitation of iron as monosulphides (e.g. mackinawite, FeS), siderite ($FeCO_3$), gypsum, or organic sulphide compounds (Benner et al., 1999; Herbert et al., 2000). After flowing through the PRB, potential acidity of the groundwater decreased from 8 to 46 meq/L (net acid producing) to -16 to -45 meq/L (net acid consuming). At the time of construction, a lifetime of >15 years was suggested, based on comparison with column experiments (Benner et al., 1997). However, after just 3 years, it was observed that sulphate reduction had decreased by 30% and iron removal by 50% (Benner et al., 2002). Possible factors in temporal and spatial variations in PRB performance were preferential flow through the PRB with faster movement through the center than at the top or bottom observed, depletion of organic carbon, and temperature fluctuations that may have limited SRB activity. Benner et al. (2002) suggested as a rule of thumb that a water temperature increase of 10°C will cause the sulphate reduction rate to double. Temperature changes may also affect iron concentrations, owing to the resulting changes in iron sulphide solubility.

Case study: In situ bioreactor, lilly/orphan boy mine, montana, USA

An in situ bioreactor with a substrate of cow manure (~70%), wood chips (~20%), and straw (~10%) was used to treat AMD-contaminated groundwater emanating from the abandoned Lilly/Orphan Boy Mine in Montana (Nordwick and Bless, 2008). The bioreactor consisted of a platform suspended on cables 30 feet below the water level in a mineshaft that was open to the surface (Figure 7). The platform held the organic substrate, and AMD flowed upwards through the substrate before exiting the mine portal/adit. Additional organic substrate was injected into the main portal tunnel through holes that were drilled from the surface.

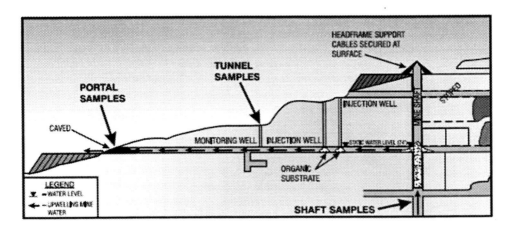

Figure 7. Cross section of underground mine subsurface SRB bioreactor, Lilly/Orphan Boy mine, Montana (reproduced from Nordwick and Bless (2008) with permission of the US EPA)

Table 6. Lilly/Orphan Boy AMD chemistry before and after installation of in situ bioreactor (reproduced from Nordwick and Bless (2008) with permission of the US EPA)

	Fe (mg/L)	Zn (mg/L)	Al (mg/L)	Mn (mg/L)	As (mg/L)	Cd (mg/L)	Cu (mg/L)	SO_4 (mg/L)	pH
Baseline (Average, September 1993 to August 1994)	13.8	19.4	7.36	5.46	0.08	0.24	0.33	213	3.4
Tunnel (May 2002)	9.7	<0.01	<0.02	1.51	0.04	<0.005	<0.002	21.0	6.6
Adit (May 2002)	28.4	12.5	0.51	5.44	3.66	0.064	0.041	223	5.2

During 11 years of field demonstration (1994-2005), the bioreactor effectively removed metals, decreased the sulphate concentration and raised the pH of the effluent (Table 6), indicating strong SRB activity. Aluminum, copper, zinc, and cadmium were removed more effectively than iron, arsenic, or manganese. Nordwick and Bless (2008) recommended the use of an oxidizing step downstream of the reducing process to remove metals, such as iron and manganese, which are more mobile and likely to remain in the effluent.

Seasonal effects of spring runoff events were evident in the fluctuations in effluent water quality and increases in metal concentrations in the portal effluent. These fluctuations were probably owing to infiltration of oxygenated runoff water through the ground above the portal as well as increased AMD infiltration into the shaft through fractures in the shaft walls. Longevity of similar SRB designs remain unknown. The Lilly/Orphan Boy Mine field demonstration, which is one of the longest-running in-situ hard rock mine treatment systems in the US, indicates that SRB-based systems are effective in removing metals and acidity from AMD. However, the upward trend in dissolved metal concentrations indicates that the bioreactor is becoming less effective, probably owing to depletion of organic matter in the substrate. Similar studies are needed to optimize SRB system design and determine the effective lifespan of various substrates.

Successive alkalinity producing system (SAPS)

The successive alkalinity producing system (SAPS) (Kepler and McCleary, 1994, 1997) is also known as the reducing and alkalinity producing system (RAPS) (Younger et al., 2003). Sufficient hydraulic head is required to drive vertical flow of AMD downward through the system. Typically, AMD is loaded onto the top layer of a constructed lagoon or cell at a depth of 0.5–1.5 m. The AMD first flows vertically downward through a reducing environment, which is typically an unvegetated layer of alkalinity-generating media and/or compost 0.15-0.5 m thick, where DO is removed and ferric iron is reduced to the ferrous form (Figure 8) (Demchak et al., 2001; Gusek and Wildeman, 2002). Alkalinity is increased during this stage of the process. Metals may be retained in the organic substrate by exchange processes and filtering (Demchak et al., 2001), and metal sulphides precipitate out of solution. The anoxic AMD then flows down through a layer of limestone gravel (0.5-2 m thick), where bicarbonate alkalinity is generated. Metals may precipitate as hydroxides in the limestone layer (Kepler and McCleary, 1997). After treatment, the effluent flows to a sedimentation pond or aerobic wetland, where iron (as ferric hydroxides) and other metals are precipitated. If the AMD is still net-acidic, it may be diverted through another SAPS for further treatment. The SAPS may also be used in series with oxidizing ponds (Ziemkiewicz et al., 2003).

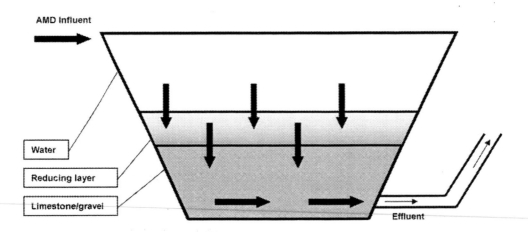

Figure 8. Schematic diagram of a successive alkalinity producing system (SAPS)

The SAPS is used to treat net-acidic AMD, and is recommended for treating AMD with pH 4.0-5.5 (Kleinmann et al., 1998; Mayes et al., 2008). The process is often used upstream of an aerobic wetland, or for reducing ferric iron to the ferrous form prior to flowing into an ALD. Because SAPS is based on the same chemical and biological processes as compost wetlands, it is sometimes called a vertical flow wetland. However, Demchak et al. (2001) reported that water interaction with the compost layer and the limestone is much higher in the SAPS than in compost wetlands. The SAPS may be more efficient at removing acidity than compost wetlands with removal rates of 20-62 g acidity/m^2/day reported (Watzlaf et al., 2000; Ziemkiewicz et al., 2003; Watzlaf et al., 2004).

Demchak et al. (2001) examined the long-term performance of four SAPS treating AMD (Table 7). Metal removal, decreased acidity and increased alkalinity were observed in all four of the SAPS, although treatment efficiencies varied. Acidity removal rates were 17.4 g/m^2/day for Howe Bridge, 51.5 g/m^2/day for Filson 1, 18.2 g/m^2/day for Sommerville, and 11.2 g/m^2/day for McKinley. Sulphate concentrations were increased in the effluent at all sites. They observed that SAPS efficiency to be dependent on influent water chemistry, system age, flow rates through the system, and water residence time. The nature of the compost layer is critical and must be thick enough to create reducing conditions but any compaction or short-circuiting in this layer can lead to reduced system performance. A minimum thickness of 50 cm is recommended for the compost layer, which should be composed of materials such as mushroom compost mixed with wood chips to provide a short- and long-term carbon source. Also recommended are periodic replacement of compost or addition of fresh material, and increasing the number of drainage pipes underlying the limestone to reduce the risk of preferential flow through the system. Flushing the system may increase longevity and efficiency in cases where aluminum concentration is high (Kepler and McCleary, 1997).

Table 7. Influent and effluent water chemistry for SAPS (reproduced from Demchak et al. (2001) with permission of the Geological Society of London and the authors)

Site Flow rate (L/min)	pH Inflow	pH Outflow	Acidity* Inflow (mg/L)	Acidity* Outflow (mg/L)	Alkalinity* Inflow (mg/L)	Alkalinity* Outflow (mg/L)	Iron Inflow (mg/L)	Iron Outflow (mg/L)	Aluminum Inflow (mg/L)	Aluminum Outflow (mg/L)	Sulphate Inflow (mg/L)	Sulphate Outflow (mg/L)	Calcium Inflow (mg/L)	Calcium Outflow (mg/L)
Howe Bridge														
High (240)	6.1	6.2	332	206	54	64	154	78	<1	<1	947	782	146	174
Low (80)	4.5	5.5	324	192	10	28	191	59	<1	<1	890	1010	170	202
Filson 1														
High (225)	3.9	4.9	240	194	0	17	21	2	17	9	1603	1770	128	173
Low (20)	3.4	4.4	254	164	0	10	30	7	9	7	649	824	109	141
Sommerville														
High (350)	3.4	4.4	296	224	0	15	1	3	37	30	444	2350	75	115
Low (1)	3.5	4.6	398	246	0	15	3	7	48	29	678	736	108	163
McKinley														
High (185)	4.4	6.1	42	7	8	46	6	<1	4	0.3	312	571	64	121
Low (20)	4.0	6.4	72	0	3	58	4	<1	2	0.2	556	478	135	143

*Acidity and alkalinity are in mg/L as $CaCO_3$

4.5. Other Passive Systems

Kalin et al. (2006) described a novel system called acid reduction using microbiology (ARUM), which minimizes the problem of depletion of available organic carbon by using products of photosynthesis in the water column as the source of organic matter. Primary productivity in the water column provides electron donors for microbial processes in the sediment, as well as generating alkalinity. The system consists of aerobic and anaerobic "wetlands" consisting of two iron oxidation/ precipitation cells and two reducing (ARUM) cells (Kalin et al., 1991) containing organic material (typically floating macrophytes) in which alkalinity and sulphides are generated. The system may be used for AMD treatment or pre-treatment for conventional systems.

A pilot ARUM system comprised of two oxidation/precipitation cells and two reducing cells was used to treat AMD containing between 300 - 400 mg/L of iron (Kalin and Smith, 1997). Using the high concentrations of ferrous iron in the AMD, the system generated iron oxyhydroxides which co-precipitated and adsorbed other metals. Seepage flow into the oxidation/precipitation cells was approximately 1 L/min with a pH of between 5 and 6. The pH decreased to between 3 and 4 after the oxidation stage. The reducing cells formed the ARUM stage in which alkalinity and sulphide were generated through reductive processes. The ARUM cells contained sediments made of organic materials while floating vegetation mats provided organic matter to the sediment, provided microbial habitats, and protected the system from wind-induced turbulence (Kalin, 1993). The ARUM system demonstrates that organic matter can decompose in AMD, thus providing biodegradable carbon to the sediments where anaerobic microbial reduction generates alkalinity. ARUM systems have been used in Canada (Fyson et al., 1995; Kalin et al., 1999) and in Brazil (Kalin and Caetano-Chaves, 2003).

In an effort to identify possible long-term passive treatment options for Wheal Jane (Cornwall, UK) AMD (and other similar drainages), a pilot passive treatment plant was constructed in 1994 (Whitehead and Prior, 2005; Whitehead et al., 2005). The plant consisted of three integrated steps: (i) constructed aerobic wetlands (to remove iron and arsenic), (ii) an anaerobic cell to encourage sulphate reduction and metal sulphide precipitation (to remove zinc, copper, cadmium, and the remaining iron), and (iii) aerobic rock filters to promote algal growth (to precipitate manganese oxides). Each system included a different pre-treatment stage to modify influent pH: (i) dosing with calcium carbonate, (ii) an anoxic cell followed by an ALD, or (iii) no modification. Maximum removal of iron, aluminum, cadmium, copper, manganese, and zinc during 1999 to 2001 was achieved using the ALD pre-treatment stage (Whitehead et al., 2005).

As described above, manganese removal is difficult in both active and passive treatment systems (Hallberg and Johnson, 2005a). Johnson and Younger (2005) have described a passively-aerated subsurface gravel bed with catalytic substrates designed for manganese removal. The system, an alternative to existing passive treatment systems such as open-air gravel filters that require large areas of land, can be used as a final polishing step in a series of passive treatment systems. The system removed over 95% manganese when influent manganese concentration was ~20 mg/L, and functioned at low temperatures and in darkness. Manganese oxyhydroxides recovered from this system may be used to enhance removal of other dissolved metals, such as zinc. A similarly novel method for manganese removal is a passive in-situ bioreactor that employs biofilms for remediating Mn^{2+}-contaminated waters

(Mariner et al., 2008). The black-stained biofilms formed by manganese-oxidizing microorganisms contain manganese oxides that can scavenge Mn^{2+} (Tessier et al., 1996; Hoover and Rightnour, 2002). Microorganisms such as fungi and bacteria catalyze the oxidation of Mn^{2+} to Mn^{4+}, increasing oxidation rates by several orders of magnitude compared to spontaneous abiotic oxidation (Nealson, 1983). Mariner et al. (2008) operated a pilot reactor containing stones 2-5 cm in diameter and covered with Mn^{4+}-containing biofilm. The concentration of Mn^{2+} was reduced from 10 mg/L to <0.25 mg/L, with attenuation most effective at low pH (pH 4-6) although Mn^{2+} removal slowed at pH <3.5. The efficiency of the system improved with addition of Mn^{2+}, possibly owing to increased numbers of Mn-oxidizing microorganisms and increased abiotic uptake of Mn^{2+} by the biofilm.

Surface-catalyzed oxidation of ferrous iron (SCOOFI) systems can treat net-alkaline AMD containing iron and manganese in situations where land availability is limited (Younger, 2000b; Jarvis and Younger, 2001). The SCOOFI reactors are packed with high surface area media such as blast furnace slag or PVC trickling filter media (Younger, 2000b) and designed to encourage ferric hydroxide precipitation to remove iron from AMD. As AMD flows down through the media, ferrous iron is removed from solution and sorbed onto the highly polar ochre surface. With ferric hydroxide acting as a catalyst, ferrous iron is then rapidly oxidized in situ by dissolved oxygen (Younger, 2000b). Alkalinity in the mine drainage buffers the acidity generated during iron oxidation and precipitation.

Whereas aerobic wetlands must be sized up to treat influent with low iron and manganese concentrations (Tarutis et al., 1999), SCOOFI reactors require less land area and shorter residence times. However, secondary wastes are an issue, and the system requires regular ochre removal and disposal (Younger, 2000b). Jarvis and Younger (2001) investigated iron oxidation and accretion onto high surface area plastic trickling filter media in two down-flow pilot-scale reactors treating ferruginous, net-alkaline coal mine drainage with pH ~6.95. Concentrations of iron and manganese were significantly decreased (Table 8), and residence times were significantly shorter than those of aerobic wetlands.

Sapsford et al. (2008) described a similar passive system for treating net-alkaline, circum-neutral pH ferruginous mine drainage, the vertical flow reactor (VFR). The VFR differs from the SCOOFI reactor in that no high surface area media for enhancing ochre accretion are included. Instead, the mine water passes through a sandstone gravel bed that supports the accreting layer of ochre. The ochre bed filters iron and catalyzes iron oxidation, leading to further ochre precipitation. Sapsford and Williams (2009) formulated sizing criteria for the VFR based on the performance of a pilot system at a colliery in South Wales. Influent iron concentrations were typically <10 mg/L, and treatment efficiency by normal systems is normally poor at such low concentrations. However, the VFR very effectively removed iron, with percentage removal ranging from 42%-100% and removal rates many times the standard sizing criteria for aerobic wetlands of 10 $g/m^2/day$. Manganese was also removed, which was unexpected given the slow kinetics of manganese oxidation at the circum-neutral pH of the system, and the presence of ferrous iron at concentrations >1 mg/L (Sapsford and Williams, 2009). Removal efficiency ranged from 0% when influent manganese concentration was at its lowest (0.04 mg/L) to 100% at the higher concentrations (0.7-0.8 mg/L) (Sapsford et al., 2008).

Table 8. Metal concentrations of coal mine AMD before and after treatment by a SCOOFI system, based on 5 months of monitoring (adapted from Jarvis and Younger (2001))

Parameter (metal concentrations in mg/L)	Influent (average)	Reactor A^a effluent (average)	Reactor B^b effluent (average)
Fe (total)	1.43	0.41	0.38
Mn	0.67	0.53	0.47

a: total media surface area 124 m^2, b: total media surface area 178 m^2

5. CONCLUSION

The task of preventing and treating acid mine drainage will continue to challenge engineers and scientists in the search for treatment systems that are both sustainable and efficient. There are advantages and disadvantages associated with all treatment systems described in this review, and these are summarized in Table 9. In situations where acidity and metal loading are extreme, active systems are normally the only practical treatment solution. Active systems can be optimized to treat large volumes of highly contaminated AMD efficiently and rapidly. The technologies commonly used are now well established, and with several decades of practical experience available to both designers and operators, new active treatment systems offer even better treatment capabilities and reliability. However, the systems are energy-intensive and require continual operation and maintenance. Chemical reagents must also be purchased, and large volumes of secondary wastes and toxic by-products must be contained and disposed of in designated disposal sites. Therefore, active systems are ideal solutions to AMD contamination in the short-term, but the associated high costs and energy demand make these systems unsuitable for long-term treatment. Metal recovery by selective precipitation is possible with active systems, offering the opportunity to recapture and reuse valuable metals, which can also partially offset operating costs. As world metal prices continue to rise, recovery of metals will become an increasingly important factor in treatment plant design.

In contrast, passive systems offer the best long-term solution to AMD treatment. Although they are not as readily optimized as active systems, passive systems offer many advantages. For example, no toxic by-products are generated, requirements for external energy supply, chemical reagents, and maintenance are all minimal, and they require infrequent disposal of secondary wastes. The major drawback to passive systems is that the remediative processes are slower, necessitating longer residence times that often translate into large footprints (e.g. wetlands). However, in situations where land availability is not limiting and acidity and metal loading are not extreme, passive systems are generally the most practical and cost-effective option. Passive systems are also preferred when decommissioning active treatment plants, or after the initial flush of highly contaminated AMD has dissipated. Microbial metabolic processes are the basis of most passive systems, and in order to achieve maximum performance, optimum operating conditions must be maintained. One major problem is an unforeseen increase in contaminant loading due to increased influent volumes or changes in AMD chemistry, either of which can severely reduce performance or even destroy passive systems. The lack of long-term performance data of passive treatment systems remains an obstacle to their widespread adoption. To overcome this, careful analysis of performance data from a wide range of scenarios must be performed, with the results used to refine design guidelines. In addition, as understanding of microbial tolerance to acidity and metals is improved, optimization based on microbial selection, acclimatization of species to specific AMD conditions and deliberate system inoculation may become feasible.

Table 9. Summary of advantages and disadvantages of treatment systems

Treatment system	Advantages	Disadvantages
Oxidation and chemical neutralization	+ Relatively low land area required + Simple system with low cost reagents + Metal recovery by selective precipitation	- Voluminous sludge with low chemical stability, requiring disposal in specially designated landfills to avoid redissolution/migration - Poor Mn removal - Limestone may become armored if treating net-acidic, Fe-rich AMD - Poor sulphate removal - High pH required to precipitate some metals as hydroxides
Proprietary reagents		- Sole source availability - Toxic by-products - High cost
Flotation	+ Metal recovery	- Energy intensive
Ultrafiltration	+ Metal recovery	- Energy intensive - Membrane durability, fouling
Nanofiltration	+ Metal recovery	- Energy intensive - Membrane durability, fouling
Reverse osmosis	+ Metal recovery	- Energy intensive - Membrane durability, fouling
Ion exchange	+ Metal recovery	- High costs of chemicals and adsorbents
SRB bioreactors	+ Precipitate metals as sulphides at lower pH than that required for hydroxide precipitation + Improved sludge characteristics: lower volume and better thickening and dewatering compared to hydroxide sludges + Metal recovery by selective precipitation + Sulphate removal	- Require continuous supply of organic carbon in the substrate - SRB must be protected from toxic metal concentrations and high acidity - Mn precipitation as sulphide difficult
Sorptive flotation	+ Metal recovery + Sludge volume lower than hydroxide sludge	- Toxic sludge
Sorption onto inorganic materials	+ Metal recovery + Use low-cost waste products as sorbents + Sludge volume lower than hydroxide sludge	
Biosorption	+ Metal recovery + Use low-cost waste products as sorbents + Sludge volume lower than hydroxide sludge + Use as pretreatment for SRB bioreactor	- pH sensitive
Aerobic wetlands	+ Effective for treating alkaline, neutral, and weakly acidic metal and coal mine drainages + Precipitates retained within wetland + Little/ no external energy input required	- Long-term stability of deposits is uncertain - Large land requirement - Poor removal of Al, Zn, Mn as hydroxides at low pH - If low pH or high metal loading, must incorporate other treatment step(s) eg ALD or alkali dosing - Disposal of biomass -Removal of metals such as Zn and Mn at pH <8.0 difficult

Table 8. (Continued)

Treatment system	Advantages	Disadvantages
Compost wetlands	+ Little/ no external energy input required + Precipitates retained within wetland + Treat net-acidic (pH < 5.5) AMD with high Fe^{3+}, Al, DO	- Long-term stability of deposits is uncertain - Large land requirement - Require continuous supply of organic carbon in the substrate - Temperature may affect SRB - Long-term stability of deposits is uncertain - Disposal of biomass -Removal of metals such as Zn and Mn at pH <8.0 difficult - Mn precipitation as sulphide difficult
Anoxic limestone drains	+ Little/ no external energy input required + Treat net-acidic AMD + Treat Zn-rich net-alkaline AMD using aeration step to precipitate Zn as carbonate	- Limestone may become armored if treating oxic, Fe-rich AMD
Open limestone channels, limestone leach beds, limestone ponds	+ Little/ no external energy input required + Treat net-acidic AMD + Open limestone channel may be best option in mountainous areas with steep slopes (>10%)	-Limestone may become armored if treating Fe-rich AMD
SRB bioreactor (passive)	+ Little/ no external energy input required + Treat net-acidic AMD + Metal recovery possible + Improved sludge thickening and dewatering compared to hydroxide sludge	- Require continuous supply of organic carbon in the substrate - Temperature may affect SRB - SRB must be protected from toxic metal concentrations and high acidity - Mn precipitation as sulphide difficult
Permeable reactive barrier	+ Little/ no external energy input required + Treat acidic, metalliferous groundwater	- Require continuous supply of organic carbon in the substrate - Temperature may affect SRB - SRB must be protected from toxic metal concentrations and high acidity - Mn precipitation as sulphide difficult
Successive alkalinity producing system (SAPS)	+ Little/ no external energy input required + More compact passive system than constructed wetland + Increased water and substrate contact compared to compost wetland + Treat net-acidic (pH <5.5) AMD with high Fe^{3+}, Al, DO +Use to raise pH and remove Fe^{3+} and Al prior to ALD	- Require continuous supply of organic carbon in the substrate - Temperature may affect SRB - SRB must be protected from toxic metal concentrations and high acidity

REFERENCES

Acero, P; Ayora, C; Carrera, J. Coupled thermal, hydraulic, and geochemical evolution of pyritic tailings in unsaturated column experiments. *Geochim. Cosmochim. Acta*, 2007, 71(22), 5325-5338.

Adams, M; Lawrence, R; Bratty, M. Biogenic sulphide for cyanide recycle and copper recovery in gold-copper ore processing. *Miner. Eng.*, 2008, 21, 509-517.

Ahluwalia, SS; Goyal, D. Microbial and plant derived biomass for removal of heavy metals from wastewater. *Bioresour. Technol.*, 2007, 98(12), 2243-2257.

Ahn, KH; Song, KG; Cha, HY; Yeom, IT. Removal of ions in nickel electroplating rinse water using low-pressure nanofiltration. *Desalination*, 1999, 122, 77-84.

Akabzaa, TM; Armah, TEK; Baneong-Yakubo, BK. Prediction of acid mine drainage generation potential in selected mines in the Ashanti metallogenic belt using static geochemical methods. *Environ. Geol*, 2007, 52(5), 957-964.

Akita, S; Castillo, LP; Nii, S; Takahashi, K; Takeuchi, H. Separation of Co(II)/Ni(II) via micellar-enhanced ultrafiltration using organophosphorus acid extractant solubilized by nonionic surfactant. *J. Membr. Sci.*, 1999, 162, 111-117.

Alpers, CN; Nordstrom, DK; Thompson, JM. Seasonal variations of Zn/Cu ratios in acid mine water from Iron Mountain, California. In *Environmental Geochemistry of Sulphide Oxidation*, American Chemical Society: New York, 1994, 323-344.

Álvarez-Ayuso, E; García-Sánchez, A; Querol, X. Purification of metal electroplating wastewaters using zeolites. *Water Res.*, 2003, 37(20), 4855-4862.

Amos, PW; Younger, PL. Substrate characterisation for a subsurface reactive barrier to treat colliery spoil leachate. *Water Res.*, 2003, 37, 108-120.

Andres, NF; Francisco, MS. Effects of sewage sludge application on heavy metal leaching from mine tailings impoundments. *Bioresour. Technol.*, 2008, 99(16), 7521-7530.

Arnesen, RT; Nygaard, K; Iversen, ER; Christensen, B; Slørdahl, A; Bollingmo, Å. *Water filling of Wallenberg mine* (in Norwegian); Norsk Institutt for Vannforskning (NIVA) Report No. O-92174, NIVA: Oslo, Norway, 1994.

Arnesen, RT; Iversen, ER. *Transport of heavy metals from Norwegian sulphide mines* (in Norwegian); Norsk Institutt for Vannforskning (NIVA) Report No. O-94021, NIVA: *Oslo*, Norway, 1995.

Aubé, B; Zinck, JM. Lime treatment of acid mine drainage in Canada. In *Brazil-Canada Seminar on mine rehabilitation technological innovations*; JP; Barbosa, PS; Moreira Soares, B; Dixon, B. Tisch, Eds; *Desktop Publishing: Rio de Janeiro*, Brazil, 2003, 23-40.

Babich, H; Stotzky, G. Heavy metal toxicity to microbe-mediated ecologic processes: A review and potential application to regulatory policies. *Environ. Res.*, 1985, 36(1), 111-137.

Bain, JG; Blowes, DW; Robertson, WD. Hydrogeochemistry of a sand aquifer affected by drainage from the Nickel Rim tailings. *J. Contam. Hydrol*, 1999, 41, 23-47.

Banks, D; Younger, PL; Arnesen, RT; Iversen, ER; Banks, SB. Mine-water chemistry: the good, the bad, and the ugly. *Environ. Geol*, 1997, 32(3), 157-174.

Batty, LC; Younger, PL. Critical role of macrophytes in achieving low iron concentrations in mine water treatment wetlands. *Environ. Sci. Technol.* 2002, 36(18), 3997-4002.

Batty, LC. Wetland plants - more than just a pretty face? *Land Contamination & Reclamation*, 2003, 11(2), 173-180.

Batty, LC; Younger, PL. Growth of *Phragmites australis* (Cav.) Trin ex. Steudel in mine water treatment wetlands: effects of metal and nutrient uptake. *Environ. Pollut*, 2004, 132, 85-93.

Batty, LC. Wetland systems associated with mine sites as a source of biodiversity. In *Proceedings of the 9th International Mine Water Association Symposium*, Oviedo, Spain, September 5-7, 2005.

Batty, LC; Younger, PL. The effect of pH on plant litter decomposition and metal cycling in wetland mesocosms supplied with mine drainage. *Chemosphere*, 2007, 66, 158-164.

Batty, L; Hooley, D; Younger, P. Iron and manganese removal in wetland treatment systems: Rates, processes and implications for management. *Sci. Total Environ*, 2008, 394, 1-8.

Benner, SG; Blowes, DW; Ptacek, CJ. A full-scale porous reactive wall for prevention of acid mine drainage. *Ground Water Monit. Rem*, 1997, 17(4), 99-107.

Benner, SG; Blowes, DW; Gould, WD; Herbert Jr., RB; Ptacek, CJ. Geochemistry of a permeable reactive barrier for metals and acid mine drainage. *Environ. Sci. Technol.*, 1999, 33, 2793-2799.

Benner, SG; Blowes, DW; Ptacek, CJ; Mayer, KU. Rates of sulfate reduction and metal sulfide precipitation in a permeable reactive barrier. *Appl. Geochem*, 2002, 17, 301-320.

Bhowal, A; Dhatta, S. Studies on transport mechanism of Cr(VI) extraction from an acidic solution using liquid surfactant membranes. *J. Membr. Sci.*, 2001, 188(1), 1-8.

Black, C; Ziemkiewicz, P; Skousen, J. Construction of a limestone leach bed and preliminary water quality results in Beaver Creek. In *Proceedings of the 20th Annual West Virginia Surface Mine Drainage Task Force Symposium, Morgantown*, WV, April 13-14, 1999.

Bloom, JE; Yuretich, RF; Gál, NE. Environmental consequences of acid mine-drainage from Davis Pyrite Mine, Rowe, Massachusetts. *Northeast. Geol. Environ. Sci.*, 2007, 29(2), 107-120.

Blowes, DW; Ptacek, CJ; Benner, SG; McRae, CWT; Bennett, TA; Puls, RW. Treatment of inorganic contaminants using permeable reactive barriers. *J. Contam. Hydrol*, 2000, 45, 123-137.

Blue, LY; Van Aelstyn, MA; Matlock, M; Atwood, DA. Low-level mercury removal from groundwater using a synthetic chelating ligand. *Water. Res.*, 2008, 42, 2025-2028.

Bouazza, A; Rahman, F. Oxygen diffusion through partially hydrated geosynthetic clay liners. *Geotechnique*, 2007, 57(9), 767-772.

Brezonik, PL. *Chemical Kinetics and Process Dynamics in Aquatic Systems*; Lewis Publishers: Boca Raton, 1994.

Brix, H; Dyhr-Jensen, K; Lorenzen, B. Root-zone acidity and nitrogen source affects *Typha latifolia* L. growth and uptake kinetics of ammonium and nitrate. *J. Exp. Bot*, 2002, 53(379), 2441-2450.

Brodie, GA; Britt, CR; Taylor, H; Tomaszewski, T; Turner, D. Passive anoxic limestone drains to increase effectiveness of wetlands acid drainage treatment systems. In *Proceedings of the 12th Annual National Association of Abandoned Mine Land Programs Conference*, Breckenridge, CO, September 16-20, 1990.

Brown, H; Skousen, J; Renton, J. Floc generation by chemical neutralization of acid mine drainage. In *Acid Mine Drainage Control and Treatment*; J; Skousen, P. Ziemkiewicz, Eds; *West Virginia University and the National Mine Land Reclamation Center: Morgantown*, WV, 1995, Chapter 19.

Brown, ME. The amelioration of contaminated mine water by wetlands. PhD thesis, *Camborne School of Mines*, University of Exeter, UK, 1997.

Bulusu, S; Aydilek, AH; Rustagi, N. CCB-based encapsulation of pyrite for remediation of acid mine drainage. *J. Hazard. Mater*, 2007, 143(3), 606-619.

Bussière, B. Colloquium 2004: Hydrogeotechnical properties of hard rock tailings from metal mines and emerging geoenvironmental disposal approaches. *Canadian Geotech. J*, 2007, 44(9), 1019-1052.

Canovas, CR; Olias, M; Nieto, JM; Sarmiento, AM; Ceron, JC. Hydrogeochemical characteristics of the Tinto and Odiel Rivers (SW Spain): Factors controlling metal contents. *Sci. Total Environ*, 2007, 373(1), 363-382.

Chakraborty, M; Murthy, Z; Bhattacharya, C; Datta, S. Process intensification: extraction of chromium (VI) by emulsion liquid membrane. *Sep. Sci. Technol.*, 2005, 40, 2353-2364.

Chang, JS; Chen, CC. Quantitative analysis and equilibrium models of selective adsorption in multimetal systems using a bacterial biosorbent. *Sep. Sci. Technol.*, 1998, 33, 611-632.

Charerntanyarak, L. Heavy metals removal by chemical coagulation and precipitation. *Water Sci. Technol.*, 1999, 39(10/11), 135-138.

Chen, F; Finch, JA; Distin, PA; Gomez, CO. Air assisted solvent extraction. *Can. Metall. Q*, 2003, 42, 277-280.

Cheng, S; Grosse, W; Karrenbrock, F; Thoennessen, M. Efficiency of constructed wetlands in decontamination of water polluted by heavy metals. *Ecol. Eng.*, 2002, 18, 317-325.

Choi, H; Zhang, K; Dionysiou, DD; Oerther, DB; Sorial, GA. Effect of permeate flux and tangential flow on membrane fouling for wastewater treatment *Sep. Purif. Technol.*, 2005, 45, 68-78.

Code of Federal Regulations (CFR) 40, 141, 261, 268.40. *U.S. Government Printing Office*, Superintendent of Documents: Washington, DC, 1994.

Cohen, RRH; Staub, MW. *Technical Manual for the Design and Operation of a Passive Mine Drainage System*, prepared for the U.S. Bureau of Reclamation by the Colorado School of Mines: Golden, CO, 1992.

Cohen, RRH. Use of microbes for cost reduction of metal removal from metals and mining industry waste streams. *Journal of Cleaner Production*, 2006, 14, 1146-1157.

Colleran, E; Finnegan, S; Lens, P. Anaerobic treatment of sulfate-containing waste streams. *Antonie van Leeuwenhoek Int. J. Gen. Mol. Microbiol*, 1995, 67, 29-46.

Collins, BS; Sharitz, RR; Coughlin, DP. Elemental composition of native wetland plants in constructed mesocosm treatment wetlands. *Bioresour. Technol.*, 2005, 96(8), 937-948.

Collon, P; Fabriol, R; Bues, M. Modelling the evolution of water quality in abandoned mines of the Lorraine Iron Basin. *J. Hydrol*, 2006, 328(3-4), 620-634.

Cooke, JG. Nutrient transformations in a natural wetland receiving sewage effluent and the implications for waste treatment. *Water Sci. Technol.*, 1994, 29(4), 209-217.

Coulton, R; Bullen, C; Hallett, C. The design and optimisation of active mine water treatment plants. *Land Contamination & Reclamation*, 2003a, 11(2), 273-279.

Coulton, R; Bullen, C; Dolan, J; Hallett, C; Wright, J; Marsden, C. Wheal Jane mine water active treatment plant – design, construction, and operation.*Land Contamination & Reclamation*, 2003b, 11(2), 245-252.

Cravotta, CA; Trahan, MK. Limestone drains to increase pH and remove dissolved metals from acidic mine drainage. *Appl. Geochem*, 1999, 14, 581-606.

Cravotta, CA. Size and performance of anoxic limestone drains to neutralize acidic mine drainage. *J. Environ. Qual*, 2003, 32, 1277-1289.

Dąbrowski, A; Hubicki, Z; Podkościelny, P; Robens, E. Selective removal of the heavy metals from waters and industrial wastewaters by ion-exchange method. *Chemosphere*, 2004, 56(2), 91-106.

Davis, TA; Volesky, B; Mucci, A. A review of the biochemistry of heavy metal biosorption by brown algae. *Water Res.*, 2003, 37(18), 4311-4330.

Demchak, J; Morrow, T; Skousen, J. Treatment of acid mine drainage by four vertical flow wetlands in Pennsylvania. *Geochem. Explor. Environ. Anal*, 2001, 1, 71-80.

Demchak, J; Skousen, J; McDonald, LM. Longevity of acid discharges from underground mines located above the regional water table. *J. Environ. Qual*, 2004, 33, 656-668.

Demin, OA; Dudeney, AWL; Tarasova, II. Remediation of ammonia-rich mine water in constructed wetlands. *Environ. Technol.*, 2002, 23(5), 497-514.

Demirbas, A. Heavy metal adsorption onto agro-based waste materials: A review. *J. Hazard. Mater*, 2008, 157(2-3), 220-229.

Dempsey, BA; Jeon, B. Characteristics of sludge produced from passive treatment of mine drainage. *Geochem: Explor. Environ. Anal*, 2001, 1, 89-94.

Desbarats, AJ; Dirom, GC. Temporal variations in the chemistry of circum-neutral drainage from the 10-level portal, Myra Mine, Vancouver Island, British Columbia. *Appl. Geochem*, 2007, 22(2), 415-435.

Diehl, D; Stumm, W. Is dissolved Mn^{2+} being oxidised by O_2 in absence of Mn-bacteria or surface catalysts? *Geochem. Cosmochim. Acta*, 1984, 48, 1571-1573.

Dolenec, T; Serafimovski, T; Tasev, G; Dobnikar, M; Dolenec, M; Rogan, N. Major and trace elements in paddy soil contaminated by Pb-Zn mining: a case study of Kočani Field, Macedonia. *Environ. Geochem. Health*, 2007, 29(1), 21-32.

Drizo, A. Physico-chemical screening of phosphate-removing substrates for use in constructed wetland systems. *Water Res.*, 1999, 33(17), 3595-3602.

Dvorak, DH; Hedin, RS; Edenborn, HM; McIntyre, PE. Treatment of metal-contaminated water using bacterial sulfate reduction: Results from pilot-scale reactors. *Biotechnol. Bioeng*, 1992, 40(5), 609-616.

Eger, P; Wagner, JR; Kassa, Z; Melchert, GD. Metal removal in wetland treatment systems, In Proceedings of the International Land Reclamation and Mine Drainage Conference and the Third International Conference on the Abatement of Acidic Drainage, *United States Department of the Interior*, Bureau of Mines Special Publication SP 06A-94: Washington, DC, 1994.

Eger, P; Wagner, J. Sulfate reduction: designing systems for long term treatment. *Trans. Soc. Min. Metall. And Expl.*, 2001, 310, 118-124.

Eger, P; Wagner, J. Wetland treatment systems—How long will they really work? In Proceedings of Sudbury 2003, *Mining and Environment, Sudbury*, Ontario, May 25-28, 2003.

Elliott, P; Ragusa, S; Catcheside, D. Growth of sulfate-reducing bacteria under acidic conditions in an upflow anaerobic bioreactor as a treatment system for acid mine drainage. *Water Res.*, 1998, 32(12), 3724-3730.

Ettner, DC. Pilot scale constructed wetland for the removal of nickel from tailings drainage, southern Norway. *In Proceedings of the Congress of the International Mine Water Association, Sevilla*, Spain, September 13-17, 1999.

Eusden, Jr., JD; Gallagher, L; Eighmy, TT; Crannell, BS; Krzanowski, JR; Butler, LG; Cartledge, FK; Emery, EF; Shaw, EL; Francis, CA. Petrographic and spectroscopic characterization of phosphate-stabilized mine tailings from Leadville, Colorado. *Waste Manag*, 2002, 22(2), 117-135.

Evangelou, VP. *Pyrite Oxidation and its Control*, CRC Press: Boca Raton, FL, 1995.

Evangelou, VP. Pyrite chemistry: the key for abatement of acid mine drainage. In *Acidic Mining Lakes: Acid Mine Drainage, Limnology and Reclamation*, A; Geller, H; Klapper, W. Salomons, Eds; Springer: Berlin, 1998, 197-222.

Feng, D; Aldrich, C; Tan, H. Treatment of acid mine water by use of heavy metal precipitation and ion exchange. *Miner. Eng.*, 2000, 13(6), 623-642.

Fyson, A; Kalin, M; Smith, M. Microbially-mediated metal removal from acid mine drainage. In *Environmental Biotechnology: Principles and Applications*, M; Moo-Young, WA; Anderson, AM. Chakrabarty, Eds; Kluwer Academic Publishers: *Dordrecht*, 1995.

Gazea, B; Adam, K; Kontopoulos, A. A review of passive systems for the treatment of acid mine drainage. *Miner. Eng.*, 1996, 9(1), 23-42.

Geremias, R; Laus, R; Macan, JM; Pedrosa, RC; Laranjeira, MCM; Silvan, J; Favere, FV. Use of coal mining waste for the removal of acidity and metal ions Al(III), Fe(III) and Mn(II) in acid mine drainage. *Environ. Technol.*, 2008, 29(8), 863-869.

Gibert, O; de Pablo, J; Cortina, JL; Ayora, C. Chemical characterization of natural organic substrates for biological mitigation of acid mine drainage. *Water Res.*, 2004, 38(19), 4186-4196.

Glover, HG. Mine water pollution – an overview of problems and control strategies in the United Kingdom. *Water Sci. Technol.*, 1983, 15, 59-70.

Gouzinis, A; Kosmidis, N; Vayenas, DV; Lyberatos, G. Removal of Mn and simultaneous removal of NH_3, Fe and Mn from potable water using a trickling filter. *Water Res.*, 1998, 32(8), 2442-2450.

Gray, NF. Environmental impact and remediation of acid mine drainage: a management problem. *Environ. Geol*, 1997, 30, 62-71.

Gray, NF. Acid mine drainage composition and the implications for its impact on lotic systems. *Water Res.*, 1998, 32, 2122-2134.

Gray, NF. *Biology of wastewater treatment*, 2nd ed; Imperial College Press: London, 2004.

Gray, NF; Delaney, E. Comparison of benthic macroinvertebrate indices for the assessment of the impact of acid mine drainage on an Irish river below an abandoned Cu-S mine. *Environ. Pollut*, 2008, 155, 31-40.

Groudev, S; Kontopoulos, A; Spasova, I; Komnitas, K; Angelov, A; Georgiev, P. In situ treatment of groundwater at Burgas Copper Mines, Bulgaria, by enhancing microbial sulphate reduction. In *Groundwater Quality: Remediation and Protection*, Proceedings of the GQ '98 Conference, Tübingen, Germany, September 21-25, 1998; M; Herbert, K. Kovar, Eds; IAHS Publication, No. 250, 1998.

Gusek, JJ; Wildeman, TR. Passive Treatment of Aluminum-Bearing Acid Rock Drainage, *In Proceedings of the 23rd Annual West Virginia Surface Mine Drainage Task Force Symposium*, April 16-17, 2002, Morgantown, West Virginia.

Gusek, JJ. Sulfate-reducing bioreactor design and operating issues: is this the passive treatment technology for your mine drainage? *In Proceedings of the 2002 National Association of Abandoned Mine Land Programs*: Reclamation 2002, September 15-18, 2002, Park City, UT.

Gusek, JJ. Scaling up design challenges for large scale sulfate reducing bioreactors, *In Proceedings of the 2004 National Meeting of the American Society of Mining and Reclamation and the 25th West Virginia Surface Mine Drainage Task Force*, April 18-24, 2004, Morgantown, WV.

Gusek, JJ; Rutkowski, T; Blumenstein, E; Shipley, B. Two-year sulfate reducing bioreactor pilot test results at the Golinski Mine, California. *In Proceedings of the 25th Annual Meeting of the American Society of Mining and Reclamation*, June 14-19, 2008, Richmond, VA.

Haferburg, G; Merten, D; Büchel, G; Kothe, E. Biosorption of metal and salt tolerant microbial isolates from a forming uranium mining area – Their impact on changes in rare earth element patterns in acid mine drainage. *J. Basic Microbiol*, 2007a, 47(6), 474-484.

Haferburg, G; Reinicke, M; Merten, D; Büchel, G; Kothe, E. Microbes adapted to acid mine drainage as source for strains active in retention of aluminum or uranium. *J. Geochem. Explor*, 2007b, 92(2-3), 196-204.

Hall, GH; Puhlmann, T. Spatial distribution of iron oxidation in the aerobic cells of the Wheal Jane Pilot Passive Treatment Plant. *Sci. Tot. Environ*, 2005, 338, 73-80.

Hallberg, KB; Johnson, DB. Biological manganese removal from acid mine drainage in constructed wetlands and prototype bioreactors. *Sci. Total Environ*, 2005a, 338, 115-124.

Hallberg, KB; Johnson, DB. Microbiology of a wetland ecosystem constructed to remediate mine drainage from a heavy metal mine. *Sci. Total Environ.*, 2005b 338, 53-66.

Hammarstrom, JM; Sibrell, PL; Belkin, HE. Characterization of limestone reacted with acid-mine drainage in a pulsed limestone bed treatment system at the Friendship Hill National Historical Site, Pennsylvania, USA. *Appl. Geochem*, 2003, 18, 1705-1721.

Hao, OJ; Huang, L; Chen, JM; Buglass, RL. Effects of metal additions on sulfate reduction activity in wastewaters. *Toxicol. Environ. Chem.*, 1994, 46, 197-212.

Hao, OJ., Chen, JM., Huang, L., Buglass, RL. Sulfate-reducing bacteria. *Crit. Rev. Environ. Sci. Technol.*, 1996, 26, 155-187.

Heal, KV; Salt, CA. Treatment of acidic metal-rich drainage from reclaimed ironstone mine spoil. *Water Sci. Technol.*, 1999, 39(12), 141-148.

Hedin, RS; Nairn, RW. Contaminant removal capabilities of wetlands constructed to treat coal mine drainage. In *Constructed Wetlands for Water Quality Improvement*; GA. Moshiri, Ed; Lewis Publishers: Boca Raton; 1993, 187-195.

Hedin, RS; Nairn, RW; Kleinmann, RLP. *Passive treatment of coal mine drainage.* U.S. Bureau of Mines Information Circular 9389: Washington, DC, 1994.

Hedin, RS; Watzlaf, GR. The effects of anoxic limestone drains on mine water chemistry, *In Proceedings of the International Land Reclamation and Mine Drainage Conference and the Third International Conference on the Abatement of Acidic Drainage, United States Department of the Interior*, Bureau of Mines Special Publication SP 06A-94: Washington, DC, 1994.

Hedin, RS. Environmental engineering forum: long-term effects of wetland treatment of mine drainage. *J. Environ. Eng.*, 1996, 122(1), 83-86.

Hellier, WW; Giovannitti, EF; Slack, PT. Best professional judgment analysis for constructed wetlands as a best available technology for the treatment of post-mining groundwater seeps, *In Proceedings of the International Land Reclamation and Mine Drainage Conference and the Third International Conference on the Abatement of Acidic Drainage*, United States Department of the Interior, Bureau of Mines Special Publication SP 06A-94: Washington, DC, 1994.

Henke, KR. Chemistry of heavy metal precipitates resulting from reactions with Thio-Red®. *Water Environ. Res.*, 1998, 70(6), 1178-1185.

Henke, KR; Robertson, D; Krepps, M; Atwood, DA. Chemistry and stability of precipitates from aqueous solutions of 2,4,6-trimercaptotriazine, trisodium salt, nonahydrate (TMT-55) and mercury (II) chloride. *Water Res.*, 2000, 34 (11), 3005-3013.

Henrot, J; Weider, RK. Processes of iron and manganese retention in laboratory peat microcosms subjected to acid mine drainage. *J. Env. Qual*, 1990, 19, 312-320.

Herbert Jr., RB; Benner, SG; Blowes, DW. Solid phase iron-sulfur geochemistry of a reactive barrier for treatment of mine drainage. *Appl. Geochem*, 2000, 15, 1331-1343.

Herrera, PS; Uchiyama, H; Igarashi, T; Asakura, K; Ochi, Y; Ishizuka, F; Kawada, S. Acid mine drainage treatment through a two-step neutralization ferrite-formation process in northern Japan: Physical and chemical characterization of the sludge. *Miner. Eng.*, 2007, 20, 1309-1314

Hetland, MD; Gallagher, JR; Daly, DJ; Hassett, DJ; Heebink, LV. Processing of plants used to phytoremediate lead-contaminated sites. In *Phytoremediation, Wetlands, and Sediments*, Proceedings of the Sixth International In Situ and On-Site Bioremediation Symposium, San Diego, CA, June 4-7, 2001, A; Leeson, EA; Foote, MK; Banks, VS. Magar, Eds; Battelle Press: Columbus, Richland, 2001.

Hoover, KL; Rightnour, TA. Design approaches for passive treatment of coal combustion byproduct leachate – project experience within the utility industry. In *Environmental Challenges and Greenhouse Gas Control for Fossil Fuel Utilization in the 21st Century*, MM; Maroto-Valer C; Song, Y. Soong, Eds; Kluwer Academic/ Plenum Publishers: New York, 2002, 417-429.

Huisman, JL; Schouten, G; Schultz, C. Biologically produced sulphide for purification of process streams, effluent treatment and recovery of metals in the metal and mining industry. *Hydrometallurgy*, 2006, 83, 106-113.

Igarashi, T; Asakura, K; Yoshida, T; Miyamae, H; Iyatomi, N; Hashimoto, K. Ferrite formation using precipitate in the treatment of acid mine drainage for reducing its volume. *In Proceedings of the 5th International Congress on Environmental Geotechnics*, Cardiff, Wales, UK, 2006.

Jacob, DL; Otte, ML. Conflicting processes in the wetland plant rhizosphere: Metal retention or mobilization? *Water Air Soil Pollut. Focus*, 2003, 3(1), 91-104.

Jarvis, AP; Younger, PL. Design, construction, and performance of a full-scale compost wetland for mine-spoil drainage treatment at Quaking Houses, UK. *Water Environ. J*, 1999, 13(5), 313-318.

Jarvis, AP; Younger, PL. Passive treatment of ferruginous mine waters using high surface area media. *Water Res.*, 2001, 35(15), 3643-3648.

Johnson, DB; Hallberg, KB. Acid mine drainage remediation options: a review. *Sci. Total Environ.*, 2005, 338, 3-14.

Johnson, DB; Sen, AM; Kimura, S; Rowe, OF; Hallberg, KH. Novel biosulfidogenic system for the selective recovery of metals from acidic leach liquors and waste streams. *Mineral Process. Extr. Metall*, 2006, 115, 19-24.

Johnson, KL; Younger, PL. Rapid manganese removal from mine waters using an aerated packed-bed bioreactor. *J. Environ. Qual*, 2005, 34(3), 987-993.

Johnson, KL; Younger, PL. The co-treatment of sewage and mine waters in aerobic wetlands. *Eng. Geol*, 2006, 85, 53-61.

Jones, PM; Mulvay, SM; Fish, D. The role of sulphate and ionic strength on the shift from acid to alkaline mine drainage in southwest Pennsylvania, In *Proceedings of the International Land Reclamation and Mine Drainage Conference and the Third International Conference on the Abatement of Acidic Drainage, United States Department of the Interior*, Bureau of Mines Special Publication SP 06A-94: Washington, DC, 1994.

Juang, RS; Shiau, RC. Metal removal from aqueous solutions using chitosan-enhanced membrane filtration, *J. Membr. Sci.*, 2000, 165, 159-167.

Kaewsarn, P. Biosorption of copper(II) from aqueous solutions by pre-treated biomass of marine algae *Padina* sp. *Chemosphere*, 2002, 47, 1081-1085.

Kaksonen, AH; Puhakka, JA. Sulfate reduction based bioprocesses for the treatment of acid mine drainage and the recovery of metals. *Eng. Life Sci.*, 2007, 7(6), 541-564.

Kalin, M; Cairns, J; McCready, R. Ecological engineering methods for acid-mine drainage treatment of coal wastes. *Resour. Conserv. Recycl*, 1991, 5, 265-275.

Kalin, M. Treatment of acidic seepages using wetland ecology and microbiology: Overall program assessment. MEND project 3.11.1, DSS File #015SQ.23440-2-9217, Contract serial number 23440-2-9217/01-SQ, *Canmet*, Canada, 1993.

Kalin, M; Smith, MP. Microbial acid reduction in sediments—concepts and application. *In Proceedings of the 4th International Conference on Acid Rock Drainage*, May 31–June 6, 1997, Vancouver, BC, Canada.

Kalin, M. Biological polishing of zinc in a mine waste management area. In *Acidic Mining Lakes: acid mine drainage,limnology, and reclamation*, W; Geller, H; Klapper, W. Salomons, Eds; Springer-Verlag: Berlin, 1998, 321-324.

Kalin, M; Fyson, A; Smith, MP. Maintenance-free microbial treatment of AMD: four years of performance. *In Proceedings of the Sudbury '99 Mining and the Environment Conference*, Sudbury, Ontario, September 13-15, 1999.

Kalin, M; Caetano Chaves, WL. Acid reduction using microbiology: treating AMD effluent emerging from an abandoned mine portal. *Hydrometallurgy*, 2003, 71, 217-225.

Kalin, M. Passive mine water treatment: the correct approach? *Ecol. Eng.*, 2004, 22, 299-304.

Kalin, M; Fyson, A; Wheeler, W. The chemistry of conventional and alternative treatment systems for the neutralization of acid mine drainage. *Sci. Total Environ.*, 2006, 366, 395-408.

Karate, VD; Marathe, KV. Simultaneous removal of nickel and cobalt from aqueous stream by cross flow micellar enhanced ultrafiltration. *J. Hazard. Mater*, 2008, 157, 464-471.

Kargari, A; Kaghazchi, T; Soleimani, M. Role of emulsifier in the extraction of gold (III) ions from aqueous solutions using the emulsion liquid membrane technique. *Desalination*, 2004, 162, 237-247.

Kasan, HC. The role of waste activated sludge and bacteria in metal-ion removal from solution. *Crit. Rev. Env. Sci. Technol.*, 1993, 23, 79-117.

Kelly, MG. *Mining and the freshwater environment*. Elsevier Applied Science: London, 1988.

Kepler, DA; McCleary, EC. Successive alkalinity-producing systems (SAPS) for the treatment of acidic mine drainage. *In Proceedings of the International Land Reclamation and Mine Drainage Conference and the Third International Conference on the Abatement of Acidic Drainage, United States Department of the Interior*, Bureau of Mines Special Publication SP 06A-94: Washington, DC, 1994.

Kepler, DA; McCleary, EC. Passive aluminum treatment successes. *In Proceedings of the 18th Annual West Virginia Surface Mine Drainage Task Force Symposium*, April 15-16, 1997, Morgantown, WV.

Kerndorf, H; Schnitzer, M. Sorption of metals onto humic acid. *Geochim. Cosmochim. Acta*, 1980, 44, 1701-1708.

Kittle, DL; McGraw, JB; Garbutt, K. Plant litter decomposition in wetlands receiving acid mine drainage. *J. Environ. Qual*, 1995, 24, 301-306.

Kleinmann, RLP; Crerar, D. *Thiobacillus ferrooxidans* and the formation of acidity in simulated coal mine environments. *Geomicrobiol. J*, 1979, 1, 373-388.

Kleinmann, RLP; Crerar, P; Pacelli, R. Biogeochemistry of acid mine drainage and a method to control acid formation. *Min. Eng.*, 1981, 33, 300-304.

Kleinmann, RLP. Acid mine drainage. *Eng. Min. J*, 1989, 190, 16I-16N.

Kleinmann, RLP; Hedin, RS; Nairn, RW. Treatment of mine drainage by anoxic limestone drains and constructed wetlands. In: A; Geller, H; Klapper, W. Salomons. Eds; *Acidic Mining Lakes: Acid Mine Drainage, Limnology, and Reclamation*. Springer: Berlin, 1998, 303-319.

Kosolapov, DB; Kuschk, P; Vainshtein, MB; Vatsourina, AV; Wiebner, A; Kastner, M; Muller, RA. Microbial processes of heavy metal removal from carbon-deficient effluents in constructed wetlands. *Eng. Life Sci.*, 2004, 4(5), 403-411.

Kostenbader, PD; Haines, GF. High Density Sludge treats acid mine drainage. *Coal Age*, 1970, (September), 90-97.

Kratochvil, D; Volesky, B. Biosorption of Cu from ferruginous wastewater by algal biomass. *Water Res.*, 1998, 32(9), 2760-2768.

Kumbasar, R; Tutkun, O. Separation and concentration of gallium from acidic leach solutions containing various metal ions by emulsion type of liquid membranes using TOPO as mobile carrier. *Hydrometallurgy*, 2004, 75, 111-121.

Kumpulainen, S; Carlson, L; Raisanen, ML. Seasonal variations of ochreous precipitates in mine effluents in Finland. *Appl. Geochem*, 2007, 22(4), 760-777.

Kurniawan, TA; Chan, GYS; Lo, WH; Babel, S. Physico-chemical treatment techniques for wastewater laden with heavy metals. *Chem. Eng. J*, 2006, 118, 83-98.

Kuyucak, N. Mining, the environment and the treatment of mine effluents. *Int. J. Environ. Pollut*, 1998, 10(2), 315-325.

Laîné, JM; Vial, D; Moulart, P. Status after 10 years of operation – overview of UF technology today. *Desalination*, 2000, 131, 17-25.

Laine, SM. The treatment of pumped mine water at Woolley Colliery, West Yorkshire. *J. Chart. Inst. Water Environ. Manag*, 1999, 13, 127-130.

Lange, K; Rowe, RK; Jamieson, H. Metal retention in geosynthetic clay liners following permeation by different mining solutions. *Geosynth. Int.*, 2007, 14(3), 178-187.

Lazaridis, NK; Matis, KA; Webb, M. Flotation of metal-loaded clay anion exchangers. Part I: the case of chromates. *Chemosphere*, 2001, 42, 373-378.

Lee, G; Bigham, JM; Faure, G. Removal of trace metals by coprecipitation with Fe, Al, and Mn from natural waters contaminated with acid mine drainage in the Ducktown Mining District, Tennessee. *Appl. Geochem*, 2002, 17(5), 569-581.

Lee, G; Faure, G. Processes controlling trace-metal transport in surface water contaminated by acid-mine drainage in the Ducktown Mining District, Tennessee. *Water Air Soil Pollut*, 2007, 186(1-4), 221-232.

Lee, MK; Saunders, JA. Effects of pH on metals precipitation and sorption: field Bioremediation and geochemical modelling approaches. *Vadose Zone J*, 2003, 2, 177-185.

Li, X; Tang, Y; Cao, X; Lu, D; Luo, F; Shao, W. Preparation and evaluation of orange peel cellulose adsorbents for effective removal of cadmium, zinc, cobalt, and nickel. *Colloids Surf. A*, 2008, 317(1-3), 512-521.

Liang, HC; Thomson, BM. Minerals and mine drainage. *Water Environ. Res.*, 2008, 80(10), 1481-1509.

Liao, B; Huang, LN; Ye, ZH; Lan, CY; Shu, WS. Cut-off net acid generation pH in predicting acid-forming potential in mine spoils. *J. Environ. Qual*, 2007, 36(3), 887-891.

Logan, MV; Ahmann, D; Figueroa, L. Assessment of microbial activity in anaerobic columns treating synthetic mine drainage. *In Proceedings of the 2003 National Meeting of the American Society of Mining and Reclamation and the 9th Billings Land Reclamation Symposium*, Billings, MT, June 3-6, 2003.

Logan, MV; Reardon, KF; Figueroa, LA; McLain, JET; Ahmann, DM. Microbial community activities during establishment, performance, and decline of bench-scale passive treatment systems for mine drainage. *Water Res.*, 2005, 39(18), 4537-4551.

Lottermoser, BG. *Mine Wastes: Characterization, Treatment, and Environmental Impacts*, 2nd ed; Springer: Berlin, 2007.

Ludwig, RD; McGregor, RG; Blowes, DW; Benner, SG; Mountjoy, K. A permeable reactive barrier for treatment of heavy metals. *Ground Water*, 2002, 40(1), 59-66.

Luptakova, A; Kusnierova, M. Bioremediation of acid mine drainage contaminated by SRB. *Hydrometallurgy*, 2005, 77, 97-102.

Macek, T; Macková, M; Káš, J. Exploitation of plants for the removal of organics in environmental remediation. *Biotechnology Advances*, 2000, 18, 23-24.

Malmström, ME; Berglund, S; Jarsjo, J. Combined effects of spatially variable flow and mineralogy on the attenuation of acid mine drainage groundwater. *Appl. Geochem*, 2008, 23(6), 1419-1436.

Manyin, T; Williams, FM; Stark, LR. Effects of iron concentration and flow rate on coal mine drainage in wetland mesocosms: an experimental approach to sizing of constructed wetlands. *Ecol. Eng.*, 1997, 9, 171-185.

Mapanda, F; Nyamadzawo, G; Nyamangara, J; Wuta, M. Effects of discharging acid-mine drainage into evaporation ponds lined with clay on chemical quality of the surrounding soil and water. *Phys. Chem. Earth*, 2007, 32(15-18), 1366-1375.

Mariner, R; Johnson, DB; Hallberg, KB. Characterisation of an attenuation system for the remediation of Mn(II) contaminated waters. *Hydrometallurgy*, 2008, 94, 100-104.

Matis, KA; Zouboulis, AI; Lazaridis, NK; Hancock, IC. Sorptive flotation for metal ions recovery. *Int. J. Miner. Process*, 2003, 70, 99-108.

Matlock, MM; Henke, KR; Atwood, DA; Robertson, JD. Aqueous leaching properties and environmental implications of cadmium, lead, and zinc trimercaptotriazine (TMT) compounds. *Water Resour*, 2001, 35(15), 3649-3655.

Matlock, MM; Henke, KR; Atwood, DA. Effectiveness of commercial reagents for heavy metal removal from water with new insights for future chelate designs. *J. Hazard. Mater*, 2002, B92, 129-142.

Mavrov, V; Erwe, T; Blöcher, C; Chmiel, H. Study of new integrated processes combining adsorption, membrane separation and flotation for heavy metal removal from wastewater. *Desalination*, 2003, 157, 97-104.

Mayes, WM; Batty, LC; Younger, PL; Jarvis, AP; Kõiv, M; Vohla, C; Mander, U. Wetland treatment at extremes of pH: A review. *Sci. Total Environ*, 2008 doi:10.1016/j.scitotenv.2008.06.045.

Mays, PA; Edwards, GS. Comparison of heavy metal accumulation in a natural wetland and constructed wetlands receiving acid mine drainage. *Ecol. Eng.*, 2001, 16, 487-500.

McCauley, CA; O'Sullivan, AD; Milke, MW; Weber, PA; Trumm, DA. Sulfate and metal removal in bioreactors treating acid mine drainage dominated with iron and aluminum. *Water Res.*, 2009 doi:10.1016/j.watres.2008.11.029.

Meek, FA. Evaluation of acid prevention techniques used in surface mining. In *Acid Mine Drainage Control and Treatment*, JG; Skousen, PF. Ziemkiewicz, Eds., West Virginia University and the National Mine Land Reclamation Center: Morgantown, WV, 1996, Chapter 11.

Mendez-Ortiz, B; Carrillo-Chavez, A; Monroy-Fernandez, MG. Acid rock drainage and metal leaching from mine waste material (tailings) of a Pb-Zn-Ag skarn deposit: environmental assessment through static and kinetic laboratory tests. *Rev. Mex. Cienc. Geol.*, 2007, 24(2), 161-169.

Mohan, D; Chander, S. Removal and recovery of metal ions from acid mine drainage using lignite – a low cost sorbent. *J. Hazard. Mater*, 2006, 137(3), 1545-1553.

Munk, L; Faure, G; Pride, DE; Bigham, JM. Sorption of trace metals to an aluminum precipitate in a stream receiving acid rock-drainage; Snake River, Summit County, Colorado. *Appl. Geochem*, 2002, 17, 421-430.

Murdock, DJ; Fox, JRW; Bensley, JG. Treatment of acid mine drainage by the high density sludge process, In *Proceedings of the International Land Reclamation and Mine Drainage Conference and the Third International Conference on the Abatement of Acidic Drainage*, United States Department of the Interior, Bureau of Mines Special Publication SP 06A-94: Washington, DC, 1994.

Nairn, RW; Hedin, RS; Watzlaf, GR. A preliminary review of the use of anoxic limestone drains in the passive treatment of acid mine drainage. In *Proceedings of the 12th West Virginia Surface Mine Drainage Task Force* Symposium, Morgantown, WV, April 3-4, 1991.

Nairn, RW; Hedin, RS; Watzlaf, GR. Generation of alkalinity in an anoxic limestone drain. In *Proceedings of the 9th Annual National Meeting of the American Society for Surface Mining and Reclamation*, Duluth, MN, June 14-18, 1992.

Nealson, KH. The microbial manganese cycle. In *Microbial Geochemistry*; WE. Krumbein, Ed; Blackwell Science Publishing: Oxford, 1983, 191-221.

Neculita, CM; Zagury, GJ; Bussiere, B. Passive treatment of acid mine drainage in bioreactors using sulfate-reducing bacteria: critical review and research needs. *J. Environ. Qual*, 2007, 36, 1-16.

Nehdi, M; Tariq, A. Stabilization of sulphidic mine tailings for prevention of metal release and acid drainage using cementitious materials: a review. *J. Environ. Eng. Sci.*, 2007, 6(4), 423-436.

Nelson, PO; Chung, AK; Hudson, MC. Factors affecting the fate of heavy metals in the activated sludge process. *J. Water Pollution Control Federation*, 1981, 53(8), 1323-1333.

Nordstrom, DK. Aqueous pyrite oxidation and the consequent formation of secondary iron minerals. In *Acid Sulphate Weathering*; JS; Kittrick, DS; Fanning, LR. Hosser, Eds; Special Publication, Vol. 10, Soil Science of America: Madison, 1982, 37-56.

Nordstrom, DK; Alpers, CN. Negative pH, efflorescent mineralogy, and consequences for environmental restoration at the Iron Mountain Superfund site, California. *In Proceedings of the National Academy of Sciences Colloquium*, USA, 1999, 96, 3455-3462.

Nordstrom, DK; Alpers, CN; Ptacek, CJ; Blowes, DW. Negative pH and extremely acidic mine waters from Iron Mountain, California. *Environ. Sci. Technol.*, 2000, 34, 254-258.

Nordwick, S; Bless, D. *In-situ source control of acid generation using sulfate-reducing bacteria*. Mine Waste Technology Program Activity III, Project 3, Contract No. DE-AC09-96EW96405. EPA/600/R-08/096; National Risk Management Research Laboratory, Office of Research and Development, US EPA: Cincinnati, OH, 2008.

Nuttall, CA; Younger, PL. Zinc removal from hard, circum-neutral mine waters using a novel, closed-bed limestone reactor. *Water Res.*, 2000, 34(4), 1262-1268.

Nuttall, CA. Testing and performance of a newly constructed full-scale passive treatment system at Whittle Colliery, Northumberland. *Land Contam. Reclam.*, 2003, 11, 105-112.

Odom, JM; Singleton, R. *The sulphate reducing bacteria. Contemporary perspectives*, Springer Verlag: New York, 1993.

O'Flaherty, V; Mahony, T; O'Kennedy, R; Colleran, E. Effect of pH on growth kinetics and sulphide toxicity thresholds of a range of methanogenic, syntrophic and sulphate-reducing bacteria. *Process Biochem*, 1998, 33, 555-569.

Otte, ML. What is stress to a wetland plant? *Environ. Exp. Bot*, 2001, 46, 195-202.

Ozaki, H; Sharma, K; Saktaywin, W. Performance of an ultra-low-pressure reverse osmosis membrane (ULPROM) for separating heavy metal: effects of interference parameters, *Desalination*, 2002, 144, 287-294.

Pagnanelli, F; Luigi, M; Mainelli, S; Toro, L. Use of natural materials for the inhibition of iron oxidizing bacteria involved in the generation of acid mine drainage. *Hydrometallurgy*, 2007, 87(1-2), 27-35.

Papadopoulos, A; Fatta, D; Parperis, K; Mentzis, A; Harambous, KJ; Loizidou, M. Nickel uptake from a wastewater stream produced in a metal finishing industry by combination of ion-exchange and precipitation methods. *Sep. Purif. Technol.*, 2004, 39(3), 181-188.

Pavlović, J; Stopić, S; Friedrich, B; Kamberović, Z. Selective removal of heavy metals from metal-bearing wastewater in a cascade line reactor. *Environ. Sci. Pollut. Res.*, 2007, 14(7), 518-522.

Pearson, FH; McDonnell, AJ. Use of crushed limestone to neutralize acid wastes. *J. Env. Eng. Div. ASCE*, 1975, 101, 139-158.

Pinetown, KL; Ward, CR; van der Westhuizen, WA. Quantitative evaluation of minerals in coal deposits in the Witbank and Highveld coalfields, and the potential impact on acid mine drainage. *Int. J. Coal Geol.*, 2007, 70(1-3), 166-183.

PIRAMID Consortium. *Engineering guidelines for the passive remediation of acidic and/or metalliferous mine drainage and similar wastewaters*. European Commission 5th Framework RTD Project No. EVK1-CT-1999-000021 "Passive in-situ remediation of acidic mine/ industrial drainage" (PIRAMID); University of Newcastle upon Tyne: Newcastle upon Tyne, UK, 2003.

Postgate, JR. *The Sulfate-Reducing Bacteria*, 2nd ed., Cambridge University Press: New York, 1984.

Pruden, A; Hong, HH; Inman, LY; Logan, MV; Sans, C; Ahmann, D; Figueroa, LA; Reardon, KF. Microbial characterization of sulfate-reducing columns remediating acid mine drainage. *In Proceedings of the 2005 National Meeting of the American Society of Mining and Reclamation, Breckenridge*, CO, June 19-23, 2005.

Pruden, A; Pereyra, LP; Hiibel, SR; Inman, LY; Kashani, N; Reardon, KF; Reisman, D. Microbiology of sulphate-reducing passive treatment systems. *In Proceedings of the 7th International Conference on Acid Rock Drainage (ICARD)*, St. Louis, MO, March 26-30, 2006.

Qdais, HA; Moussa, H. Removal of heavy metals from wastewater by membrane processes: a comparative study. *Desalination*, 2004, 164, 105-110.

Qin, JJ; Wai, MN; Oo, MH; Wong, FS. A feasibility study on the treatment and recycling of a wastewater from metal plating. *J. Membr. Sci.*, 2002, 208, 213-221.

Rai, PK. Heavy metal pollution in aquatic ecosystems and its phytoremediation using wetland plants: an ecosustainable approach. *International Journal of Phytoremediation*, 2008, 10(2), 133-160.

Ramsay, IJ. South African patent application number 98/4724, 1998.

Rao, SR; Finch, JA; Kuyucak, N. Ferrous-ferric oxidation in acidic mineral processing effluents – comparison of methods. *Miner. Eng.*, 1995, 8(8), 905-911.

Reisman, DJ; Sundaram, V; Al-Abed, SR; Allen, D. Statistical validation of sulphate quantification methods used for analysis of acid mine drainage. *Talanta*, 2007, 71(1), 303-311.

Richardson, JP; Nicklow, JW. In situ permeable reactive barriers for groundwater contamination. *Soil Sediment Contam. Int. J*, 2002, 11, 241-268.

Riekkola-Vanhanen, M. In situ bioreclamation of acid mine drainage, *In Proceedings of the 4th Finnish Conference of Environmental Sciences*, Tampere, Finland, May 21-22, 1999.

Rios, CA; Williams, CD; Roberts, CL. Removal of heavy metals from acid mine drainage (AMD) using coal fly ash, natural clinker and synthetic zeolites. *J. Hazard. Mater*, 2008, 156(1-3), 23-35.

Romero, FM; Armienta, MA; Gonzalez-Hernandez, G. Solid-phase control on the mobility of potentially toxic elements in an abandoned lead/zinc mine tailings impoundment. *Appl. Geochem*, 2007, 22(1), 109-127.

Ross, SM. Overview of the hydrochemistry and solute processes in British wetlands. In *Hydrology and Hydrochemistry of British Wetlands*, JMR; Hughes, AL. Heathwaite, Eds; John Wiley: Chichester, UK, 1996, 133-181.

Rowley, M; Warkentin, DD; Sicotte, V. Site demonstration of the biosulfide process at the former Britannia mine. *In Proceedings of the Fourth International Conference on Acid Rock Drainage*, May 30–June 6, 1997, Vancouver, BC, vol. IV, American Society of Surface Mining and Reclamation, 4, 1997.

Rubio, J; Tessele, F. Removal of heavy metal ions by adsorptive particulate flotation, *Min. Eng.*, 1997, 10(7), 671-679.

Saffaj, N; Loukil, H; Younssi, SA; Albizane, A; Bouhria, M; Persin, M; Larbot, A. Filtration of solution containing heavy metals and dyes by means of ultrafiltration membranes deposited on support made of Moroccan clay. *Desalination*, 2004, 168, 301-306.

Salomons, W. Environmental impact of metals derived from mining activities: Processes, predictions, prevention. *J. Geochem. Explor.*, 1995, 52, 5-23.

Sand, W; Jozsa, PG; Kovacs, ZM; Sasaran, N; Schippers, A. Long-term evaluation of acid rock drainage mitigation measures in large lysimeters. *J. Geochem. Explor.*, 2007, 92(2-3), 205-211.

Santomartino, S; Webb, JA. Estimating the longevity of limestone drains in treating acid mine drainage containing high concentrations of iron. *Appl. Geochem*, 2007, 22(11), 2344-2361.

Sapsford, DJ; Barnes, A; Dey, M; Williams, KP; Jarvis, A; Younger, PL. Low footprint mine water treatment: field demonstration and application. *Mine Water Environ.*, 2008, 26, 243-250.

Sapsford, DJ; Williams, KP. Sizing criteria for a low footprint passive mine water treatment system. *Water Res.*, 2009, 43, 423-432.

Saunders, JA; Lee, MK; Shamsudduha, M; Dhakal, P; Uddin, A; Chowdury, MT; Ahmed, K.M. Geochemistry and mineralogy of arsenic in (natural) anaerobic groundwaters. *Appl. Geochem*, 2008, 23(11), 3205-3214.

Scheeren, PJH; Koch, RO; Buisman, CJN. Geohydrological containment system and microbial water treatment plant for metal-contaminated groundwater at Budelco, *In Proceedings of the International Symposium – World Zinc '93, Hobart*, Tasmania, Australia, October 10-13, 1993, GM. Ian, Ed; Australian Institute of Mining and Metallurgy, Parkville, Victoria, Australia, 1993.

Schiewer, S; Volesky, B. Modeling of the proton-metal ion exchange in biosorption. *Environ. Sci. Technol.*, 1995, 29(12), 3049-3058.

Seki, H; Suzuki, A. Biosorption of heavy metal ions to brown algae, Macrocystis pyrifera, Kjellmaniella crassiforia, and Undaria pinnatifida. *Journal of Colloid and Interface Science*. 1998, 206(1), 297-301.

Sheoran, AS; Sheoran, V. Heavy metal removal mechanism of acid mine drainage in wetlands: A critical review. *Miner. Eng.*, 2006, 19, 105-116.

Sibrell, PL; Watten, BJ; Friedrich, AE; Vinci, BJ. ARD remediation with limestone in a CO_2 pressurized reactor. In Proceedings of the 5th International Conference on Acid Rock Drainage, Denver, CO, May 21-24, 2000, *Society for Mining, Metallurgy, and Exploration, Inc.*, 2000.

Sibrell, PL; Watten, BJ. Evaluation of sludge produced by limestone neutralization of AMD at the Friendship Hill National Historic Site. *In 2003 National Meeting of the American Society of Mining and Reclamation and the 9th Billings Land Reclamation Symposium*, Billings, MT, June 3-6, 2003; ASMR: Lexington, KY.

Sibrell, PL; Chambers, MA; Deaguero, AL; Wildeman, TR; Reisman, DJ. An innovative carbonate coprecipitation process for the removal of zinc and manganese from mining impacted waters. *Environ. Eng. Sci.*, 2007, 24(7), 881-895.

Sierra-Alvarez, R; Hollingsworth, J; Zhou, MS. Removal of copper in an integrated sulfate reducing bioreactor-crystallization reactor system. *Environ. Sci. Technol.*, 2007, 41(4), 1426-1431.

Simmons, J; Ziemkiewicz, P; Black, DC. Use of steel slag leach beds for the treatment of acid mine drainage. *Mine Water and the Environment*, 2002, 21, 91-99.

Singer, PC; Stumm, W. Acidic mine drainage: the rate-determining step. *Science*, 1970, 167, 1121-1123.

Skousen, J; Politan, K; Hilton, T; Meek, A. Acid mine drainage treatment systems: chemicals and costs. In *Acid mine drainage control and treatment*; JG; Skousen, PF. Ziemkiewicz, Eds; *West Virginia University and the National Mine Land Reclamation Center*, Morgantown, WV, 1995, Chapter 14.

Skousen, J; Simmons, J; McDonald, LM; Ziemkiewicz, P. Acid-base accounting to predict post-mining drainage quality on surface mines. *J. Environ. Qual*, 2002, 31(6), 2034-2044.

Smith, RM; Sobek, AA; Arkle, T; Sencindiver, JC; Freeman, JR. *Extensive overburden potentials for soil and water quality*. EPA-600/2-76-184; US EPA: Cincinnati, OH, 1976.

Smuda, J; Dold, B; Friese, K; Morgenstern, P; Glaesser, W. Mineralogical and geochemical study of element mobility at the sulfide-rich Excelsior waste rock dump from the polymetallic Zn-Pb-(Ag-Bi-Cu) deposit, Cerro de Pasco, Peru. *J. Geochem. Explor*, 2007, 92, 97-110.

Snoeyink, VL; Jenkins, D. *Water Chemistry*, John Wiley and Sons: New York, 1980.

Solari, P; Zouboulis, AI; Matis, KA; Stalidis, GA. Removal of toxic metals by biosorption onto nonliving sewage biomass. *Sep. Sci. Technol.*, 1996, 31, 1075-1092.

Søndergaard, J; Elberling, B; Asmund, G; Gudum, C; Iversen, KM. Temporal trends of dissolved weathering products released from a high Arctic coal mine waste rock pile in Svalbard (78 degrees N). *Appl. Geochem.*, 2007, 22(5), 1025-1038.

Spinti, M; Zhuang, HN; Trujillo, EM. Evaluation of immobilized biomass beads for removing heavy metals from wastewaters. *Water Environ. Res.*, 1995, 67(6), 943-952.

Stark, PC; Rayson, GD; Darnall, DW. Survey of non-viable biosorbents for toxic metal removal. *Adv. Environ. Res.*, 1999, 3, 74-82.

Starr, RC; Cherry, JA. In situ remediation of contaminated ground water: the funnel-and-gate system. *Ground Water*, 1994, 32(3), 465-476.

Stoveland, S; Lester, JN. A study of the factors which influence metal removal in the activated sludge process. *Sci. Total Environ.*, 1980, 16, 37-54.

Stumm, W; Morgan, JJ. *Aquatic Chemistry: An Introduction Emphasizing Chemical Equilibria in Natural Waters*, 2nd ed; Wiley: New York, 1981.

Subudhi, E; Kar, RN. Decontamination of metals from metallurgical effluent utilising *Rhizopus arrhizus* biomass. *Int. J. Environ. Stud.*, 1996, 50, 111-116.

Sud, D; Mahajan, G; Kaur, MP. Agricultural waste material as potential adsorbent for sequestering heavy metal ions from aqueous solutions – A review. *Bioresour. Technol.*, 2008, 99(14), 6017-6027.

Sung, W; Morgan, JJ. Kinetics and product of ferrous iron oxygenation in aqueous systems. *Environ. Sci. Technol.*, 1980, 14, 561-568.

Tabak, HH; Scharp, R; Burckle, J; Kawahara, FK; Govind, R. Advances in biotreatment of acid mine drainage and biorecovery of metals: 1. Metal precipitation for recovery and recycle. *Biodegradation*, 2003, 14, 423-436.

Tang, XW; Li, ZZ; Chen, YM. Behaviour and mechanism of Zn(II) adsorption on Chinese loess at dilute slurry concentrations. *J. Chem. Technol. Biotechnol*, 2008, 83(5), 673-682.

Tarkan, HM; Finch, JA. Air-assisted solvent extraction: towards a novel extraction process. *Miner. Eng.*, 2005, 18, 83-88.

Tarutis, WJ; Stark, LR; Williams, FM. Sizing and performance estimation of coal mine drainage wetlands. *Ecol. Eng.*, 1999, 12, 353-372.

Temple, KL; LeRoux, NW. Syngenesis of sulfidic ores: sulfate reducing bacteria and copper toxicity. *Econ. Geol.*, 1964, 59, 271-278.

Tessier, A; Fortin, ID; Belzile, N; Devitre, RR; Leppard, GG. Metal sorption to diagenetic iron and manganese oxyhydroxides and associated organic matter: narrowing the gap between field and laboratory measurements. *Geochim. Cosmochim. Acta*, 1996, 60(3), 387-404.

Tien, CT; Huang, CP. Adsorption behavior of Cu(II) onto biomass particulate surfaces. *J. Environ. Eng.*, 1987, 113, 285-299.

Trettin, R; Glaser, HR; Schultze, M; Strauch, G. Sulfur isotope studies to quantify sulfate components in water of flooded lignite open pits – Lake Goitsche, Germany. *Appl. Geochem*, 2007, 22(1), 69-89.

Tünay, O; Kabdasli, NI. Hydroxide precipitation of complexed metals. *Water Res.*, 1994, 28(10), 2117-2124.

Turner, D; McCoy, D. Anoxic alkaline drain treatment system, a low cost acid mine drainage treatment alternative. In *Proceedings of the 1990 National Symposium on Mining*, May 14-15, 1990, TN; Knoxville, DH; Graves, RW. DeVore, Eds; OES Publishing: University of Kentucky, Lexington, KY, 1990.

Ujang, Z; Anderson, GK. Application of low-pressure reverse osmosis membrane for Zn^{2+} and Cu^{2+} removal from wastewater. *Water Sci. Technol.*, 1996, 34(9), 247-253.

Ueshima, M; Fortin, D; Kalin, M. Development of iron-phosphate biofilms on pyritic mine waste rock surfaces previously treated with natural phosphate rocks. *Geomicrobiol. J*, 2004, 21(5), 313-323.

Utgikar, V; Chen, BY; Tabak, HH; Bishop, DF; Govind, R. Treatment of acid mine drainage: 1. Equilibrium biosorption of zinc and copper on non-viable activated sludge. *Int. Biodeterior. Biodegrad*, 2000, 46(1), 19-28.

Valenzuela, F; Cabrera, J; Basualto, C; Sapag-Hagar, J. Kinetics of copper removal from acidic mine drainage by a liquid emulsion membrane. *Miner. Eng.*, 2005a, 18, 1224-1232.

Valenzuela, F; Fonseca, C; Basualto, C; Correa, O; Tapia, C; Sapag, J. Removal of copper ions from a waste mine water by a liquid emulsion membrane. *Miner. Eng.*, 2005b, 18, 33-40.

Valenzuela, F; Auspont, J; Basualto, C; Tapia, C; Sapag, J. Use of a surfactant liquid membrane contactor for zinc uptake from an acid aqueous effluent. *Chem. Eng. Res. Des.*, 2005c, 83, 247-255.

Valenzuela, F; Cabrera, J; Basualto, C; Sapag, J. Separation of zinc ions from an acidic mine drainage using a stirred transfer cell-type emulsion liquid membrane contactor. *Sep. Sci. Technol*, 2007, 42(2), 363-377.

Valenzuela, F; Araneda, C; Vargas, F; Basualto, C; Sapag, J. Liquid membrane emulsion process for recovering the copper content of a mine drainage. *Chem. Eng. Res. Des.*, 2009, 87(1), 102-108.

Villinski, JE; O'Day, PA; Corley, TL; Conklin, MH. In situ spectroscopic and solution analyses of the reductive dissolution of MnO_2 by Fe(II). *Environ. Sci. Technol.*, 2001, 35, 1157-1163.

Visser, A; Hulshoff-Pol, L; Lettinga, G. Competition of methanogenic and sulfidogenic bacteria. *Water Sci. Technol.*, 1996, 33, 99-110.

Walker, DJ; Hurl, S. The reduction of heavy metals in a stormwater wetland. *Ecol. Eng.*, 2001, 18, 407-414.

Wallace, S; Parkin, G; Cross, C. Cold climate wetlands: design and performance. *Water Sci. Technol.*, 2001, 44(11/12), 259-266.

Walton-Day, K. Geochemistry of the processes that attenuate acid mine drainage in wetlands. In *The environmental geochemistry of mineral deposits. Part A. Processes, techniques, and health issues. Reviews in Economic Geology*; GS; Plumlee, MJ. Longsdon, Eds; Society of Economic Geologists: Littleton, CO, 1999, 215-228.

Wang, J; Huang, CP; Allen, HE; Poesponegoro, I; Poesponegoro, H; Takiyama, LR. Effects of dissolved organic matter and pH on heavy metal uptake by biomass particulates exemplified by copper (II) and nickel (II): three-variable model. *Water Environ. Res.*, 1999, 71, 139-147.

Watten, BJ. Process and Apparatus for Carbon Dioxide Pretreatment and Accelerated Limestone Dissolution for Treatment of Acidified Water. US Patent No. 5,914,046, *US Department of Commerce*, Washington, DC, 1999.

Watzlaf, GR; Hedin, RS. A method for predicting the alkalinity generated by anoxic limestone drains. In *Proceedings of the 14th Annual West Virginia Surface Mine Drainage Task Force Symposium, Morgantown*, WV, April 27-28, 1993.

Watzlaf, GR; Schroeder, KT; Kairies, CL. Long-term performance of anoxic limestone drains. *Mine Water Environ*, 2000, 19, 98-110.

Watzlaf, GR; Schroeder, KT; Kleinmann, RLP; Kairies, CL; Nairn, RW. *The Passive Treatment of Coal Mine Drainage*, National Energy Technology Laboratory, U.S. Department of Energy and University of Oklahoma: OK, 2004.

Waybrant, KR; Blowes, DW; Ptacek, CJ. Selection of reactive mixtures for use in permeable reactive walls for treatment of mine drainage. *Environ. Sci. Technol.*, 1998, 32(13), 1972-1979.

Weber, KP; Gehder, M; Legge, RL. Assessment of the changes in microbial community of constructed wetland mesocosms in response to acid mine drainage exposure. *Water Res.*, 2008, 42, 180-188.

Wehrli, B. Redox reactions of metal ions at mineral surfaces. In *Aquatic Chemical Kinetics: Reaction Rates of Processes in Natural Waters*, Stumm, W., Ed; John Wiley: New York, 1990, 311-336.

White, C; Sayer, GM; Gadd, GM. Microbial solubilization and immobilization of toxic metals: key biogeochemical process of treatment of contamination. *FEMS Microbiol. Rev.*, 1997, 20, 503-516.

Whitehead, PG; Prior, H. Bioremediation of acid mine drainage: an introduction to the Wheal Jane wetlands project. *Sci. Total Environ.*, 2005, 338(1-2), 15-21.

Whitehead, PG., Hall, G., Neal, C., Prior, H. Chemical behaviour of the Wheal Jane bioremediation system. *Sci. Total Environ.*, 2005, 338(1-2), 41-51.

Wieder, RK. Ion input/output budgets for five wetlands constructed for acid coal mine drainage treatment. *Water Air Soil Poll*, 1993, 71, 231-270.

Wildeman, TR; Gusek, JJ; Higgins, J. Passive treatment of mine influenced waters. *In Course material for the ARD Treatment Short Course presented at the 7th International Conference on Acid Rock Drainage (ICARD) St. Louis*, MO, March 26-30, 2006.

Willow, MA; Cohen, RRH. pH, dissolved oxygen, and adsorption effects on metal removal in anaerobic bioreactors. *J. Environ. Qual.*, 2003, 32, 1212-1221.

Wingenfelder, U; Hansen, C; Furrer, G; Schulin, R. Removal of heavy metals from mine water by natural zeolites. *Environ. Sci. Technol.*, 2005, 39, 4606-4613.

Wolkersdorfer, C. *Water Management at Abandoned Flooded Underground Mines: Fundamentals, Tracer tests, Modelling, Water Treatment*; Springer: Berlin, 2008.

Wood, SC., Younger, PL., Robins, NS. Long-term changes in the quality of polluted minewater discharges from abandoned underground coal workings in Scotland. *Q. J. Eng. Geol. Hydrogeol*, 1999, 32, 69-79.

Woulds, C; Ngwenya, BT. Geochemical processes governing the performance of a constructed wetland treating acid mine drainage, Central Scotland. *Appl. Geochem*, 2004, 19(11), 1773-1783.

Yang, B; Lan, CY; Yang, CS; Liao, WB; Chang, H; Shu, WS. Long-term efficiency and stability of wetlands for treating wastewater of a lead/zinc mine and the concurrent ecosystem development. *Environ. Pollut*, 2006, 143, 499-512.

Yong Gan, W; Selomulya, C; Tapsell, G; Amal, R. Densification of iron(III) sludge in neutralization. *Int. J. Miner. Process*, 2005, 76, 149-162.

Younger, PL. Possible environmental impact of the closure of two collieries in County Durham. *J. Inst. Water Environ. Manage*, 1993, 7(5), 521-531.

Younger, PL; Sherwood, JM. The cost of decommissioning a coalfield: potential environmental problems in County Durham. *Mineral Planning*, 1993, 57, 26-29.

Younger, PL; Barbour, MH; Sherwood, JM. Predicting the consequences of ceasing pumping from the Frances and Michael collieries, Fife. In *Proceedings of the 5th National Hydrology Symposium*, Edinburgh, September 4-7, 1995; AR; Black, RC. Johnson, Eds; British Hydrological Society: London, 1995.

Younger, PL; Harbourne, KJ. To pump or not to pump: cost-benefit analysis of future environmental management options for the abandoned Durham Coalfield. *J. Chart. Inst. Water Environ. Manag*, 1995, 9(4), 405-415.

Younger, PL; Curtis, TP; Jarvis, AP; Pennell, R. Effective passive treatment of aluminium-rich, acidic, colliery spoil drainage using a compost wetland at Quaking Houses, County Durham. *J. Chart. Inst. Water Environ. Manag*, 1997, 11, 200-208.

Younger, PL. The longevity of minewater pollution: a basis for decision-making. *Sci. Total Environ*, 1997, 194/195, 457-466.

Younger, PL. Predicting temporal changes in total iron concentrations in groundwaters flowing from abandoned deep mines: a first approximation. *J. Contam. Hydrol*, 2000a, 44, 47-69.

Younger, PL. The adoption and adaptation of passive treatment technologies for mine waters in the United Kingdom. *Mine Water Environ.*, 2000b, 19, 84-97.

Younger, PL., Banwart, SA., Hedin, RS. *Mine Water: Hydrology, Pollution, Remediation*, Kluwer Academic Publishers: Dordrecht, 2002.

Younger, PL; Jayaweera, A; Elliot, A; Wood, R; Amos, P; Daugherty, AJ; Martin, A; Bowden, L; Aplin, AC; Johnson, DB. Passive treatment of acidic mine waters in subsurface flow systems: exploring RAPS and permeable reactive barriers. *Land Contam. Reclam*, 2003, 11, 127-135.

Yu, Q; Matheickal, JT; Yin, P; Kaewsar, N. Heavy metal uptake capacities of common marine macro algal biomass. *Water Res.*, 1999, 33, 1534-1537.

Yurlova, L; Kryvoruchko, A; Kornilovich, B. Removal of Ni(II) ions from wastewater by micellar-enhanced ultrafiltration. *Desalination*, 2002, 144, 255-260.

Zamboulis, D; Pataroudi, SI; Zouboulis, AI; Matis, KA. The application of sorptive flotation for the removal of metal ions, *Desalination*, 2004, 162, 159-168.

Zhong, CM; Xu, ZL; Fang, XH; Cheng, L. Treatment of acid mine drainage (AMD) by ultra-low-pressure reverse osmosis and nanofiltration. *Environ. Eng. Sci.*, 2007, 24(9), 1297-1306.

Zhou, J; Dang, Z; Cai, M; Liu, C. Soil heavy metal pollution around the Dabaoshan Mine, Guangdong Province, China. *Pedosphere*, 2007, 17(5), 588-594.

Ziemkiewicz, PF; Skousen, JG; Lovett, R. Open limestone channels for treating acid mine drainage: a new look at an old idea. *Green Lands*, 1994, 24(4), 36-41.

Ziemkiewicz, PF; Skousen, JG; Brant, DL; Sterner, PL; Lovett, RJ. Acid mine drainage treatment with armored limestone in open limestone channels. *J. Env. Qual*, 1997, 26, 1017-1024.

Ziemkiewicz, PF; Skousen, JG; Simmons, J. Long-term performance of passive acid mine drainage treatment systems. *Mine Water Environ*, 2003, 22, 118-129.

Zinck, JM. Sludge management—an update on technologies. Proceedings of the 2004 Ontario MEND Workshop, *Sludge Management and Treatment of Weak Acid or Neutral pH Drainage, Sudbury*, Canada, May 26-27, 2004, CANMET: Ottawa, Canada.

Zouboulis, AI; Matis, KA. Removal of metal ions from dilute solutions by sorptive flotation. *Crit. Rev. Env. Sci. Technol.*, 1997, 27(3), 195-235.

In: Mine Drainage and Related Problems
Editor: Brock C. Robinson, pp. 71-118
ISBN: 978-1-60741-285-4
© 2010 Nova Science Publishers, Inc.

Chapter 2

ACID MINE DRAINAGE IN THE PORTUGUESE IBERIAN PYRITE BELT

Maria Manuela Abreu[1,], Maria João Batista[2], Maria Clara F. Magalhães[3] and João Xavier Matos[4]*

[1] Instituto Superior de Agronomia, Universidade Técnica de Lisboa (TULisbon), Tapada da Ajuda, P-1349-017 Lisboa, Portugal
[2] Laboratório Nacional de Energia e Geologia, UI. Rec. Minerais e Geofísica, Ap. 7586, P-2720-866 Alfragide, Portugal
[3] Dep. Química and CICECO, Universidade de Aveiro, P-3810-193 Aveiro, Portugal
[4] Laboratório Nacional de Energia e Geologia, UI. Rec. Minerais e Geofísica, R. Frei Amador Arrais 39 rc., Ap. 104, P-7801-902 Beja, Portugal

ABSTRACT

The Iberian Pyrite Belt (IPB), a major Variscan basement unit, located in the SW sector of the Iberian Peninsula, is one of the oldest mining districts in Europe. This world-class volcanic-hosted massive sulfide province extends for nearly 250 km. Pyrite ore exploitation in the Portuguese IPB sector was developed intensively in the XIXth and XXth centuries at São Domingos, Caveira, Lousal, Aljustrel, Montinho and Chança mines which are responsible for significant acid mine drainage (AMD). Outcropping Aljustrel, São Domingos and Caveira deposits were mined during Roman times, especially the gossan and supergene enrichment zones, up to 118 m depth at Aljustrel, promoting two thousand years of AMD. At present, AMD is related with unprotected open pit and underground mining, exposed mining wastes and host mineralization volcanic and sedimentary rocks of Devonian and Carboniferous ages. The São Domingos, Aljustrel, Lousal and Caveira main mining areas present lagoons, dams and stream waters with very low pH and high electric conductivity. The minimum most recent (July 2008) measured pH was 1.04 in Caveira main stream, and the maximum conductivity of 59.5 mS cm^{-1} was measured in the São Domingos open pit in July 2000.

---

[*] E-mail: manuelaabreu@isa.utl.pt

Electric reduction potential also reflects the occurrence of AMD, being values as high as 677 mV measured in the Caveira main stream. At São Domingos and Aljustrel, different materials deposited along the main streams and in the dump areas show the ore processing, while seepage waters indicate different pH-E stabilities. From these AMD waters different mineral precipitations occur, such as copiapite, jarosite, melanterite, and iron oxides. Dissolved elements such as copper, zinc, molybdenum, selenium, cadmium, iron and arsenic attained high concentrations in Lousal and Caveira mines. In Aljustrel and São Domingos mines the dissolved metals and metalloids in AMD areas are mainly iron, aluminium, manganese, cobalt, copper, arsenic, antimony and nickel.

In the Portuguese IPB, especially in São Domingos, Aljustrel, Lousal and Caveira mines, extensive areas are affected by AMD, where different mining practices (mineral processing, period of exploitation, abandon conditions) create different impacts on superficial environments. These situations need different approaches and requirements in case of decommissioning and rehabilitation. Modern mining is implemented in Neves Corvo mine where no significant AMD impact is witnessed.

INTRODUCTION

Sulfide ore bodies are sources of economically very important elements (copper, gold, lead, iron, silver, sulfur, tin, zinc, etc.). However, these exploitations, specially developed during the XIXth and XXth centuries, have also been associated with long-term environmental impact, and their abandon is more environmentally hazardous than a sustainable exploitation. Sulfides are very easily weathered by oxidation releasing potentially hazardous trace elements to the environment and generating acidity. Drainage from an acid generation source, in association with mining activity, containing high concentrations of trace elements and low pH constitute acid mine drainage (AMD). This situation can be observed in the Iberian Pyrite Belt (IPB) causing long time changes in biogeochemical conditions of the region (Abreu et al., 2008; Cánovas et al., 2007; Luís et al., 2009; Matos & Martins, 2006; Nieto et al., 2007; Olías et al., 2004; Pérez-López et al., 2008; Silva et al., 2006).

Sulfide ore bodies are formed in the geochemical primary environment, from magmatic and/or hydrothermal origins at high temperature and pressures when compared with those occurring in the secondary environment where superficial processes of weathering, sedimentation and soil formation occur. Those environments, where mineral deposits containing sulfides are formed, are reducing rather than oxidizing being characterized by low concentrations of oxygen and high concentrations of reduced sulfur species (Williams, 1990). The IPB massive sulfide ores are related with hydrothermal systems associated with submarine volcanism and sea floor environments (Barriga et al., 1997; Carvalho et al., 1999; Relvas et al., 2002; 2006). As a consequence, sulfide ore bodies are very unstable when in contact with the Earth surface conditions. The geochemical secondary environment is characterized by low temperature and pressures, relatively free flow of solutions, and abundant free oxygen, carbon dioxide and water. In general, Earth surface environments are more oxidants than the primary environments and, the contact of the sulfides with these conditions originates a set of chemical reactions in order to form more thermodynamically stable chemical species and solid phases under the new environments. Completely different chemical species (higher oxidation states species) are found for sulfide type ore bodies when they naturally outcrop at the suface of the Earth (Williams, 1990) or when, by mining, they are placed in contact with the Earth surface. The input of trace elements in the environment, mainly soils and waters, from anthropogenic activities is greater than those from natural weathering of geologic materials, contributing to the alteration of the global biogeochemical cycles of trace elements (Adriano, 2001). Mining, especially before the 1970's, was a major contributor for landscape degradation by producing hudge quantities of waste materials, still rich in solid phases containing several hazardous trace elements (antimony, arsenic, cadmium, copper, lead, mercury, tungsten, zinc, etc.) and, often susceptible of acid generation. The pH of these systems (frequently < 3) increases not only the weathering rate of other minerals present in the waste materials (e.g. silicates from ore host rocks) but also the mobility of the majority of the trace elements and its subsequent spreading in the environment.

Depending on the nature, availability and potential effects of the chemical species the hazardous trace elements in mine areas can rise in the food chain affecting different species of animals and plants (Adriano, 2001; Adriano *et al.*, 2004; Krieger *et al.*, 1999). Changes in the physical, chemical and biological conditions of these degraded areas result sometimes in the colonization with new species, as a response to the new biogeochemical conditions. This is the case of *Erica andevalensis* Cabezudo & Rivera an endemic species from Andévalo (Spain) whose distribution is limited to the IPB mining areas affected by AMD (Abreu *et al.*, 2008; Cabezudo & Rivera, 1980).

In Europe, mining dates back ten thousands years as attested by discoveries in France (Wolkersdorfer & Bowell, 2005), whereas in North America the Native Americans erected mounds that testimony early mining activities dating back to 3000 B.C. (Caldwell, 2005). In the territory that corresponds nowadays to the place of Portugal there is evidence that mineral exploration comes from prehistorical times. Large uncertainty exists on dating, however it is accepted that there was some mining activity in the Palaeolithic age (600000 – 8000 B.C.) likewise to other places of the Iberian Peninsula (Abreu & Magalhães, 2009; Geirinhas *et al.*, 2008; Matos *et al.*, 2008a; Oliveira & Matos, 2002). During Roman Era, mining increase significantly in the main European mining regions like the Iberian Pyrite Belt. Here, important AMD was certainly observed along the Rio Tinto, Odiel, Sado and Guadiana basins (Davis *et al.*, 2000; Domergue, 2002; Matos & Martins, 2006; Matos *et al.*, 2008a; 2008b).

Old mining technologies did not take into account the environmental impacts, and the result was a gradual degradation of those sites. Old mining infrastructures are also subjected to degradation that can occur as a consequence of the chemical reactions associated with the presence of sulfur rich materials remaining in mining facilities. These chemical reactions and the subsequent by-products can destroy, for instance, the cement of consolidated dams causing geotechnical problems that may end in accidents as the example of Aznalcollar mine (Domènech & Ayora, 2002; Grimalt *et al.*, 1999; Lacal *et al.*, 2003). Most of the IPB mining areas are in abandon status, with significant degradation of the mining infra-structures (e.g. mine dams and channels). Locally, mining wastes are used as road pavement material contributing to the dispersion of the geochemical halo (Oliveira, 1997; Matos & Rosa, 2001; Oliveira *et al.*, 2002; Matos & Martins, 2006). Mining waste dams are critical high risk areas of the European mining regions like IPB; risk scenarios can be predicted to prevent possible disasters in the future (e-Ecorisk and Tailsafe EU projects).

IBERIAN PYRITE BELT

The Iberian Pyrite Belt (IPB) is a European mining region and a world-class volcanic-hosted massive sulfide belt. The IPB is integrated in the South Portuguese Zone geological terrain, a major Variscan basement unit. The IPB territory extends 250 km WNW–ESE direction, from the Portuguese Atlantic coast Marateca area (Alentejo region), south of Lisbon, to the Spanish Seville area (Andalusia region) (Matos & Martins, 2006; Matos et al., 2008a; Oliveira et al., 2006; Pereira et al., 2008; Relvas et al., 2006; Tornos, 2006). The IPB landscape is dominated by large gently waved plains in Portugal, and plains and local mountains in the Spanish sector. This area is a region with typical Mediterranean climate, according to the Thornthwaite classification, is semiarid mesothermic with no excess water and small thermal efficiency in the hot season, and it can be divided in two distinct periods, a wet period from November to March and a dry period from May to September. The IPB territory is crossed by the Odiel, Tinto, Guadiana and Sado main rivers, with drainage north to south, except the Sado River that flows from southeast to west. The Guadiana shows the largest drainage basin and the river marks, in several regions, the border between the two Iberian countries.

The IPB consists of two major units (Barriga et al., 1997; Oliveira, 1990; Oliveira et al., 1992; 1997; 2005; 2006; Pereira et al., 2007; 2008; Silva et al., 1997; Sáez et al., 1999): the Phyllite Quartzite Group (PQG) dated as Upper Devonian age and is formed by phyllites, quartzites, quartzwackes and shales with intercalations of limestone lenses and nodules in the upper sequences, and the Volcano-Sedimentary Complex (VSC) dated as Upper Devonian to Upper Viséan age. The PQG forms the detritic IPB basement forming a marine siliciclastic platform. The base of the unit is not known. The VSC incorporates several submarine volcanic episodes with dominant rhyolites, dacites, basalts and minor andesites, and intercalations of black shales, siltstones, minor quatzwackes, siliceous shales, jaspers and cherts, and purple shales. The massive sulfide deposits and the manganese IPB deposits are all most related with the VSC units (Figure 1). Locally, stockwork type structures occur in the PQG (Leistel et al., 1998; Matos et al., 2008a; Oliveira et al., 2006; Pereira et al., 2008; Relvas et al., 2006). Above the IPB, VSC and PQG units a thick flysch turbidite succession occurs (Oliveira, 1990; Oliveira et al., 1984; 1992; 2006; 2008; Pereira et al., 2007; 2008) represented by the Baixo Alentejo Flysch Group (BAFG), with > 5 km thickness. The BAFG is represented by Mértola (upper Viséan), Mira (upper Viséan to lower Bashkirian), and Brejeira (Bashkirian to upper Moscovian) formations.

The IPB is characterized by >200 Tg giant massive sulfide deposits like Neves Corvo, Aljustrel, Rio Tinto, and >100 Tg deposits like Tharsis, Aznalcollar-Los Frailes, Sotiel-Migollas and La Zarza (Barriga et al., 1997; Leistel et al., 1998; Matos & Martins, 2003; Sáez et al., 1999; Tornos, 2006). In the IPB, around 90 pyrite deposits are known among abandoned, still in exploration and unexploited. Mining activity at IPB was developed since Calcolithic period, 5000 years ago (Matos et al., 2008a; Oliveira & Matos, 2002), related with pyrite gossans, and copper veins structures mined at small scale, certainly with minor environmental impact. Some deposits were mined in Roman times but there was an increase in mining exploration especially in the XIXth and XXth centuries (Barriga, 1990; Carvalho et al., 1999; Leistel et al., 1998; Matos et al., 2008a; Relvas et al., 2002; 2006; Sáez et al., 1999). The common IPB massive sulfide ore deposits present (Tornos, 2006): giant deposits

(> 100 Tg) containing 60 g Cu/kg, 200 g Zn/kg, 88 g Pb/kg, 0.5 g Ag/kg, 9 mg Au/kg; medium deposits (3.4 Tg) containing 8.5 g Cu/kg, 11.3 g Zn/kg, 5.3 g Pb/kg, 38.5 mg Ag/kg, 0.8 mg Au/kg.

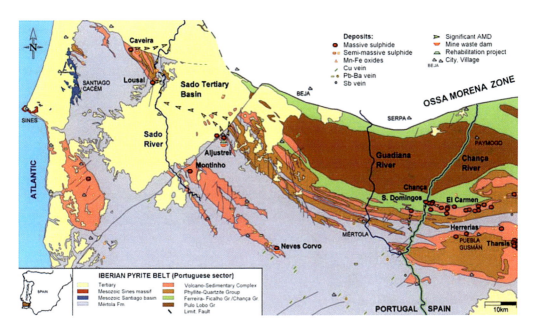

Figure 1. Main acid mine drainage (AMD) occurrences in the Iberian Pyrite Belt Portuguese sector. Main mine waste dams and rehabilitation projects. No AMD is observed downstream of the Neves Corvo active mine due to the excellence of the mining project related with the environment protection. AMD in the IPB Spanish sector is not indicated. Geology adapted from: Barriga et al., 1997; Leistel et al., 1998; Matos & Martins, 2006; Matos et al., 2008a; Oliveira, 1990.

The southern area of the IPB Portuguese sector is represented in the Figure 1. All massive sulfide deposits are associated with the Volcano-Sedimentary Complex. Nowadays, active mining is observed at Neves Corvo. Aljustrel mine is in stand by production since 2008, after a stage of non production between 1993 and 2007. AMD occurences are observed in the Sado basin associated with the Aljustrel, Caveira and Lousal pyrite mines (Feliciano et al., 2008; Luís et al., 2009; Matos & Martins, 2006; Matos et al., 2003b; Oliveira, 1997; Silva et al., 2006). In the Guadiana basin AMD is linked with the São Domingos and Chança mines (Alvarenga et al., 2002; Martins et al., 2007; Matos et al., 2006). Montinho pyrite mine do not present significant AMD owing to the small exploitation dimensions (Matos & Rosa, 2001). Salgadinho and Lagoa Salgada (not represented) are in exploration phase (Matos & Sousa, 2008). Small manganese–iron oxides and copper, antimony and lead–barium veins are also present in the Portuguese IPB sector. Environmental impacts in Portugal are much smaller comparing with the Spanish IPB mines where intense AMD is observed (e.g. Río Tinto, Tharsis, Aznalcollar and Sotiel mine areas) (Matos & Martins, 2006; Sarmiento, 2007).

PORTUGUESE IBERIAN PYRITE BELT EXPLOITATION HISTORICAL VIEW

Several mines were exploited in the Portuguese IPB (Matos & Martins, 2006): i) massive sulfide deposits: giant deposits (> 200 Tg) – Neves Corvo and Aljustrel; medium deposits (5 – 50 Tg) – Lousal, São Domingos and Caveira; minor deposits (< 5 Tg) – Montinho and Chança; ii) manganese (iron) oxides, carbonates and silicates – stratiform and veins – nearly 100 abandoned mines, e.g. Cercal, Ferragudo, Balança and Lagoas do Paço; iii) copper, lead, barium, antimony vein structures – tens of abandoned mines, e.g. Juliana, Barrigão, Brancanes, Ferrarias, Furnazinhas (copper), Castro Verde (barium, lead) and Cortes Pereiras (antimony).

During Roman times, several outcropping massive sulfide deposits were intensely exploited (Domergue, 2002; Gaspar, 1998; Martins *et al.*, 2003; Matos *et al.*, 2008a; 2008b). In the IPB Portuguese sector from NW to SE, the Roman mines were represented by the Caveira, Lousal, Montinho, Aljustrel, São Domingos and Chança deposits. Maximum known depth of Roman exploitation is 118 m at Algares, Aljustrel mine area (the Roman *Vipasca*). These pyrite mines and other similar sites located in Spain (e.g. Río Tinto, Tharsis and La Zarza) were the most important base metal sources of the Roman Empire (Domergue, 2002). Roman mining practice was aimed at the exploitation of secondary ores (oxidation zone – gossan and supergene alteration zone) and, accessory, of primary pyrite ore. The ore was roasted in place to extract the metals for export. Related with mining activities, the acid mine drainage was probably significant at the main mines like Aljustrel, and minor at São Domingos and Caveira sites, considering the exploitation dimension and the presence of interpreted Roman cementation tanks at Aljustrel (Macias *et al.*, 2008; Martins *et al.*, 2003; Matos & Martins, 2006; Matos *et al.*, 2008b). Roman slags cover now a total area of 200,736 m^2: 78,852 m^2 at Caveira, 79,809 m^2 at Aljustrel, 28,173 m^2 at São Domingos, and 13,902 m^2 at Fortes (Matos *et al.*, 2008b). Some of these Roman slags are covered by modern mine wastes produced during the XIXth-XXth centuries (Mateus *et al.*, in press; Matos *et al.*, 2006). In the XIXth century, Roman mining wastes occupied areas of 442,424 m^2 at Aljustrel and 92,410 m^2 at São Domingos (Matos *et al.*, 2008b; Pinto *et al.*, 2007). Considering the average slag density (3.1 g cm^{-3}) and thickness (2 m at Fortes mine and 3 m for the other mines), as suggested by field data, a total volume of 1.87 Mm^3 (5.79 Tg) of Roman slags is known to have existed in the IPB Portuguese sector. No remnants of Roman slag piles were so far found at Chança, Lousal and Montinho, also exploited in Roman times. Considering the efficiency values reported in literature as typical for Roman metallurgical processes and making use of the slag mass estimated, a minimum of 370 Gg of copper, 139 Mg of silver and 9 Mg of gold were produced in the Roman *Lusitania* Province if conservative ore grades are assumed (Matos *et al.*, 2008b).

The modern exploitation of the IPB pyrite deposits started at significant scale in the XIXth century, according to the dates of the first exploitation permits: Aljustrel (1849), Caveira (1854) and São Domingos (1854) (Matos *et al.*, 2006; 2008a). The outcropping pyrite ore deposits, with large oxidation zones (e.g. gossans structures at São Domingos and Algares (Aljustrel)), were exploited underground by shafts and addicts, and in open pit. The gossanized ore and the supergene zone of the deposits provided significant copper, lead, zinc, gold and silver resources. The open pit reached 20 – 30 m at Aljustrel and Caveira, and 120 m depth at São Domingos. At the small Montinho and Chança pyrite mines extractive works were limited to shafts and galleries. In the second part of the XXth century, these mines and dozens of other small mines related with the copper vein were exploited (e.g. Juliana, Brancanes, Porteirinhos, Barrigão, Martinlongo, Ferrarias, Alcaria Queimada, Furnazinhas and Fortes abandoned mines) (Matos et al., 2003b; 2003c; Oliveira et al., 2002). During the beginning of the XXth century the Lousal mine started the pyrite exploitation in open pit and underground. Until 1972 at São Domingos and until the 1980's at Aljustrel copper concentrates were produced by cementation methods, with discharge of extreme acid water flowing down stream. However, the area affected by acid mine drainage was much smaller in Portugal comparing with the Spanish IPB sector (e.g. Río Tinto and Odiel Rivers). The Neves Corvo mine represents the new European mining extractive industry with no significant environmental impacts and no acid drainage observed in the downstream area (Batista et al., 2007; Matos & Oliveira, 2003; Matos & Martins, 2006; Matos et al., 2008a; Relvas et al., 2006). Neves Corvo is a unique case in the IPB that, by the extreme high ore grades, permits a continuous activity since the beginning of the project (1987). The Neves Corvo ore concentrate production is near 2 Tg/year which is significantly more than the usual production in the 1960's, e.g. ~150 Gg at São Domingos, ~240 Gg at Aljustrel and Lousal and < 50 Gg at Caveira (Gaspar, 1998; Matos & Martins, 2003; Matos & Oliveira, 2003; Matos et al., 2008a).

ACID MINE DRAINAGE IN THE PORTUGUESE IBERIAN PYRITE BELT

Since the end of the 1990's the Portuguese IPB is a typical European post mining phase region (Matos & Martins, 2006; Matos et al., 2008a), with significant problems related with AMD, metal dispersion, mine waste management and unsafe mining infra-structures. This scenario is common in the main pyrite exploitations, but also in smaller scale in the IPB manganese and iron oxides stratiform, and copper, lead-barium and antimony vein deposits (Matos & Rosa, 2001; Oliveira, 1997). The last manganese exploitation in Portugal was the Cercal mine, closed in 2001. The Portuguese Geological Survey (Laboratório Nacional de Energia e Geologia – LNEG, former Instituto Geológico e Mineiro (IGM)) performed a national ranking of abandoned mines related with their environment damage (Oliveira et al., 2002). In the Portuguese IPB context the Caveira, Lousal, Aljustrel, and São Domingos mines are considered extreme impact cases, regarding the dimension of the areas affected by AMD (Matos & Martins, 2006) (Figure 2). The detailed geology and mining waste mapping of the Caveira, Lousal, Aljustrel and São Domingos mines were also performed by the Portuguese Geological Survey (Matos, 2004; 2005; 2006a; 2006b). Table 1 indicates the identified total area values (m^2) affected by mining activity in the IPB Portuguese pyrite ore exploitations, including mining wastes, contaminated landfills and areas affected by intense leaching with acid mine water, associated with the control of water volumes by evaporation process. This survey was performed in large scale at São Domingos and Aljustrel mines where complex systems, of acid water pump and transport, were built. In São Domingos, a ~30 km channel network and associated dams permit the store of acid water during the summer season. In the wet season, water discharges to downstream areas, when the Guadiana and Chança rivers presented high flow rates (Cabral *et al.*, 1889; Oliveira & Matos, 2004; Matos *et al.*, 2006; 2008a).

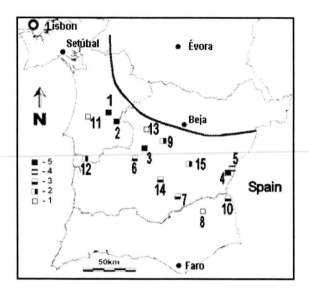

Figure 2. Mining areas with environmental impact in the IPB Portuguese sector, (adapted from Matos & Martins, 2006 and Oliveira *et al.*, 2002): maximum (5) to minimum (1). **Massive sulfides**: 1 – Caveira, 2 – Lousal, 3 – Aljustrel, 4 – São Domingos, 5 – Chança, 6 – Montinho; **Veins**: 7 – Barrigão, 8 – Ferrarias, 9 – Juliana (Cu); 10 – Cortes Pereiras (Sb); **Manganese, iron oxides**: 11 – Saramaga, 12 – Cercal, 13 – Lagoas do Paço, 14 – Ferragudo, 15 – Balança.

Table 1. Mining areas of the IPB Portuguese pyrite ore exploitations (area values in m^2)

Mine	Caveira	Lousal	Aljustrel	São Domingos
Mining wastes	217103	66294	675046	1696384
Contaminated landfill	16917	52025	254749	845551
Acid water dams	1239	21383	284085	273250
Total	**235259**	**139702**	**1213880**	**2815185**[a]
Downstream AMD	3.3 km	5 km	17 km	10 km

[a] 3156225 m^2 considering the artificial AMD slope areas related with the São Domingos water mine evaporation system. Downstream AMD confirmed by stream sediment geochemistry main halo. Sources: Batista *et al.*, 2007; Luís *et al.*, 2009; Martins *et al.*, 2007; Matos, 2004; 2005; 2006a; 2006b; Matos & Rosa, 2001; Matos *et al.*, 2003b; 2006; Silva *et al.*, 2009.

Acid Mine Drainage

Acid mine drainage is a process that results from the air-water-sulfide containing rock/wastes interactions (the primary ingredients of acid generation) releasing large amounts of sulfate and trace elements into the aqueous environment, and decreasing its pH. The exposure of sulfide bearing materials to the atmospheric conditions can occur naturally with the consequent acid drainage generation that is termed Acid Rock Drainage (ARD). Its increase by human activities, such as mining, changes the denomination to Acid Mine Drainage (AMD). Acid mine drainage can be underground mine drainage, runoff and seepage from the waste-rock dumps surface, and from beneath in the waste-bedrock interface, respectively (Walder & Schuster, 1997).

Pyrite is the most abundant mineral in polymetallic sulfide deposits but other metal sulfide minerals may also produce AMD. The extent of environmental trace elements contamination associated with AMD is related with the extent of the oxidation of the sulfide containing materials, the type of gangue material, ore host rocks, mine waste dispersion, mining infrastructures (e.g. dams, channels, ore mills, mine landfills), the hydrology and the hydrogeology of the area. Several geochemical aspects of massive sulfide ores – type (pyrrhotite, chalcopyrite, arsenopyrite, sulfosalts, etc.) and amount of sulfide minerals – make them especially sensitive to weathering and the subsequent generation of AMD.

There are many factors that affect the oxidation rate of sulfide minerals and hence, the rate of acid generation, increasing the rate with:

the increase of temperature;
the increase of the specific surface area of exposed metal sulfides, which is related with the grain size;
a decrease in the chemical activation energy necessary to initiate the acid generation that can, for instance, be decreased in sites with high excess surface energy such as grain edges and corners, defects, cleavages and fractures (Akcil & Kodas, 2006; McKibben & Barnes, 1986);
the increase of iron(III) concentration and decreasing with the increase of iron(II) concentration, once the iron(III) ions can react with pyrite as oxidant to produce more iron(II) ions (Abassi *et al.*, 2009; Holmes & Crundwell, 2000; Williams, 1990);
pH (in general the oxidation rate of sulfides increases with the decrease of hydron ions concentration (Holmes & Crundwell, 2000)); and
oxygen concentration in water and gas phase (Jerz & Rimstidt, 2004; Williamson & Rimstidt, 1994).

The oxidation of iron(II) and sulfide can be catalysed by microbiological activity (e.g. *Acidithiobacillus thiooxidans* and *Acidithiobacillus ferrooxidans*) (Garcia Jr. *et al.*, 2007; Leduc *et al.*, 2002), but bacteria can also reduce sulfate to sulfide, slowing down the acid generation (Leduc *et al.*, 2002; Walder & Schuster, 1997).

The mineralogy of the wastes and weathered mined ores of the Caveira, Lousal, Aljustrel and São Domingos mining sites has been done by some researchers (Africano, 2007; Álvarez-Valero *et al.*, 2008; Durães *et al.*, 2008; Pérez-López *et al.*, 2008; Pinto *et al.*, 2007; Rosado *et al.*, 2008), being also object of research work developed by the Portuguese Geological Survey. Table 2 lists the minerals referred in the literature to Caveira, Lousal, Aljustrel, and São Domingos mining areas.

The weathering of pyrite, that is the most abundant sulfide mineral in the IPB, can be used as an example of a global reaction that explains the low pH, and the high concentrations of, at least, iron and sulfate ions in the oxidizing environments in the neighbourhood of sites containing sulfide minerals:

$$FeS_2(s) + 7/2 O_2(g) + H_2O(l) \rightarrow Fe^{2+}(aq) + 2SO_4^{2-}(aq) + 2H^+(aq) \qquad (1).$$

The iron(II) sulfates are soluble, so both ions can exist in rather acidic aqueous solutions in more or less high concentrations. The highly acidic aqueous solutions can promote the dissolution of both ore and host rock materials originating solutions with high specific electric conductivity. In the presence of oxygen and water the iron(II) ions are also oxidized to iron(III) leading to the precipitation of various insoluble iron containing minerals. In very acidic solutions the generated iron(III) ions can also oxidize the sulfides contributing to a more extent and faster oxidation of the sulfide containing materials. A possible pathway to explain the formation of several iron(III) minerals is shown in the following scheme based on Williams (1990) and Hammarstrom et al. (2005).

$$4Fe^{2+}(aq) + O_2(g) + 4H^+(aq) \rightarrow 4Fe^{3+}(aq) + 2H_2O(l) \tag{2}$$

$$Fe^{3+}(aq) + H_2O(l) \rightleftharpoons Fe(OH)^{2+}(aq) + H^+(aq) \tag{3}$$

$$Fe(OH)^{2+}(aq) + H_2O(l) \rightleftharpoons Fe(OH)_2^+(aq) + H^+(aq) \tag{4}$$

$$Fe(OH)_2^+(aq) + H_2O(l) \rightleftharpoons Fe(OH)_3^0(aq) + H^+(aq) \tag{5}$$

$$nFe(OH)_3^0(aq) + mH_2O(l) \rightarrow \text{ferrihydrite} \\ (5Fe_2O_3 \cdot 9H_2O \text{ or } Fe_2O_3 \cdot 2FeOOH \cdot 2.6H_2O) \tag{6}$$

$$\text{ferrihydrite} \rightarrow FeOOH(\text{goethite, akaganeite, lepidocrocite,} \\ \text{feroxyhyte}) + zH_2O(l) \tag{7}$$

$$2FeOOH(s) \rightarrow Fe_2O_3(\text{hematite, maghemite}) + H_2O(l) \tag{8}$$

The formation of the different solid phases seems to be related with ageing processes, besides the aqueous solutions saturation. Nevertheless, the formation of the polymorphs, that must have similar stabilities, is deeply related with the composition of the aqueous solutions (the pH and the presence or absence of some ions), the atmospheric relative humidity and the temperature.

To establish the relationship between the average composition of the waters percolating in the mining areas (Fonseca & Silva, 2000; Luís et al., 2009; Pinto, 2007; Silva et al., 2006; 2009) and the main solid phases observed there (Table 2), the diagrams of the Figure 3 were drawn, using the thermodynamic data of Table 3. These diagrams were drawn considering the most common percolating surface waters compositions of $c_{sulphur,T}/c^\circ = 10^{-2}$, and $c_{iron,T}/c^\circ = 10^{-4}$ that corresponds, to the total concentration of dissolved sulfate of approximately 1 g/L, and 5 mg/L of total dissolved iron, respectively. For higher concentrations of these elements similar phase diagrams can be found in the literature (Nordstrom & Alpers, 1999).

The direct oxidation of the ion disulfide to the sulfate ion is a process that would involve fourteen electrons, what turns it very unlikely to occur. This is a stepwise process involving many intermediate stable phases existing many of them in natural waters only in a transient way (Williams, 1990). However, some of them as elemental sulfur can be identified in many mining areas (e.g. Caveira, Aljustrel and São Domingos), and even the odour of sulfur dioxide can also be detected in the sulfide mining areas. Phase diagram of Figure 3a wants to emphazise this point and to show that a stability field can be drawn for some species, under particular conditions.

Table 2. List of the primary and secondary ore, and supergene minerals described from the Iberian Pyrite Belt Portuguese mines (Caveira, Lousal, Aljustrel and São Domingos)

Mineral	Ideal chemical formula*	Occurrence with reference
Pyrite	FeS_2	Caveira[a], Lousal[c,d,e], Aljustrel[f], São Domingos[g,i]
Chalcopyrite	$CuFeS_2$	Caveira[a], Lousal[c,d,e], Aljustrel[f], São Domingos[g,i]
Galena	PbS	Caveira[a], Lousal[c,d], Aljustrel[f], São Domingos[g,i]
Sphalerite	ZnS	Caveira[a], Lousal[c,d], Aljustrel[f], São Domingos[g,i]
Arsenopyrite	$FeAsS$	Caveira[a], Lousal[c,e], Aljustrel[f], São Domingos[g]
Bornite	Cu_5FeS_4	Caveira[a]
Pyrrhotite	$Fe_{1-x}S$ ($x = 0.1 - 0.2$)	Caveira[a], Lousal[c,e], São Domingos[g]
Marcasite	FeS_2	Caveira[a], Lousal[c], Aljustrel[f]
Tetrahedrite	$Cu_6Cu_4(Fe,Zn)_2(Sb,As)_4S_{13}$	Caveira[a], Lousal[c]
Cobaltite	$CoAsS$	Caveira[a], Lousal[c]
Bournonite	$CuPbSbS_3$	Caveira[c], Lousal[c]
Goethite	$FeOOH$	Caveira[b], São Domingos[g,h,i]
Hematite	Fe_2O_3	Lousal[d], São Domingos[g,h,i]
Magnetite	Fe_3O_4	Caveira[c], Lousal[c], São Domingos[h,i]
Ferrihydrite	Poorly crystallized iron oxides $5Fe_2O_3 \cdot 9H_2O$ or $Fe_2O_3 \cdot 2FeOOH \cdot 2.6\ H_2O$	São Domingos[h]
Sulphur	S (orthorhombic)	Caveira[b], São Domingos[h]
Gold	Au	Caveira[a], Lousal[c]
Copiapite	$Fe^{2+}Fe_4^{3+}(SO_4)_6(OH)_2 \cdot 20H_2O$	Caveira[b], Aljustrel[f], São Domingos[g,l,k]
Melanterite	$FeSO_4 \cdot 7H_2O$	Caveira[k], Lousal[d,e], Aljustrel[l], São Domingos[k]
Siderotil	$FeSO_4 \cdot 5H_2O$	Aljustrel[f], São Domingos[k]
Rozenite	$FeSO_4 \cdot 4H_2O$	Lousal[d], Aljustrel[L], São Domingos[k]
Szomolnokite	$FeSO_4 \cdot H_2O$	Aljustrel[L]
Coquimbite	$Fe_2^{3+}(SO_4)_3 \cdot 9H_2O$	São Domingos[k]
Rhomboclase	$Fe^{3+}H(SO_4)_2 \cdot 4H_2O$	Aljustrel[f]
Schwertmannite	$Fe_{16}^{3+}O_{16}(OH)_{12}(SO_4)_2$	Aljustrel[f]
Jarosite	$KFe_3(SO_4)_2(OH)_6$	Lousal[c], São Domingos[g,h,i]
Natrojarosite	$NaFe_3(SO_4)_2(OH)_6$	Aljustrel[f,m], São Domingos[g,m]
Anglesite	$PbSO_4$	São Domingos[g]
Scorodite	$Fe^{3+}AsO_4 \cdot 2H_2O$	Lousal[c]
Plumbogummite	$PbAl_3(PO_4)_2(OH,H_2O)_6$	São Domingos[g]
Beudantite	$PbFe_3(AsO_4,SO_4)_2(OH,H_2O)_6$	São Domingos[i]
Poitevinite	$(Cu,Fe)SO_4 \cdot H_2O$	São Domingos[i]
Alunite	$KAl_3(SO_4)_2(OH)_6$	São Domingos[g,m]
Natroalunite	$NaAl_3(SO_4)_2(OH)_6$	Aljustrel[m]
Minamiite	$(Na,Ca,V)_2Al_6(SO_4)_4(OH)_{12}$	Aljustrel[m]
Gypsum	$CaSO_4 \cdot 2H_2O$	Lousal[d], Aljustrel[f], São Domingos[h,i]
Epsomite	$MgSO_4 \cdot 7H_2O$	Lousal[d]
Hexahydrite	$MgSO_4 \cdot 6H_2O$	Lousal[d]
Wavellite	$Al_3(PO_4)_2(OH)_3 \cdot 5H_2O$	Aljustrel[m]

* Ideal mineral chemical formula as presented in Back & Mandarino (2008).

[a]Mateus et al., 2008; [b]http://www.mindat.org/loc-126108.html; [c]Strauss, 1970; [d]Silva et al., 2006; [e]Silva et al., 2005; [f]Bobos et al., 2006; [g]Quental et al., 2002; [h]Rosado et al., 2008; [i]Álvarez-Valero et al., 2008; [j]http://www.mindat.org/gallery.php?min=2633; [k]authors unpublished data; [L]Durães et al., 2008; [m]Matos et al., 2003a.

Table 3. Standard Gibbs energies of formation, at 298.15 K, of relevant species

Species	$\Delta_f G°/\text{kJ mol}^{-1}$	References
FeS_2 (pyrite)	−160.229	Robie et al., 1978
FeS(s)	−101.333	Robie et al., 1978
Fe_2O_3 (hematite)	−742.683	Robie et al., 1978
FeOOH (goethite)	−488.55	Robie et al., 1978
Fe_3O_4 (magnetite)	−1012.566	Robie et al., 1978
$FeSO_4 \cdot 7H_2O$ (melanterite)	−2509.87	Wagman et al., 1982
$FeSO_4 \cdot 4H_2O$ (rozenite)	−1769.216	Wagman et al., 1982 + Chou et al., 2002
$KFe_3(SO_4)_2(OH)_6$ (jarosite)	−3270.99	Robie et al., 1978 + Baron & Palmer, 1996
PbS (galena)	−96.075	Robie et al., 1978
$PbSO_4$ (anglesite)	−813.026	Robie et al., 1978
ZnS (sphalerite)	−202.496	Robie et al., 1978
S (rhombic)	0	Wagman et al., 1982
$H_2O(l)$	−237.141	Robie et al., 1978
Fe^{2+}(aq)	−78.87	Robie et al., 1978
Fe^{3+}(aq)	−4.60	Robie et al., 1978
K^+(aq)	−282.49	Robie et al., 1978
Pb^{2+}(aq)	−24.40	Robie et al., 1978
Zn^{2+}(aq)	−147.26	Robie et al., 1978
HSO_4^-(aq)	−755.91	Wagman et al., 1982
SO_4^{2-}(aq)	−744.53	Wagman et al., 1982
HSO_3^-(aq)	−527.73	Wagman et al., 1982
H_2S(aq)	−27.83	Wagman et al., 1982
HS^-(aq)	12.08	Wagman et al., 1982
SO_2(g)	−300.194	Wagman et al., 1982

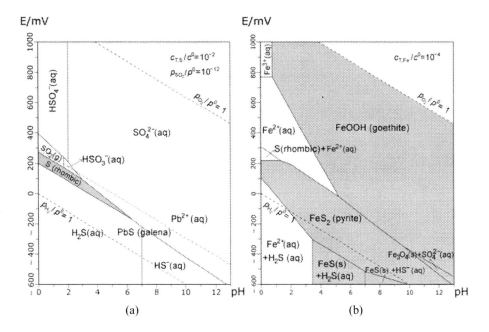

Figure 3. Phase diagrams, at 298.15 K, of (a) sulfur containing species and (b) iron containing species for the average concentration of both elements found in the percolating surface waters in Lousal, Caveira, Aljustrel, and São Domingos mine areas. The total concentration of sulfur is the same in both diagrams.

Other than iron oxides and hydroxide oxides, Table 2 shows that many other iron containing solid phases, mainly sulfates, are being found in the Portuguese IPB mine areas. The complexity and diversity of the acid mine drainage solutions composition is translated to the variety of solid phases that can be formed under different conditions. It is not easy to explain the formation of the secondary solid phases, but some authors (Hammarstrom et al., 2005; Montero et al., 2005), based on previous works, present general possible reaction pathways indicating the conditions (solutions composition, temperature and atmospheric relative humidity) for the crystallization of some secondary sulfate minerals from the oxidation of the primary metal sulfides. For instance, jarosite is expected to crystallize from solutions with pH bewean 1.5 and 3 and sulfate concentrations higher than 3 g/L, while schwertmannite will crystallize from solutions with pH between 3 and 4 and sulfate concentrations between 1 and 3 g/L (Hammarstrom et al., 2005; Montero et al., 2005).

The occurrence of solid phases under particular situations can also be analysed from a possible equilibrium viewpoint, and give rise to the application of the thermodynamic model. The common association of the iron sulfates melanterite (bluish-green), rozenite (white) and coquimbite (brownish-yellow), in São Domingos mines (Table 2), shown in Figure 4a can be translated by the equations:

$$FeSO_4 \cdot 7H_2O(s) \rightarrow FeSO_4 \cdot 4H_2O(s) + 3H_2O(g) \qquad (9)$$

$$4FeSO_4 \cdot 7H_2O(s) + O_2(g) + 2HSO_4^-(aq) + 2H^+(aq) \rightarrow$$

$$\rightarrow 2Fe_2^{3+}(SO_4)_3 \cdot 9H_2O(s) + 12H_2O(l) \qquad (10)$$

or

$$6FeSO_4 \cdot 7H_2O(s) + O_2(g) + 4H^+(aq) \rightarrow 2Fe_2^{3+}(SO_4)_3 \cdot 9H_2O(s) +$$

$$2Fe^{2+}(aq) + 26H_2O(l) \qquad (11)$$

(a) (b)

Figure 4. São Domingos mine área – Achada do Gamo sulfur plants. **(a)** Melanterite (bluish-green) covered by whitish rozenite, and coquimbite (brownish-yellow). **(b)** Extreme saturated acid waters (pH <1.5) and bluish cuprianmelanterite in the bank of a mine waste dam.

The presence of rozenite indicates that the relative humidity of the mineral environments has to be lower than 59%, if the system was at the temperature of 20 °C (Hammarstrom et al., 2005). Melanterite–rozenite transformation is a reversible and fast process in both directions (Chou et al., 2002). At temperatures lower than 20 °C and/or high relative humidity rozenite changes to melanterite. The reported presence of siderotil in Aljustrel (Bobos et al., 2006) that was also identified in São Domingos indicates that in both locations occur copper containing melanterites with different amounts of copper (possibly poitevinite (Álvarez-Valero et al., 2008)). Siderotil may be formed from the dehydration of richer cuprian melanterite than that that originate rozenite (Hammarstrom et al., 2005).

The possible transformation of melanterite to coquimbite, shown in Figure 4, and observed also in several other places in São Domingos mine area, needs the presence of aqueous acid media as can be seen in equations (10) and (11). If the solid is in contact with the saturated very acidic aqueous mining drainage the reaction can be considered as written in equation (10), but if the reaction occurs under other environments, where there is no direct contact of melanterite and aqueous solutions, the reaction must end with the formation of brownish orange iron(III) hydroxide oxides (Hammarstrom et al., 2005). Besides the formation of copiapite, equation (11) shows that melanterite can also produce aqueous iron(II) ions, which in contact with oxidizing environments can undergo the reactions translated by the equations (2) to (8). The precipitation of these minerals occurs seasonally at São Domingos mine according to local humidity conditions along the year (e.g. in the Achada do Gamo sulfur plants area, Figure 4). The increase of non–saturated water levels in the mine dams contributes to the mineral dissolution.

The equilibrium between melanterite and acid aqueous solutions represented in Figure 4b from São Domingos mine, is also very frequently observed in Caveira, Lousal, and Aljustrel mining areas. In São Domingos mine it is also found large amounts of melanterite showing that the waters present in some lagoons/dams besides being saturated in relation to melanterite have a moderate electric reduction potential. The occurrence of melanterite indicates that the system has an electric reduction potential that allows the existence of iron(II) salts (Figure 3b). In fact, the water of the dam shown in Figure 4b (sample São Domingos ASD4) had pH = 1.65, T = 16.6 °C, and an electric reduction potential of 544 mV. For this water sample there is no experimental data for the total concentrations of iron and sulphate, but their concentrations can be estimated by an iterative method from the values of the pH, solution ionic strength and the solubility constant of melanterite.

To estimate the solubility of melanterite in the natural water of pH = 1.65, it is necessary to consider only the following chemical equilibria

$$FeSO_4 \cdot 7H_2O(s) \rightleftharpoons Fe^{2+}(aq) + SO_4^{2-}(aq) + 7H_2O(l) \tag{12}$$

and

$$HSO_4^-(aq) \rightleftharpoons H^+(aq) + SO_4^{2-}(aq) \tag{13}$$

Using the thermodynamic approach the solubility can be calculated from the relevant thermodynamic constants – the solubility constant, K_{s0}, in the case of equilibrium (12), and the acid ionization constant, K_a, for equilibrium (13). Thermodynamic equilibrium constants (K), in general, are defined by the expression from mining wastes, related with Roman and XIXth exploitation periods, which were rich in copper and lead. Mine wastes were leached by acid mine water (pumped from underground galleries) and metal recovered by local precipitation in evaporation and cementation tanks.

The mineral processing included size reduction and/or classification, physico-chemical separation, chemical treatment and smelting (predominant in the Roman period and minor in the XIXth century). The late metallurgic activities occurred in the northern sector of the mine, near the abandoned ruins of the miner's quarters, linked to open air pyrite ore roasting, probably using *teleras* method. This roasted method was commonly used in Spain (e.g. Rio Tinto) and Aljustrel Portuguese mine (Cabral et al., 1889; Matos & Martins, 2006; Vara, 1963) and is defined by slow roasting of the pyrite using wood and small metric and circular infrastructures with air tunnels. This mine wasting with barren pyrite was dedicated to metal recovering after leaching process. A small dam located in the tailing downstream area collected the enriched water to the cementation process (Matos et al., 2008a).

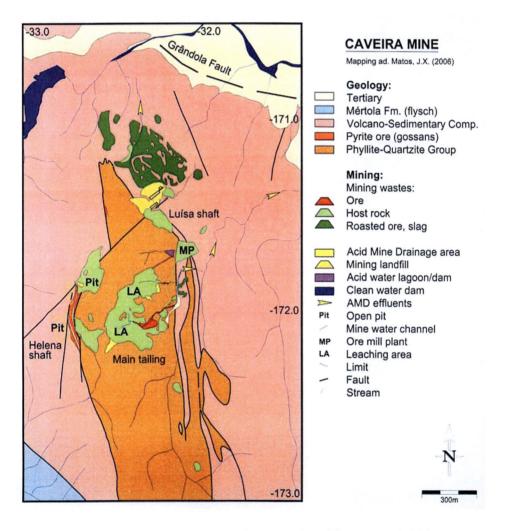

Figure 6. Geological and mining map of Caveira mine area (adapted from Matos, 2006a).

The geomorphology of the mining area is strongly controlled by the major WNW–ESE Grândola active fault, present in the northern mine sector (Figure 6). The relief close to the principal streams is moderate (slopes 2 – 5%), but the neighbouring hills are dominated by long steep slopes (10 – 35%).

Acid mine drainage characterization

The Caveira mining site shows significant downstream contamination and unsafe mining infrastructures like unprotected shafts and open pits (Fonseca & Silva, 2000; Mateus *et al.*, 2008; Matos, 2006a; Matos & Martins, 2006; Oliveira, 1997). The mine is characterized by the main tailing located in a hill, the Helena and Luísa shafts, two small acid water dams, several ruins of the miner's quarters, power plant, mill plant and the roasting area (Figure 6). The mine wastes cover an area of approximately 217,103 m^2 (Table 1). Globally, the waste distribution is characterized by large dispersion of the main waste classification (Matos, 2006a; Matos *et al.*, 2008a; 2008b): brittle pyrite ore (pyrite, chalcopyrite, galena, sphalerite, arsenopyrite, and minor tetrahedrite) together with host rocks, represented by dark shales with quartz veins and felsic quartz-feldspar volcanics with quartz-sericite and/or chlorite hydrothermal alteration; host rocks with rare pyrite; slags and roasted pyrite; Roman slags. Roman slag wastes are common in the eastern sector of the Caveira mine and are represented by an *in situ* stable thin layer (< 2 m) (Matos *et al.*, 2008b). Other minor waste tailings are present at Caveira, with a sparse distribution located in the western and north east mine sector near the Helena and Luísa mining shafts, respectively. The use of the mining slags in road pavement, reported since the 1960´s, strongly contribute to the chemical elements spreading along the region (Matos & Martins, 2006).

The geochemical characterization of the dump materials (fraction < 80 mesh), using Principal Component Analysis statistics (14 variables of 44 individuals, where 71% of the total variance was explained in three axis) identified three groups of chemical elements: i) silver, bismuth, mercury, lead, antimony and sulfur predominant in all dump materials; ii) copper, zinc, iron, and sulfur more located in the extreme NE and SW area of the dump, and probably related with the pyrite leaching and lower depositional areas of the leached materials; iii) arsenic, lead and potassium. All these groups of chemical elements are compatible with the original mineralogy of pyrite, rich silver galena, sphalerite, chalcopyrite, pyrrhotite, marcasite, bornite, tetrahedrite, arsenopyrite, cobaltite and gold (Mateus *et al.*, 2008; Matos & Martins, 2006; Matos *et al.*, 2008a; 2008b).

The main tailing (up to 6 m thick) is situated on the top of the Caveira central hill, in unstable slope conditions, showing intense rill and gully erosion and an important visual impact in the landscape. The main dump material (composed of ore host rocks like dark grey shales with silica veins, felsic volcanics and massive and semi-massive sulfide ore) was dispersed in the high slope hills to facilitate the leaching process of the poor metal content ore and the liquors collection in a tank where copper cementation took place (Matos & Martins, 2006; Matos *et al.*, 2008a; 2008b). Nowadays, this leaching process is uncontrolled due to mine abandon and rupture of the main dam wall, located in the tailing slope toe (Figure 7a), and intense AMD is observed.

Located 25 km near the Atlantic coast the Caveira mine is the IPB Portuguese mine most affected by usual rain showers, especially during the wet season. Rainwater circulates and percolates easily over and through the tailing materials causing significant chemical elements leaching, erosion and transport of tailing debris to areas nearby and downstream (Figure 7b)

(Fonseca & Silva, 2000; Mateus et al., 2008; Matos, 2006a; Matos et al., 2003b; Silva et al., 2009). The water reacts with fine brittle pyrite rich materials in a slow sub-superficial movement of water percolating slowly through the brittle pyrite wastes. AMD is observed and its spread is increased by local topography, mine waste location and destruction of the single protection dam. In the dry season there is a small flux of transparent water from the hole of the dam with a pH around one and containing high concentrations of trace elements (Table 5).

Figure 7. Caveira mine area – **(a)** Dam wall rupture in the downstream area of the Caveira mine main tailing **(b)** Gully and rill erosion, and waste dispersion at the Caveira mine main tailing formed by fragmented pyrite ore and host rocks.

Table 5. Statistics of chemical elements concentrations in superficial water (mg/L) from samples collected in the downsteam area of the Caveira mine dam

Superficial water	As	Cu	Fe	Pb	S	Zn
Scarce rain period (March 2000)[a]						
Median	16.6	110	2516	2.17	3.85	125
Minimum	0.172	1.14	80.0	1.02	0.252	6.89
Maximum	86.1	52.6	1110	4.52	16.6	313
Intense rain period (April 2000)[a]						
Median	328	27.5	2160	1.29	3.02	162
Minimum	43.3	1.32	104	0.523	0.258	11.6
Maximum	388	55.2	3173	1.69	3.35	280
Dry period (June 2000)[a]						
Median	68.5	59.1	9686	3.19	5.07	283
Minimum	0.335	1.48	150	1.92	0.302	10.5
Maximum	86.4	64.0	11500	3.75	5.23	324
Dry period (July 2008)[b]						
Minimum	42.3	34.5	5280	3.90	10.4	173
Maximum	80.4	64.2	9130	5.01	17.6	304

[a](Fonseca & Silva, 2000). [b]Author's unpublished data.

Seepage waters collected in the main tailing toe show significant values for soluble arsenic, lead and iron, and very low pH that reflect, the long period of water-tailing materials interaction and meteorological conditions at the moment of sampling.

The water sampled near the broken wall dam during July 2008 (summer season) show pH around one, and very high electric conductivity (13.8 mS cm^{-1}) and concentrations of heavy metals and metalloids such as copper (48.7 g/L), iron (7510 g/L), lead (4.33 g/L), zinc (226 g/L), and arsenic (59.8 g/L). Downstream, the water sample collected in a < 0.5 m water column and very low flow conditions, and in the presence of neoforming minerals such as copiapite, showed physical and chemical characteristics even more extreme (pH=1.04, electric conductivity 39.4 mS cm^{-1}). These conditions are characteristic of the dry season when low dillution is observed. During the sampling program (summer 2008, author's unpublished data) the water and air temperatures were 30 °C and > 35 °C, respectivelly. The attenuation effects, that usually occur, with increased distances from the main acid drainage source, were not observed. Approximately 2 km downstream the pH of the water was still bellow two. These particular conditions of the Caveira downstream area reflect the local topograpphy and the no input of clean waters in the hydrographic basin. According to Mateus *et al.* (2008) the dispersion of the trace elements, specially copper, zinc and lead, in sediments collected in a large area of the Caveira mine (a total of 163 sediment samples) is almost confined around the main tailing.

The AMD and the mechanical dispersion of tailing materials control will be the key issues of a rehabilitation program of this mine (Martins & Carvalho, 2005; Matos & Martins, 2006).

Lousal Mine Area

Geology and mining exploration

The Lousal polymetallic massive sulfide mine is located in the NW region of the Iberian Pyrite Belt, in a lineament of the Volcano–Sedimentary Complex located NNW of Ermidas do Sado (Matos, 2006b) (Figure 1).

The Lousal pyrite ore present commonly primary banding sometimes more intense due to ore recrystalization by deformation process. Interstitial hydrothermal silica occurs, locally with chalcopyrite dissemination. Interstitial carbonates are referred by Strauss (1970). According to this author, pyrite is the main sulfide phase and variable chalcopyrite, galena, sphalerite, pirrotite, marcasite, bournonite, tetrahedrite, cobaltite, safflorite and native gold can occur.

The geomorphology of the region is conditioned at NW by the Palaeozoic basement of the South Portuguese Zone and at SE by the tertiary sediments of the Sado Basin (Matos & Oliveira, 2003; Matzke, 1971; Schermerhorn *et al.*, 1987). Close to the principal streams the relief presents a moderate development and the hills have significant slopes. The Lousal mining area is limited to the N by the Espinhaço de Cão stream and at the south by the Corona main stream, both tributaries of the Sado River.

The Lousal mine was exploited basically for pyrite between 1900 and 1988 at surface (one main open-cast pit) and by underground works reaching about 500 m depth (Shaft n° 1) (Matos & Oliveira, 2003). In 1882 the local farmer António Manuel discovered the gossans of the *Sul* and *Extremo Sul* orebodies, located near the Corona stream (Figure 8). In 1901 several shafts were on exploitation, from NW to SE: Miguel (*Oeste* orebody), Mendonça, Santa Bárbara and Guilherme (*Central* orebody), Rui, Valdemar and Luís (*Sul* and *Extremo Sul* orebodies). The three main gossans and their respective supergene enrichment zone were explored for copper ore.

Until the 1920's the superficial sectors (gossans and/or supergene enrichment zone) of the *Sul, Extremo Sul, Central, Oeste, Miguel* and *Norte* orebodies were exploited mainly for copper (Strauss, 1970). Later, the mining works had been directed for the deeper lenses *José* and *Fernando* and the first objective of the owner company becomes the production of pyrite concentrates. In the 1930's the exploitation of the Lousal mine increased to give more volume of pyrite concentrates to the SAPEC phosphates plant located at Barreiro (near Lisbon) and built in 1928 (Silva, 1968). In the late 1950's and early 1960's large scale mechanization of the exploitation and ore treatment methods occurs at Lousal, which permitted an annual production of 230 – 250 Gg of concentrate, with 450 g S/kg and 7 g Cu/kg (Matzke, 1971). Thadeu (1989) estimated for the Lousal mine a total tonnage of 50 Tg, with 7 g Cu/kg, 8 g Pb/kg and 14 g Zn/kg. Matzke (1971) presents the following average grades for the Lousal pyrites: 390 g Fe/kg; 7 g Cu/kg; 8 g Pb/kg; 14 g Zn/kg; 0.8 g Co/kg; 0.1 g Ni/kg; 450 g S/kg; 4 g As/kg; 0.3 g Sb/kg; 0.5 g Mn/kg; 2 g Sn/kg; 0.1 g Cr/kg; 0.2 g BaO/kg; 35 mg Ag/kg; 1 mg Au/kg.

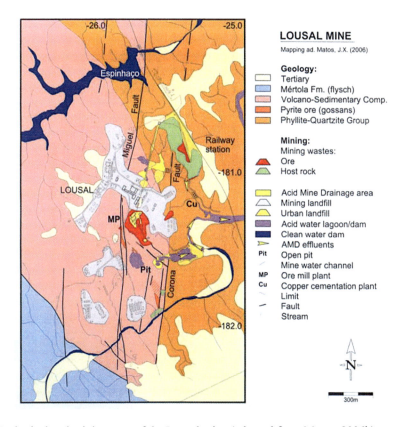

Figure 8. Geological and mining map of the Lousal mine (adapted from Matos, 2006b).

Chemical analysis of one semi–massive pyrite ore sample (Matos & Rosa, 2001) indicate the following concentrations: 2.42 g Cu/kg; 9 g Pb/kg; 3 g Zn/kg; 140 mg As/kg; 358 mg Sb/kg; 21.4 mg Ag/kg; 352 µg Au/kg; 128 g Fe/kg; 38 mg Mn/kg; 82 mg Ba/kg; 122 mg Co/kg.

Acid mine drainage characterization

In the Lousal mining site two tailing areas can be considered, one related with the mine shafts and ore mill plant with 23,907 m^2, the other related with the railway station, from where the mine ore was transported, with 59,542 m^2 (Matos, 2006b). The first area is located in the northern sector of the mine open pit, the second is located 600 m north–eastward, near the railway Lisbon–Algarve. These two areas are different in mine waste composition (Matos, 2006b; Matos & Martins, 2006; Matos & Oliveira, 2003; Silva *et al.*, 2006; 2009). The railway area is predominantly composed of brittle pyrite ore (> 60%) mixed with mineralization host rocks represented by quartz, dark grey to black shale, grey chert and felsic well cleaved quartz, rare feldspar porphyrytic volcanic. The open pit tailing group is predominantly composed of host rocks and minor massive and semi-massive pyrite ore (< 40%). This waste material is more coarse (centimetric to decimetric dimension) when compared with the railway area wastes (centimetric to milimetric dimension). Parts of these tailings are mine landfill related with the plateau areas were the main mine shafts and ore mill building were constructed. During winter times acid waters occur, especially near the base of the tailing where an acid water spring is identified, near the shaft n° 2. The Lousal tailings present high contents of copper, lead, zinc, arsenic, antimony, silver, gold, cobalt, bismuth, molybdenum, mercury and thallium (Matos & Martins, 2006).

The ore host rocks present in the mine tailings are sediments and felsic volcanics of the Volcano–Sedimentary Complex (Upper Famennian–Upper Visean age (Oliveira *et al.*, 2005; 2006; Pereira *et al.*, 2008)). These rocks were extracted from underground mining works and constitute the ore gang. In some areas of the mine, local tailing erosion is observed related with less favourable slope angle. However, Lousal does not present significant tailing instability, comparing with other Portuguese IPB mines (e.g. Aljustrel and Caveira (Matos & Martins, 2006)).

Nowadays, the large volumes of waste materials with high concentrations of copper, lead, zinc, arsenic, cadmium, antimony, sulfur and iron (Silva *et al.*, 2006) produce significant AMD that is visible in the Corona stream located downstream from the mining facilities. In fact, seepage waters flowing beneath the waste dumps drain directly into the Corona valley near the railway sector. A mine landfill isolates the Lousal open pit from the fluvial waters of the Corona stream, as a safety infrastructure related to fluvial flooding (Matos & Oliveira, 2003; Matos, 2006b). However, the high porosity of the landfill material (mostly composed of ore volcanics and shales host rocks) does not stop the continuous infill circulation of acid waters and the consequent contamination of the fluvial ecosystem. Other smaller unprotected acid water dams and tailings (e.g. north area of the Mining Museum) easily release acid water into the surroundings. As a consequence, the Corona stream is affected with acid mine drainage for several kilometres (Matos & Oliveira, 2003) (Figure 9a). However, according to Silva *et al.* (2006) the severity of contaminants generally decreases 4 km downstream of the source due to mixing with fresh waters leading to acidity neutralization.

Figure 9. Lousal mine area – (a) Corona stream affected by Lousal mine AMD. (b) Bubbles at the acid water spring located in the mine open pit.

Acid mine drainage, associated with interaction of rain water and sulfide minerals existing in the exposed tailings, shows pH values ranging from 1.9 to 2.9 and high concentrations of sulfate (9.25 – 20.7 g/L), iron (0.959 – 4.83 g/L) and aluminium (136 – 624 mg/L) (Silva et al., 2005).

The pH measured in the waters from Corona stream and collected upstream from the mine site was 7.93, with electric conductivity of 944 µS cm^{-1} and $E = 340$ mV, whereas downstream from the mine site the water had the pH = 2.8, electric conductivity 5.23 mS cm^{-1} and $E = 515$ mV (author's unpublished data).

Superficial water of the lagoon, although with some wastes and host rock in the border, did not present very high concentrations of the ore/waste chemical elements, such as, arsenic, lead, zinc and iron. However, Silva et al. (2006) found electric conductivity as high as 5.04 mS cm^{-1} and average pH of 3.0. In the dry periods the total concentrations of sulfate and iron increased, as well as for manganese, aluminium, zinc, copper, cadmium and cobalt. On the other hand, the flowing channels waters and the acid water spring located nearby the lagoon had very high concentrations of arsenic, iron, copper and zinc (Table 6, Figure 9b), which are compatible with the weathering of the sulfide minerals present in mine wastes.

Acid spring water, that refluxs from the galleries of the underground mine, reflects the continuing oxidation of the ore still *in situ*. This spring is situated in the bottom of a small lake where waters from the lexiviation of tailings are retained and gas bubbles released from the bottom (Figure 9 b) (Matos & Oliveira, 2003; Silva et al., 2005; 2006; 2009). The spring water had low pH (2.7 – 2.8), high sulfate content (1.6 – 3.9 g sulfate/L), and high electric conductivity (4.9 – 11.6 mS cm^{-1}) (Silva et al., 2006).

Silva et al. (2006) stated that the major AMD input to the Corona stream during spring season is due to the flowing channels waters originated from the waste heaps whereas during summer the major AMD input emanates from leaky dam/lagoon.

Table 6. Statistics of concentrations of chemical elements in superficial water from Lousal mine area

Superficial water	As (µg/L)	Cu (mg/L)	Fe (mg/L)	Pb (µg/L)	Zn (mg/L)
Superficial water of the lagoon[a]					
Minimum	bdl	7	14	140	66
Maximum	bdl	11	22	197	77
Superficial water in the open pit upstream of Corona stream[a]					
Minimum	2	9	395	75	90
Maximum	95	12	959	306	170
Flowing channels water from the dump near the mine facilities[a]					
Minimum	21,064	23	2,250	177	199
Maximum	36,455	111	8,215	302	269
Downstream of confluence with effluents from railway station dump[a]					
Minimum	4,574	29	1,526	bdl	130
Maximum	9,000	50	1,629	bdl	219
Dry period (July 2008)[b]					
Upstream of the mine	7.2	0.134	7.8	4.9	1.52
Downstream of the mine	5.0	5.41	69.5	38.6	54.7

[a]Silva et al., 2009; [b]Author's unpublished data; bdl - below detection limit.

Acid mine drainage influences the chemical composition of soils located near the tailing deposits, which present significant amounts of several elements: 272 – 12,930 mg Pb/kg; 597 – 6,377 mg As/kg; 1 – 130 mg Hg/kg; 292 – 7,013 mg Cu/kg; 1 – 130 mg Mo/kg; 126 – 7,481 mg Zn/kg. Also the stream sediments downstream the tailings site contain high amounts of copper (1 – 1,986 mg/kg), lead (41 – 5,981 mg/kg), zinc (17 – 1,756 mg/kg), arsenic (6 – 1,988 mg/kg) and Cd (0.2 – 5.7 mg/kg) (Silva et al., 2005).

Aljustrel Mine Area

Geology and mining exploration

The Aljustrel area is characterized by the presence of the Messejana Fault (Figure 10), with NE–SW direction, that limits a southeastern block formed by the Iberian Pyrite Belt Palaeozoic basement (Schermerhorn et al., 1987; Matos et al., 2008a; Oliveira et al., 2006) and a north-western block represented by de sedimentary overburden of the Sado Tertiary Basin. The downstream area of the Aljustrel mine is dominated by this basin, specially the Água Forte, Água Azeda, Farrobo and Roxo streams, all Sado River affluents. The IPB Volcano–Sedimentary Complex (VSC) is represented in Aljustrel by the following units: Paraíso Formation – siliceous shales, phyllites, tuffites, purple shales, jaspers and cherts; six massive sulfides ore deposits; felsitic/mine metavolcanics and megacryst/green metavolcanics sequences – sericitic felsic volcanics, felsites, felsophyres, volcanic breccias, feldspar megacryst volcanics and lavas. The lower IPB unit, the sedimentary Phyllite–Quartzite Group is not recognized in the Aljustrel area.

The VSC Aljustrel Anticlinorium (Leitão, 1998; Matos, 2005; Schermerhorn et al., 1987; Silva et al., 1997) is represented with a NW–SE direction along 4.5 km by 1.5 km across. Close to the Messejana fault, at São João area, the Palaeozoic structures are affected by its senestral movement and present a NE–SW direction. Several NW–SE thrust are identified, mainly in the short limbs of the anticlines. The Aljustrel thrust is one of the main structures and materializes the SW contact between the VSC and the Flysch Group. Late variscan faults with NNE–SSW direction produced important dextral strike-slip movements, probably conjugated faults of the Messejana fault. This system is represented from SE to NW by the Esteval, the Azinhal, the Feitais, the Represa, the Castelo and the Moinho faults. In the SE block of the Messejana fault well-developed hills occur, locally controlled by differential erosion of outcropping Palaeozoic VSC units, e.g. Algares and São João gossans, jasper and chert horizons.

The hydrographic network of the Aljustrel region is well developed and hierarchised. Locally NNE–SSW faults, like Represa, control stream development (e.g. Água Forte stream) (Luís et al., 2009; Matos et al., 2008a). The NW block of the Messejana fault is characterized by a flat relief. Here the Palaeozoic basement is covered by the sediments of the Alto Sado Tertiary Basin (up to one hundred meters thick). The Aljustrel mine is located in the upstream sector and the drainage flow goes to north to the Roxo stream, affluent of the Sado River, and to the northwest directly to Sado River.

Since prehistoric days until the end of the last millennium the mining exploitation was a preponderant factor in the social-economic structure of Aljustrel mine surrounding areas. The Aljustrel mine is one of the major IPB extractive centres where six orebodies are recognized (Moinho, Feitais, Estação, Gavião, Algares and São João) the last two were mined since the Roman times (Leitão, 1997; 1998; Matos & Martins, 2006; Silva et al., 1997), where the exploitation of the gossans and the supergene zones reached 100 m deep (Allan, 1965; Domergue, 1983; 2002; Matos et al., 2008b). Two bronze tablets containing the Roman Mining Law were discovered in the slag fields from this period. Previous mining works (Iron Age) are also recognized at Mangancha hill (Martins, 1996).

The Aljustrel mine is an underground base metal mine that includes surface infrastructure consisting of a processing plant with 1.8 Tg capacity. The modern exploitation of the Aljustrel mining started in 1847 (Leitão, 1997; Martins, 1996; Matos et al., 2008a). The Algares, São João and Moinho deposits were exploited primarily by underground "cut and fill" methods for their sulfur content. Intense exploration surveys started in the late 1950's led to the discovery of the following deposits: Moinho (1954) by Turam method, and Feitais (1964), Estação (1969) and Gavião (1970) by gravity surveys (Leitão, 1998; Queiroz et al., 1989). The owners of the exploitation permits were Sebastião Gargamala (1849–1854), the Lusitanian Mining Company (1854), the Empresa de Mineração Transtagana (1867–1895), the Companhia Fonseca, Santos e Viana (1895–1898), the Societé Anonyme Belge des Mines de Aljustrel (1898–1956), the Belgium company Mines d´Aljustrel S.A. (1956–1973), the Portuguese Pirites Alentejanas (1973–2009) and Almina (2009).

After the II World War, the mine annual production was increased to 200 – 300 Gg/year of pyrite concentrates (eg. 1,780,521 Mg (1946 – 1953), 248,068 Mg in 1953, 195,033 Mg in 1977). The Moinho deposit was intensely exploited for its copper content from 1991 to 1993, and the maximum annual production was established in 1992 (1.2 Tg). Since then the mine has been on stand-by production with a detailed care and maintenance program. A zinc project started in the end of 2007. After the metal prices decrease the mine stoped in 2008.

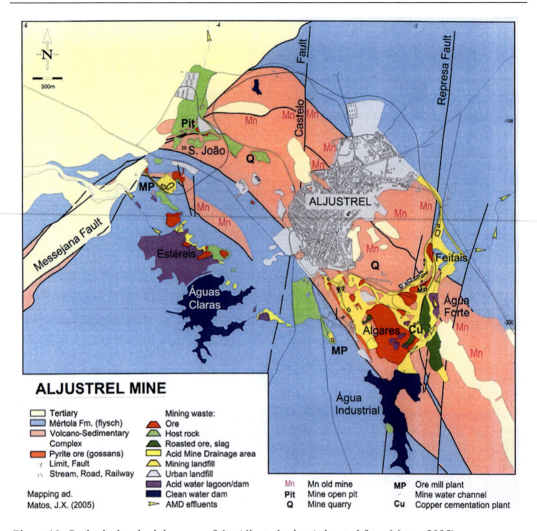

Figure 10. Geological and mining map of the Aljustrel mine (adapated from Matos, 2005).

The Aljustrel mining heritage is very significant in the IPB context (Matos & Martins, 2006; Matos et al., 2002; 2008a). The main mining infrastructures with cultural interest are the Algares and São João open pits and gossans, the Vipasca, Viana, Moinho and São João mining shafts, and the Algares copper cementation fields. Other infrastructures are also important such as, the electrical power plant, the processing plant, the railway infrastructure and equipment, the Algares mills, the Transtagana chimney and the mine laboratories and quarters. Underground mining works are fundamental for further tourist projects. The Aljustrel mine present spectacular mining chambers and a galleries network which extends from São João to Feitais orebodies (4 km long).

The manganese–iron abandoned exploitations are represented by small open pits (up to 30 m length) with a sparse distribution along several areas (Figure 10), where the VSC Paraíso sedimentary is represented (Matos, 2005; Matos & Martins, 2006; Matos et al., 2008a). The occurrence of manganese–iron mineralizations is confined to this unit and commonly associated with jaspers horizons. The manganese–iron oxides were exploited in the late XIXth century by small and probably family companies. This extractive industry occurred between 1869 and 1873 in the following mines: Cabeço do Moinho, Herdade de Monte Ruas, Serra dos Feitais, Cerro da Cabeça do Homem, Malpique, Nossa Senhora do Castelo, Moinho Velho, Penedões, Mangancha, Moinho de Vale São João, Santo Antão. These mine sites present tailings, having less than 1 m thickness. The environmental impact of these exploitations is not quantified but, certainly, is locally significant. Some of these small mines are used for illegal waste disposals.

The Aljustrel landscape is characterized by strong urban and industrial (mining) influence (Matos & Martins, 2006). The mining shafts, the evaporation and cementation fields and the large grey tailings dominate the wide industrial area of Algares. In the São João valley the open pit, the infill quarries of Moinho and the Mangancha hill outcrops (jaspers and manganese–iron oxides abandoned exploitations) expose local geology, commonly a reference in the landscape.

Acid mine drainage characterization

In general, the wastes are composed of Roman slag, pyrite ore (blocks and brittle massive pyrite ore) and Volcano Sedimentary Complex host rocks (felsic volcanic and shales). In a petrogenic study the waste materials were identified as containing chalcopyrite, sphalerite, galena, arsenopyrite and minor sulfosalts (Luís et al., 2009). The pyrite ore contain high amounts of iron, copper, lead, zinc, silver, antimony, mercury, selenium, cobalt, gold and cadmium, and the roasted pyrite ores present high concentrations of gold, lead, silver, iron, antimony, bismuth, selenium, copper, zinc and molybdenum (Luís et al., 2009).

Large volumes of acid waters are discharged in the principal streams like the Água Azeda, Farrôbo and Água Forte (Figure 5) streams. The Água Azeda stream is affected by the São João mining sector (tailings and open pit) and by the São João urban area. Farrôbo stream is affected by the Santo Antão new industrial area where the ore processing plant, the main acid water dam (Estéreis) and the Águas Claras clean water dam are located. Água Forte stream drains the Algares large brittle pyrite and slag tailings (Figures 10 and 11a). The abandoned copper cementation mining process and underground exploitation originate intense acid mine drainage along the valleys, especially in the Algares and São João sectors (Candeias, 2008; Luís et al., 2009; Matos & Martins, 2006; Matos et al., 2008a; 2008b; Salgueiro et al., 2007).

Acid mine drainage at the Feitais–Algares area can, presently, be characterised by the waters collected in July 2008 (Figure 5) at the basement of a recent big dump at Feitais. This dump is composed of ore host rocks (volcanic and shales) and is related with the recent excavation of the Feitais underground ramp more than 3 km length. The new tailing is located above old brittle pyrite ore tailings (Figure 10). The water collected in a small spring located in this area had pH 1.84, electric conductivity 30.03 mS cm^{-1} and electric reduction potential 389 mV (sample Aljustrel 4 in Table 4). Melanterite is visible in the border of the water flow showing, besides the waters saturation, the low value of electrical reduction potential as was discussed and shown (Figure 3b) in the acid mine drainage thermodynamic analysis. Another water sampling point was located in the Água Forte stream resulting from the drainage of the acid water from Água Forte dam (Figure 11a). Until the 1990's the acid water of this dam was pumped to the top of the brittle pyrite ore piles, and percolated again to the dam (Martins & Carvalho, 2005; Matos & Martins, 2006). A small decrease of water volume was verified by evaporation process, concomitant with metal enrichment in the water system. Common over flow during winter times is responsible for uncontrolled AMD discharges. The Água Forte stream water had pH 2.06, electric conductivity 13.02 mS cm^{-1}, and higher electric reduction potential (E = 672 mV) than the Feitais–Algares tailing spring water. Downstream, the mixture of the water of the Água Forte stream with the water flowing from the Feitais–Algares area had pH 1.98, electric conductivity 24.5 mS cm^{-1} and 408 mV electric reduction potential (sample Aljustrel 2 in Table 4). In this site it is also visible the presence of melanterite in the banks of the water flow, as a result of the electric reduction potential decrease. Other efflorescent salts are also present, such as, rozenite and szomolnokite (Table 2) (Durães et al., 2008). The efflorescent salts and weather conditions may be responsible for the variations in the concentration in trace elements in superficial waters between wet and dry seasons (Pinto, 2007).

Along the Água Forte stream the collected water samples had high total dissolved elemental concentrations, which slightly decrease with distance from the waste materials and Feitais mining facilities origin. This stream receives effluent water from an urban residual water treatment plant (2 km away from the tailings) with pH ≈ 8 and electric conductivity 2.25 mS cm^{-1}, which changes the water physical–chemical conditions in the vicinity of the water mixture. For instance, the pH changes from 2 to near 4 (Candeias, 2008; Gonçalves et al., 2007; Luís et al., 2009; Pinto, 2007) which is kept at least 4 km downstream (Figure 11b) where the measured pH continues to be around 4.

Superficial water samples collected in Água Forte hydrographical basin (Luís et al., 2009) in May 2005 and, March and June 2006, presented pH values ranging from 1.5 to 3.5 and trace elements concentrations as high as: 6.84 mg As/L; 0.455 mg Cd/L; 68.8 mg Cu/L; 1.26 g Fe/L; 19.5 mg Mn/L; 0.136 mg Pb/L; 264 mg Zn/L. Microbial analysis of the Água Forte waters showed the presence of the consortium of bacteria *Acidithiobacillus ferrooxidans*, *Leptospirillum ferrooxidans*, *Acidithiobacillus albertensis* and *Leptospirillum ferriphilum* (Gonçalves et al., 2007), which can catalyse the oxidation reactions of sulphur and iron.

Figure 11. Aljustrel mine area – **(a)** Água Forte acid waters dam located in the Algares mine sector. **(b)** Água Forte stream acid waters, ~ 4 km downstream of the Aljustrel mine.

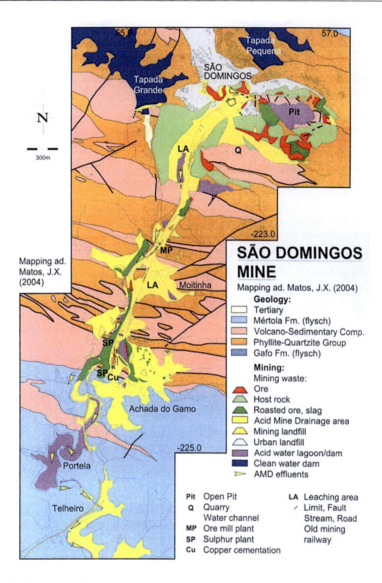

Figure 12. Geological and mining map of the São Domingos mine, between the mine open pit and the Telheiro downstream sector (adapted from Matos, 2004).

According to Luís *et al.* (2009) the sediments of Água Forte stream have very high concentrations of trace elements (maximum values: 2.49 g As/kg; 843 mg Cu/kg; 213 g Fe/kg; 2.63 g Pb/kg; 311 mg Sb/kg; 937 mg Zn/ kg) which are above the Consensus–Based Sediments Quality (CSST, 2003) for the tolerance of organisms for elements.

São Domingos Mine Area

Geology and mining exploration

São Domingos mine, one of the most emblematic Portuguese massive sulfide deposits, is located in the northern sector of the IPB, about 5 km far from the Spanish border. The São

Domingos deposit is a sub-vertical massive sulfide orebody located at the top of a Volcano–Sedimentary Complex (VSC) sequence, represented by black shales, felsic, basic and intermediate–basic volcanics (Matos et al., 2006; Pereira et al., 2008). These rocks are bounded, through thrust faults vergent to SW, by shales and quartzites of the Phyllite–Quartzite Formation and black shales and banded shales ascribed to the Represa Formation, both of late Devonian age. The VSC and the deposit are preserved in a tectonic sheet with sigmoidal geometry (Figures 1 and 12).

The São Domingos orebody is a subvertical lense shape massive sulfide deposit with the following main characteristics (Carvalho, 1971; Matos, 2004; Webb, 1958): pyrite, sphalerite, chalcopyrite, galena, arsenopyrite and sulfosalts as mineral assemblage; the northern sector (orebody footwall) is rich in copper, and the western and southern sectors are rich in zinc; grades – 12.5 g Cu/kg, 20 – 30 g Zn/kg and 450 – 480 g S/kg; >25 Tg of massive sulfides, one pyrite lense with 537 m length; and 45 to 70 m thickness.

The outcropping São Domingos orebody was mined since the Calcolithic, and extensively exploited during the Roman times. The modern mine exploitation started in 1857 and finished in 1966. The São Domingos gossan and associated Roman slags were discovered by Nicolau Biava in 1854 (Matos et al., 2006). The Mason & Barry Company, the mine owner, exploited the São Domingos deposit by open pit until 120 m depth and by underground works until 420 m depth. Part of the São Domingos village was constructed over gossan waste and over landfill of the first mine village destructed in 1867, in consequence of the open pit development (Figure 13a). Between 1867 and 1880, from the open pit, 3 Mm^3 of rock were extracted. The addict network extends 1080 m at the 150 m level, from the main water pump shaft (Malacate n° 6) to the air pump shaft (n° 7) located in the eastern sector of the mine. Ore was transported by railway to Moitinha ore mills, to the Achada do Gamo sulfur factories (Figures 12 and 13b) and to the Pomarão harbour (Guadiana River). The native copper was obtained by cementation after extensive ore leaching at Moitinha plateaus.

In a favourable high metal prices context related with the Spanish civil war two sulfur plants were constructed in 1935 by the Mason & Barry Company, using the Orkla Norwich metallurgic process. The sulfur obtained by this catalytic process becomes one of main mine products. The Achada do Gamo reservoir was destinated to store not treated waste materials, an output of the Orkla pyrite ustulation process developed between the 1930's and 1960's (Mason & Barry Company, 1937 – 1955; 1950 – 1952; Matos et al., 2006). In the Moitinha and Achada do Gamo sectors large tailings of modern slags are present, also with ash waste from the 1930's sulfur plants (Matos et al., 2002; 2006).

Mining activities at São Domingos affected a total area of 3,076,900 m^2, from the village to the Chumbeiro downstream dam, located 11 km southeast. Along the valley of the São Domingos stream, 767,659 m^2 are occupied by industrial landfill + leaching/cementation tanks and 722,737 m^2 are covered by ore + host rocks tailings. In the mine area a total of 14.7 Mm^3 of mining wastes are estimated, settled in mining landfills and tailings with a heterogeneous dimension, ranging from 14 m thick near the open pit to less than 1 m in the downstream areas (Álvarez-Valero et al., 2008; Mateus et al., in press; Matos, 2004; Matos & Martins, 2006; Pérez-Lopéz et al., 2008) (Figure 12).

The São Domingos ore was intensely leachead in the Moitinha leaching fields with underground acid water pumped from mine addicts in the main shaft n° 6 (Batista et al., 2003; Matos & Martins, 2006; Matos et al., 2006; Quental et al., 2003) (Figure 14a). The acid effluents, very rich in copper, were later transported by channel to the Achada do Gamo copper cementation plants. Between Moitinha and the downstream Chumbeiro dam, and specially in the vicinity of the Achada do Gamo sulfur factories, large and significant effects of acid mine drainage are observed (Figures 12, 13b, 14, 15).

The mine closure in the 1960's led to a social and economic decline of the region. Until 2004 the urban sanitary water was not controlled by local municipality which leads to important flow of urban waters into waste tailing with consequent downstream acid mine drainage in the São Domingos stream valley.

Acid mine drainage characterization

The mining history of the São Domingos mine is reflected in great variety of mining wastes. Twelve waste classes and three landfill classes were considered in the mine mapping, done in the 1/2500 scale (Figure 12) (Álvarez-Valero *et al.*, 2008; Matos, 2004; Matos & Martins, 2006; Matos *et al.*, 2006; Pérez-Lopéz *et al.*, 2008; Pinto *et al.*, 2007). As expected, around the open pit a large variety of waste is observed (Figure 13a). The gossan wastes are predominant, deposited locally over previous Roman slags (e.g. oriental sector of the pit). Now, *in situ* Roman slags are only present in the northern and north-eastern sectors of the pit.

Slag piles at São Domingos represent 8% of the accumulated mine residues, being a historical record of the ore smelting refuses obtained in different periods of the massive sulfide and related gossan exploitation. The major slag constituents are crystalline silicates (mostly olivine and pyroxene), magnetite and glass, accessory amounts of sulfides and metal alloys are also present, sometimes coupled with sulfosalts (Pinto *et al.*, 2007). The slag is typically iron–rich (300 – 400 g/kg) and contains up to 17 g Zn/kg, 9 g Pb/kg and 5 g Cu/kg, besides minor to trace contents in many other metals and metalloids (e.g. manganese, antimony, cobalt, arsenic, indium, silver, bismuth, germanium).

The maximum tailing thickness is observed near the open pit with up to 14 m of gossan + volcanic wastes (Álvarez-Valero *et al.*, 2008; Matos, 2004; Matos *et al.*, 2006; Pérez-Lopéz *et al.*, 2008). Three main groups of wastes are present: i) Roman slags related with gossan and supergene enrichment zone exploitation; ii) gossan + host rocks + local reworked Roman slags + roasted ore related with XIX century open pit/underground exploitation; iii) modern slags + sulfide roasted ashes + brittle pyrite ore + coarse blocks of pyrite ore related with 1930's – 1960's mine exploitation. The pyrite and other minor sulfides present in modern and Roman slags, smelting ashes and pyrite-rich waste dumps (brittle pyrite and coarse blocks) generate, by oxidation, and as a consequence of the Mediterranean climatic conditions, direct AMD during all over the year (Álvarez-Valero *et al.*, 2008). According to these authors those waste materials can contain concentrations as high as 1.7 g As/kg, 5.9 g Pb/kg and 8.5 g Zn/kg, being also a significant source of hazardous elements contamination.

Nowadays, a large negative environmental impact is observed in the mining area along the São Domingos stream valley (Álvarez-Valero *et al.*, 2008; Matos, 2004; Matos & Martins, 2006; Oliveira, 1997; Oliveira *et al.*, 2002) (Figures 14 and 15). Acid rock drainage in the Achada do Gamo, near the sulfur factories, is marked by significant lead, arsenic, antimony, copper, zinc and iron anomalies in stream sediments, soils and waters (Quental *et al.*, 2002; Tavares *et al.*, 2008).

(a) (b)

Figure 13. São Domingos mine area – **(a)** Modern slag tailings (dark colour) and brittle pyrite ore (grey colour) near the acid water (pH <2.5) flood open pit. **(b)** Dam with extreme acid waters (pH <1.5) at the Achada do Gamo 1930's sulfur plants.

(a) (b)

Figure 14. São Domingos mine area – **(a)** Pyrite roasting materials (smelting ashes in grey colour) located near leaching fields (light brown plateau). **(b)** Uncontrolled AMD in the São Domingos downstream area near the ore mill plant located at Moitinha.

(a) (b)

Figure 15. São Domingos mine area – **(a)** Telheiro dam at São Domingos stream, containing contaminated sediments and mine waste materials. In second plan, hill slope representing one of the XIX century areas dedicated to acid water evaporation. **(b)** *Erica andevalensis* plants in acid waters at the river bank, downstream area near Achada do Gamo. In the stream margins black fragmented slag were deposited until the 1960's.

After waste materials classification and taking into account the chemical and mineralogical classification, the acid mine drainage generation potencial and acid mine neutralizing capacity were calculated (Álvarez-Valero et al., 2007). The obtained results indicate that the main sources of AMD are slags (modern and Roman slags), smelting ashes and pyrite-rich materials, although the reduced volume (1,325,821 m^3) of these wastes when compared with the other wastes (industrial landfields, gossan waste, leaching tanks and country rocks – 11,167,313 m^3). This is a consequence of the iron sulfides richeness, the absence of neutralizing minerals and their large open air distribution.

Owing to the heterogeneity of the São Domingos waste materials their ability to release hazardous chemical elements was evaluated by a sequencial extraction procedure (BCR) (Pérez-Lopez et al., 2008). The results show that considerable amounts of hazardous elements are extracted in the first step of the extraction procedure (water/acid soluble and exchangeable fractions) mainly from smelting ashes and brittle pyrite. This easy mobile/available fraction of the chemical elements represents, in comparison with the total concentration of the same elements in those wastes, 26 – 58% of sulfur (16.6 – 34.4 g S/kg), 31 – 52% of zinc (105 – 494 mg Zn/kg), 24 – 41% of copper (23 – 379 mg Cu/kg), 2 – 20% of arsenic (316 – 486 mg As/kg), 0.3 – 2% of lead (116 – 130 mg Pb/kg) and 0.8 – 43% of iron (69 – 34,481 mg Fe/kg).

Despite the high concentration of arsenic (ranging from 0.5 to 16.9 g As/kg) in all type of the São Domingos mine wastes, this element is mainly present in the residual fraction (76 – 100%). A similar distribution pattern of arsenic was also observed in soils from São Domingos mine developed on gossan materials (Abreu et al., 2009). Mine wastes also have high concentrations of total lead (3.7 – 45.8 g Pb/kg). This element is mainly in the residual fraction (49 – 98 %), except in the Roman slags and iron oxides that contain in the easy mobile/available fraction 24 and 12 % of lead, respectively (Pérez-Lopez et al., 2008).

High amounts of lead are also present in the fraction accounted to the iron and manganese oxides and even in the oxidizable fraction (e.g. smelting ashes with 41% of total lead) (Pérez-López et al., 2008). Considering that the three first steps on BCR (water/acid soluble and exchangeable fractions, reducible fraction (associated with iron and manganese oxides), and oxidizable fractions (chemical elements bound to sulfides)) correspond to the fractions which, in these particular mine environments, will be responsible for chemical elements release, Pérez-López et al. (2008) ordered the waste materials as follows: brittle pyrite > pyrite blocks > smelting ashes > iron oxides > country rocks > modern slag > Roman slag ≥ gossan wastes > leaching tanks > industrial landfill. However, if the mass/volume of the waste materias was also taken into account the order would be: industrial landfill > country rocks > gossan wastes >leaching tanks > modern slag > Roman slag > brittle pyrite > smelting ashes > pyrite blocks > iron oxides.

Microbial analysis of the waste materials from São Domingos mine area, indicates the presence of microorganisms capable to survive at pH <1, iron and sulfur oxidates (*Acidithiobacillus ferrooxidans*), iron oxidates (*Leptospirillum ferrooxidans*) and sulfur oxidates (*Acidithiobacillus thiooxidans*) and also moderate acidophiles (Bryan et al., 2006), which can catalyse the oxidation reactions of sulfur and iron.

Large non-vegetated areas are the result of decades of mining activity and a common use of topography to loose volume of water by surface acid water percolation (Matos & Martins, 2006) (Figure 15a). The São Domingos stream valley (Chança river basin, Guadiana main basin) is affected by acid effluents linked to uncontrolled surface water that percolate along tailings and mine infrastructures like the Achada do Gamo lagoon which commonly present pH <1 (Batista et al., 2003; Matos & Martins, 2006; Matos et al., 2003; Quental et al., 2002) (Figures 4, 13, 14 and 15).

Several research works showed geochemical anomalies in the following materials: soils – lead, arsenic, sulfur, copper, cadmium, mercury; stream sediments – lead, arsenic, antimony, copper, zinc, iron, silver, (vanadium, chromium); tailings – zinc, lead, antimony, copper, silver, mercury, cadmium; superficial waters – iron, aluminium, manganese, cobalt, copper, arsenic, antimony, and nickel (Batista et al., 2003; Matos & Martins, 2006; Matos & Rosa, 2001; Oliveira, 1997; Pinto et al., 2007; Quental et al., 2002; 2003).

The main stream, which flows through the São Domingos mine area, beginning at the mine buildings zone and extended to the Chança river confluence, collects the AMD generated in all the area as indicated by low pH, high E and high electric conductivity. Seepage waters that result from sub-surface runoff in and under the tailings and slags in the limit between consolidated bedrock material and unconsolidated weathered material run into the main channel (São Domingos stream). This water shows low pH (mean 2.6), high E (mean 481 mV), high electric conductivity (mean 4.34 mS cm^{-1}), high concentrations of iron (mean 4.2 g/L) and sulfate ions (mean 7.7 g/L) (Abreu et al., 2008).

In the Achada do Gamo industrial area there are heterogeneous mining wastes (mainly modern slag, metallurgic ashes and pyrite ore) (Figures 12 and 13b) that release high amounts of sulfate, iron and trace elements. The artificial lake waters have dark colour (Figure 4b) and in March 2007, had pH 1.7, electric reduction potential of 544 mV, and high concentrations of trace elements (93.4 mg As/L; 976 mg Cu/L; 609 mg Zn/L; 0.71 mg Pb/L). Other dams located along the São Domingos main stream contain waters (clear or with red colour) presenting pH ranging between 2.2 and 2.8, and also high contents of hazardous chemical elements (6 – 690 µg As/L; < 50 – 316 µg Pb/L; 4 – 13 mg Cu/L; 7 – 11 mg Zn/L).

As was above mentioned, in the acid mine drainage thermodynamic analysis, when the application of the thermodynamic model to the Achada do Gamo environment, soluble iron containing sulfates (Table 2) as cuprian melanterite [(Cu,Fe)SO$_4$·7H$_2$O], siderotil, coquimbite and copiapite were identified in the lake border. The conditions for their crystallization and transformation, as shown in Figure 4, was also analysed in the above "Acid Mine drainage" section. Silva et al. (2009) described similar mineral associations from Lousal mine translating analogous geochemical conditions in the area. Similar associations also occur as efflorescent salts in other places out of the IPB under the influence of acidic drainage waters (Jamieson et al., 2005; Nordstrom & Alpers, 1999).

In Achada do Gamo, copiapite was found to crystallize after coquimbite. The change from coquimbite, an iron(III) sulfate, to copiapite, an iron(II) and iron(III) sulfate translates a decrease in the electric potential of the mineral environment and also an increase in the humidity of the place. As was reported by Hammarstrom et al. (2005) and Jamieson et al. (2005) hydrated iron sulfates undergo phase changes with changes in humidity and temperature. Coquimbite also appears when copiapite is dehydrated (Jamieson et al., 2005). The change, observed in Achada do Gama, from coquimbite to copiapite must occur when the solid, in contact with the oxidizing normal atmosphere, is submersed by the saturated lake water that has a pH and an electric potential that falls inside the iron(II) ions stability field (Figure 3). This efflorescent assemblage of minerals is only observed during the summer (dry and hot season) once they dissolve with the first rains.

Despite the AMD characteristics of the surface water in the mine area, the groundwater presents low electric reduction potential, high pH, and low concentrations in trace elements, except for manganese, and in some locations also zinc, which indicates that the water contamination is confined to the superficial São Domingos waters (Quental *et al.*, 2002; Martins *et al.*, 2007). The *E*-pH stability diagram (Figure 16) represents the classification of waters from Tapada Grande until Telheiro (Quental *et al.*, 2002; 2003).

Due to the particular environment and the different situations that can be studied in the site, São Domingos is, probably, the more studied Portuguese mine area and for the longest period of time. In Quental *et al.* (2002; 2003) hyperspectral images were used to identify mineralogical/chemical dispersion of waste material related to AMD at São Domingos mine area following two approaches: one related to AMD waste material field spectra, and the other based on sulfide–sulfate–iron oxides minerals, using standard spectral libraries. The processing techniques were mainly based on the Spectral Angle Mapper classifier and Mixture Tunned Matched Filtering. Mapping results, obtained using both approaches, were able to detect mineralogical/chemical characteristics of imaged ground data, which were validated by field data. This study demonstrated the powerfull tool that remote sensing can be for AMD mapping.

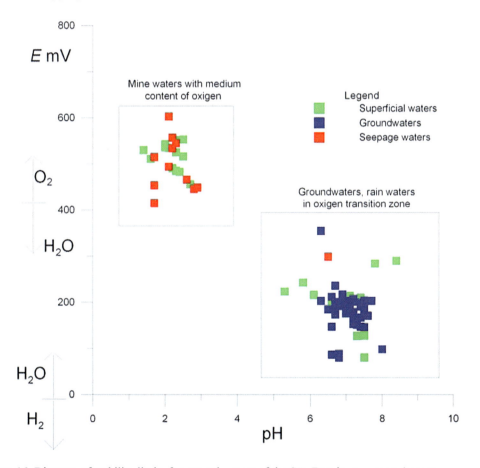

Figure 16. Diagram of stability limits for natural waters of the São Domingos waters (seepage, superficial and groundwaters), in terms of *E* and pH at 25 °C, after Garrels and Christ (1965).

Several bioassays can be used as bioindicators for aquatic risk assessment of AMD, and in São Domingos mine some studies using diatoms, *Chironomidae* larvae and *Lemna gibba* were performed (Gerhardt *et al.*, 2008; Janssens de Bisthoven *et al.*, 2005). The results showed the dominance of some diatom species in AMD waters (*Nitzschia capitellata* Hustedt and *Pinnularia sucapitata* Gregory), and morphological deformities in *Ablabesmyia monilis* and *Procladius sp.* (Chironomidae larvae). These deformities indicate the influence of low pH and high trace elements content in these organisms, which could therefore be used as AMD bioindicators. However, the reduced number of larvae can compromise its use. *L gibba*, a vascular floating hydrophyte, shows to be very sensitive to AMD-stress, as was demonstrated by necrosis and growth rate inibition (Gerhardt *et al.*, 2008).

The presence of acid water is also a biodiversity factor and *Erica andevalensis* Cabezudo & Rivera is observed in the river banks of the São Domingos stream. This species whose distribution is limited to the IPB (Cabezudo & Rivera, 1980) was only identified in the Portuguese IPB in São Domingos mine (Abreu *et al.*, 2008) (Figures 14b and 15). The implantation of this plant species may be a valuable element for mining landscape restoration. Considering the tolerant behaviour in extreme environmental conditions, this *Erica* species may be of major importance for the recovery of the sulfide mining areas of the IPB, where climatic conditions are compatible with its breeding and growing, by physical and chemical stabilization of contaminated sediments and soils and even waste materials (Abreu *et al.*, 2008). *E. andevalensis* is the only angiosperm species adapted to extreme ecological conditions (pH< 2.5, strong impoverishment in nutrients, high trace elements content) (Soldevilla *et al.*, 1992). However, it is likely that *E. andevalensis* species may become extinct as soon as its habitat changes, which illustrates the importance of the habitat in species conservation (Aparício & García-Martin, 1996). As a matter of fact, *Erica andevalensis* was classified as an endangered species by the Andalusian (southern Spain) Regional Government.

CONCLUSION

The Iberian Pyrite Belt is a European mining region in predominant post mining stage. Only the Neves Corvo mine represents the new modern, sustainable extractive industry, operating without interruption since the 1980's. In the IPB Portuguese sector all the old pyrite mining projects present negative impacts to the environment, some since the Roman times. The Caveira, Lousal, Aljustrel and São Domingos mines, presented in this chapter, show significant impacts related with AMD. Local hydrographic basins are affected promoting contamination of the Sado and the Guadiana rivers. These impacts are, however, smaller comparing with the Spanish IPB sector where large areas are intensely affected by AMD.

The Caveira, Lousal, Aljustrel and São Domingos main mining areas present lagoons, dams and streams with very low pH, high conductivity and electric reduction potential waters that allows the simultaneous presence of iron(II) and sulfate ions. The minimum most recent measured pH was 1.04 in Caveira main stream, and the maximum conductivity of 59.5 mS cm^{-1} was obtained in the São Domingos open pit. Electric reduction potential also reflects the occurrence of AMD, being values as high as 677 mV measured in the Caveira main stream. At São Domingos and Aljustrel, different materials deposited along the main streams and in the dump areas show the ore processing, while seepage waters indicate different $pH-E$ stabilities. From these AMD waters different mineral precipitations occur, such as copiapite, jarosite, melanterite, iron oxides and hydroxides. Dissolved elements such as copper, zinc, molybdenum, selenium, cadmium, iron and arsenic attained high concentrations in Lousal and Caveira mines. In Aljustrel and São Domingos mines the dissolved metals and metalloids in AMD areas are mainly iron, aluminium, manganese, cobalt, copper, arsenic, antimony and nickel.

The Portuguese Geological Survey (LNEG), local entities and universities are involved in several research programs dedicated to IPB sustainable development. One of the projects is the mining heritage Pyrite Route (Rumys project). This thematic route is based on network of minings sites (Lousal–Aljustrel–São Domingos–Neves Corvo) were sustainable programs are being developed related with mining, exploration, rehabilitation, geological mining heritage and culture.

ACKNOWLEDGMENT

To Carla Anjos for helping with the drawing of Figure 3.

REFERENCES

Abassi, R., Khan, F. & Hawboldt, K. (2009). Prediction of minerals producing acid mine drainage using a computer-assisted thermodynamic chemical equilibrium model. *Mine Water Environment, 28*, 74-78.

Abreu, M. M. & Magalhães, M. C. F. (2009). Phytostabilization of Soils in Mining Areas. Case Studies from Portugal. In L., Aachen, & P. Eichmann, (Eds), *Soil Remediation*, (297-344). NY: Nova Science Publishers, Inc.

Abreu, M. M., Tavares, M. T. & Batista, M. J. (2008). Potential use of *Erica andevalensis* and *Erica australis* in phytoremediation of sulphide mine environments: São Domingos, Portugal. *Journal of Geochemical Exploration, 96*, 210-222.

Abreu, M. M., Santos, E. S., Magalhães, M. C. F. & Nabais, C. (2009). Arsenic containing phases in soils from São Domingos mining area and in non-contaminated soils from Pomarão and Serra do Caldeirão. *Revista de Ciências Agrárias, 32*, 155-169.

Adriano, D. C. (2001). *Trace Elements in the Terrestrial Environment. Biogeochemistry, Bioavailability, and Risks of Metals*. (2nd edition). New York: Springer-Verlag.

Adriano, D. C., Wenzel, W. W., Vangronsveld, J. & Bolan, N. S. (2004). Role of assisted natural attenuation in environmental cleanup. *Geoderma, 122*, 121-142.

Africano, F. (2007). *Comportamento de elementos em traço de elevada toxicidade, de Cd como indicadores de impacto ambiental na área da Mina de São Domingos*. SFRH/BPD/237117/2005 Relatório Preliminar.

Akcil, A. & Koldas, S. (2006). Acid Mine Drainage (AMD): causes, treatment and case studies. *Journal of Cleaner Production, 14*, 1139-1145.

Allan, J. C. (1965). A mineração em Portugal na Antiguidade. *Boletim de Minas*, Lisboa, *2*, 139-175.

Alvarenga, P. M., Matos, J. X. & Fernandes, R. M. (2002). Avaliação do impacto das minas de Chança e Vuelta Falsa (Faixa Piritosa Ibérica) nas águas superficiais da bacia hidrográfica do Rio Chança. In IGM/SEDPGYM (Ed), *Actas do Congresso Internacional Sobre o Património Geológico e Mineiro* (611-620). Beja, Portugal.

Álvarez-Valero, A. M., Pérez-López, R., Matos, J., Capitán, M. A., Nieto, J. M., Sáez, R., Delgado, J. & Caraballo, M. (2008). Potencial environmental impact at São Domingos mining district (Iberian Pyrite Belt, SW Iberian Peninsula): evidence from a chemical and mineralogical characterization. *Environmental Geology, 55*, 1797-1809.

Aparício, A. & García-Martin, F. (1996). The reproductive biology and breeding system of *Erica andevalensis* Cabezudo & Rivera (Ericaceae), an endangered edaphic endemic of southwestern Spain. Implications for its conservation. *Flora, 191*, 345-351.

Back, M. E. & Mandarino, J. A. (2008). *Fleischer's Glossary of Mineral Species*. 10th Edition. Tucson: The Mineralogical Record Inc.

Banks, D., Younger, P. L., Arnesen, R. T., Iversen, E. R. & Banks, S. B. (1997). Mine-water chemistry: the good, the bad and the ugly. *Environmental Geology, 32(3)*, 157-174.

Baron, D. & Palmer, C. D. (1996). Solubility of jarosite at 4-35 °C. *Geochimica Cosmochimica Acta, 60*, 185-195.

Barriga, F. J. A. S. (1990). Metallogenesis in the Iberian Pyrite Belt. In Dallmeyer, Martinez-Garcia (Eds), *Pre-Mesozoic geology of Iberia* (369-379). Springer Verlag.

Barriga, F. J. A. S., Carvalho, D. & Ribeiro, A. (1997). Introdution to the Iberian Pyrite Belt. *SEG Neves Corvo Field Conf. Guidebook*, V.*27*, 1-20.

Batista, M. J. (2000). *Environmental State in the Portuguese Test Site S. Domingos Mine: Past and Present*. Report European Comission.

Batista, M. J., Brito, M. G., Abreu, M. M., Sousa, A. J., Quental L. & Vairinho M. (2003). Avaliação por modelação em SIG da contaminação mineira por drenagem ácida em S. Domingos (Faixa Piritosa, Alentejo). *Ciências da Terra (UNL)*, Lisboa, n°esp. V. CD-ROM, M6-M10.

Batista, M. J., Abreu, M. M. & Pinto, M. S. (2007). Biogeochemistry in Neves Corvo mining region, Iberian Pyrite Belt, Portugal. *Journal of Geochemical Exploration, 92*, 159-176.

Bobos, I., Durães, N. & Noronha, F. (2006). Mineralogy and geochemistry of mill tailings impoundments from Algares (Aljustrel), Portugal: Implications for acid sulfate mine waters formation. *Journal of Geochemical Exploration, 88*, 1-5.

Bryan, C., Hallberg, K. & Johnson, D. (2006). Mobilisation of metals in mineral tailings at the abandoned São Domingos copper mine (Portugal) by indigenous acidophilic bacteria. *Hydrometallurgy, 83*, 184-194.

Cabezudo, B. & Rivera, J. (1980). Notas taxonómicas y corológicas sobre la Flora de Andalucía Occidental. 2: *Erica andevalensis* Cabezudo & Rivera sp. Nov. *Lagascalia, 9 (2)*, 223-226.

Cabral, J., Monteiro, S. & Barata, J. (1889). *Catalogo Descriptivo da Secção de Minas Gr. I e II*. Associação Industrial Portugueza, Exposição Nacional das Industrias Fabris, Imprensa Nacional (502pp). Lisboa, Portugal.

Caldwell, J. (2005). Mining history. Proposed technology drop-down menu. *Infomine, Mining Intelligence and Technology Magazine* http://www.infomine.com/publications/docs/caldwell2005a.pdf (access July 2009)

Candeias, C. (2008). *Caracterização ambiental da zona envolvente à área mineira de Aljustrel*. Master Thesis University of Aveiro. Aveiro, Portugal, 275pp. (Unpublished).

Cánovas, C. R., Olías, M., Nieto, J. M., Sarmiento, A. M. & Cerón, J. C. (2007). Hydrogeochemical characteristics of the Tinto and Odiel Rivers (SW Spain). Factors controlling metals contents. *Science of the Total Environment*, *373*, 363-382.

Carvalho, D. (1971). Mina de S. Domingos. *Jazigos Minerais do Sul de Portugal, Livro-Guia nº4*, 59-64.

Carvalho, D., Barriga, F. J. A. S. & Munhá, J. (1999). Bimodal-siliciclastic systems – the case of the Iberian Pyrite Belt. *Reviews in Economic Geology*, *8*, 375-408.

Castelo Branco, J. M. (1994). *Sociedade Mineira Rio Artezia, Área Grândola-Alcácer*. Relatório Técnico Arquivo LNEG.

Chou, I. M., Seal II, R. R. & Hemingway, B. S. (2002). Determination of melanterite-rozenite and chalcantite-bonattite equilibria by humidity measurements at 0.1 MPa. *American Mineralogist*, *87*, 108-114.

CSST, (2003). *Consensus based sediment quality guidelines, recommendations for use & application*. Wisconsin: Dept. of Natural Resources, 35.

Davis, R. A., Welty, A. T., Borrego, J., Morales, J. A., Pendon, J. G. & Ryan, J. G. (2000). Rio Tinto estuary (Spain): 5000 years of pollution. *Environmental Geology*, *39*, 1107-1116.

Dold, B. (2005). Basic concepts of environmental geochemistry of sulphide mine-waste. Centre d'Analyse Minérale, Université de Lausanne, Switzerland. http://www.unil.ch/webdav/site/cam/users/jlavanch/public/Le_personnel/Dold_Basicconcepts.pdf (access June 2009)

Domènech, P. J. & Ayora, C. (2002). Oxidative dissolution of pyrite sludge from the Aznalcóllar mine (SW) Spain. *Chemical Geology*, *190*, 339-353.

Domergue, C. (1983). *La mine d' Aljustrel (Portugal) et les Tables de Bronze de Vipasca*, Conimbriga 20THII, Univ. Coimbra, Coimbra, Portugal, 7-194.

Domergue, C. (2002). La mina romana de Aljustrel (Portugal) y el património minero a princípios del tercer milénio. *Actas Cong. Int. Património Geológico e Mineiro*, IGM, 231-238.

Durães, N., Bobos, I. & Silva, E. F. (2008). Simple Efflorescent Sulphates from Iberian Pyrite Belt (Portugal). Resumen Workshop. *Macla*, *10*, 126-128.

e-EcoRisk project (2004-2007) – A Regional Enterprise Network Decision-Support System for Environmental Risk and Disaster Management of Large-Scale Industrial Spills (#EVGI-2002-25 0068): http://www.e-ecorisk.info/

Evangelou, V. P. & Zhang, Y. L. (1995). A Review: Pyrite oxidation mechanisms and acid mine drainage prevention. *Critical Reviews in Environmental Science Technology*, *25*, 141-199.

Feliciano, R., Mateus, A. & Matos, J. X. (2008). Distribuições espaciais das concentrações em Cu, Zn, Pb e Co apresentadas por sedimentos de corrente no domínio setentrional da Zona Sul Portuguesa., implicações para a prospecção mineral. *V Sem. Rec. Geológicos, Ambiente e Ordenamento do Território* (13-22), UTAD Vila Real, Portugal.

Fonseca, E. C. & Silva, E. F. (2000). *Projecto Estudo de Controlo Ambiental nas Áreas Mineiras Abandonadas de Lousal e Caveira*. Rel. Téc. Final Univ. Aveiro. Arq. IGM.

Garcia Jr, O., Bigham, J. M. & Tuovinen, O. H. (2007). Oxidation of isochemical FeS_2 (marcasite-pyrite) by *Acidithiobacillus thiooxidans* and *Acidithiobacillus ferrooxidans*. *Minerals Engineering*, *20*, 98-101.

Garrels, R. M. & Christ, C. L. (1965). *Solutions, Minerals and Equilibria*. New York: Harper and Row.

Gaspar, O. C. (1998). História da mineração dos depósitos de sulfuretos maciços vulcanogénicos da Faixa Piritosa Portuguesa. *Boletim de Minas, 35/4*, 401-414.

Geirinhas, F., Gaspar, M., Martinez, J. C., Figueiredo, E., Araújo, M. & Silva, R. (2008). Copper isotopes on artifacts from Fraga dos Corvos habitat site and nearby Cu occurences: an approach on metal provenance. *V Cong. Int. SEDPGYM Minería y Metalurgia Históricas SW Europeo, (16)*. León, Spain.

Gerhardt, A., Janssens de Bisthoven, L., Guhr, K., Soares, A. M. V. M. & Pereira, M. J. (2008). Phytoassessment of acid mine drainage: *Lemna gibba* bioassay and diatom community structure. *Ecotoxicology, 17*, 47-58.

Gonçalves, M. A., Figueiras, J., Pinto, C., Neng, N., Sá-Pereira, P. & Batista, M. J. (2007). Biogeochemical and mineralogical characteristics of the acid mine drainage system in Aljustrel and S. Domingos mines, Iberian Pyrite Belt. *Goldschmidt Conference Abstracts*, 2007, Cologne, A341.

Grimalt, J. O., Ferrer, M. & Macpherson, E. (1999). The mine tailing accident in Aznalcollar. *The Science of the Total Environment, 242*, 3-11.

Hammarstrom, J. M., Seal II, R. R., Meier, A. L. & Kornfeld, J. M. (2005). Secondary sulfate minerals associated with acid drainage in the eastern US: recycling of metals and acidity in surficial environments. *Chemical Geology, 215*, 407-431.

Holmes, P. R. & Crundwell, F. J. (2000). The kinetics of the oxidation of pyrite by ferric ions and dissolved oxygen: An electrochemical study. *Geochimica et Cosmochimica Acta, 64*, 263-274.

http://www.mindat.org/gallery.php?min=2633 (access October 2009).

http://www.mindat.org/loc-126108.html (access October 2009).

Jamieson, H. E., Robinson, C., Alpers, C. N., McCleskey, R. B., Nordstrom, D. K. & Peterson, R. C. (2005). Major and trace element composition of copiapite-group minerals and coexisting water from the Richmond mine, Iron Mountain, California. *Chemical Geology, 215*, 387-405.

Janssens de Bisthoven, L., Gerhardt, A. & Soares, A. M. V. M. (2005). Chironomidae larvae as bioindicators of an acid mine drainage in Portugal. *Hydrobiologia, 532*, 181-191.

Jerz, K. J. & Rimstidt, J. D. (2004). Pyrite oxidation in moist air. *Geochimica et Cosmochimica Acta, 68*, 701-714.

Krieger, G. R. & Hattermer-Frey, Kester, J. E. (1999). Bioavailability of metals in the environment: Implications for Health Risk Assessment. In The Environmental Geochemistry of Ore Deposits. (357-361). *Reviews in Economic Geology. Volume 6A*. Society of Economic Geologists, Inc.

Lacal, J., Pilar da Silva, M., Garcia, R., Sevilla, M. T., Procopioa, J. R. & Hernandeza, L. (2003). Study of fractionation and potential mobility of metal in sludge from pyrite mining and affected river sediments: changes in mobility over time and use of artificial ageing as a tool in environmental impact assessment. *Environmental Pollution, 124*, 291-305.

Leduc, D., Leduc, L. G. & Ferroni, G. D. (2002). Quantification of bacterial populations indigenous to acidic drainage streams. *Water Air and Soil Pollution, 135*, 1-21.

Leistel, J. M., Marcoux, E., Thiéblemont, D., Quesada, C., Sánchez, A., Almodóvar, G. R., Pascual, E. & Sáez, R. (1998). The volcanic-hosted massive sulphide deposits of Iberian Pyrite Belt. *Mineralium Deposita, 33*, 2-30.

Leitão, J. (1997). Geology of the Aljustrel Massive Sulfides Deposits. *SEG Guidebook Series, 27*, 82-97.

Leitão, J. (1998). Geologia dos depósitos de sulfuretos maciços de Aljustrel. *Liv.-Guia V Cong. Nac. Geologia*, IGM, 91-100.

Luís, A. T., Teixeira, P., Almeida, S. F. P., Ector, L., Matos, J. X. & Ferreira da Silva, E. A. (2009). Impact of Acid Mine Drainage (AMD) on Water Quality, Stream Sediments and Periphytic Diatom Communities in the Surrounding Streams of Aljustrel Mining Area (Portugal). *Water Air and Soil Pollution, 200*, 147-167.

Macias, J. A. P., Matos, J. X. & Martins, A. (2008). Los Scavrarii de Vipasca. *Abst. V Cong. Int. SEDPGYM Minería y Metalurgia Históricas SW Europeo*, León, Spain.

Martins, A. (1996). Aljustrel, a mina e a mineração na antiguidade. In *Mineração no Baixo Alentejo I* (94-113). Câmara Municipal de Castro Verde, Portugal.

Martins, J. & Carvalho, P. (2005). Recuperação ambiental da área mineira de Aljustrel. *Abst. III Enc. Com. Mineiras Aljustrel*, Câmara Municipal de Aljustrel, Portugal.

Martins, A., Alves, H. & Costa, T. (2003). *2000 anos de mineração em Aljustrel*. Câmara Municipal de Aljustrel, Portugal, 78.

Martins, L., Batista, M. J., Matos, J. X., Oliveira, D., Shepherd, T., Pearson, G., Abreu, M. M., Melo, T., Barrosinho, J., Neves, C., Bebianno, M. J., Serafim, A., Company, R., Cravo, A., Locutura, J., Bel-Lan, A., Nieto, J. M., Delgado, J., Sarmiento, A. M., Del Valls, T. A., Blasco, J. M., Kalman, J. & Riba, I. (2007). As origens do chumbo na parte inferior da bacia do Guadiana. *VI Cong. Geoquímica* (278-281). UTAD Vila Real, Portugal.

Mason & Barry Company, Ltd. (1937-1955). S. Domingos. Substância Cobre, (Acess.) Mina nº 11, Livro 29, Corte Pinto, Mértola. *Circunscrição Mineira do Sul* (unpublished report)

Mason & Barry Company, Ltd. (1950-1952). S. Domingos. Substância Cobre, (Pesq.) Mina nº 11, Livro 29, Corte Pinto, Mértola. *Circunscrição Mineira do Sul* (unpublished report)

Mateus, A., Figueiras, J., Matos, J. X., Gonçalves, M. A., Lopes, R., Labaredas, J. & Beleque, A. (2008). Condicionantes impostas à dispersão de metais acumulados em escombreiras: o exemplo de Caveira (Faixa Piritosa Ibérica). In Callapez (Ed), *A Terra: Conflitos e Ordem* (373-382). Coimbra: Livro de Homenagem ao Professor Ferreira Soares, Museu Mineralógico e Geológico da Universidade de Coimbra, Portugal.

Mateus, A., Pinto, A., Alves, L. C., Matos, J. X., Figueiras, J. & Neng, N. (in press) Roman and modern slag at S. Domingos mine (IPB, Portugal): compositional features and implications for their long-term stability and potential re-use. *International Journal Environment and Waste Management*.

Matos, J. X. (2004). *Carta geológica e mineira da mina de São Domingos*, esc. 1/2500. DPMM INETI.

Matos, J. X. (2005). *Carta geológica e mineira da mina de Aljustrel*, esc. 1/5000. DPMM INETI.

Matos, J. X. (2006a). *Carta geológica e mineira da Mina de Caveira* esc. 1/5000, DPMM INETI.

Matos, J. X. (2006b). *Carta geológica e mineira da mina do Lousal*, esc. 1/2500. DPMM INETI.

Matos, J. X. & Martins, L. (2003). Itinerários geo–eco–educacionais como factor de desenvolvimento sustentado do turismo temático associado à Faixa Piritosa Ibérica. *Actas IV Cong. Int. Património Geológico Y Minero* (539-557). Utrillas, Spain: *SEDPGYM*.

Matos, J. X. & Martins, L. (2006). Reabilitação ambiental de áreas mineiras do sector português da Faixa Piritosa Ibérica: estado da arte e perspectivas mineiras. *IGME, Boletin Geológico y Minero España, 117,* 289-304.

Matos, J. X. & Oliveira, V. (2003). Mina do Lousal (Faixa Piritosa Ibérica) – Percurso geológico e mineiro pelas cortas e galerias da antiga mina. *IGME, Pub. Museo Geominero, 2,* 117-128.

Matos, J. X. & Rosa, C. (2001). Diagnóstico Preliminar de Minas Abandonadas – Área Sul. *Relatório Interno IGM,* 276.

Matos, J. X. & Sousa, P. (2008). Prospecção de Sulfuretos Maciços no Sector Português da Faixa Piritosa Ibérica. *Congresso Luso-Moçambicano de Engenharia.* Maputo, Moçambique.

Matos, J. X., Oliveira, J. M. S., Farinha, J. B., Ávila, P., Rosa, C., Leite, M. R. M., Daniel, F. & Martins, L. (2002). Património mineiro português: estado actual da herança cultural de um país mineiro. *Actas do Congresso Internacional Sobre Património Geológico e Mineiro,* (539-554). IGM/SEDPGYM, Beja, Portugal.

Matos, J. X., Barriga, F. J. A. S. & Oliveira, V. (2003a). Alunite veins versus supergene kaolinite/halloysite alteration in the Lagoa Salgada, Algares and S. João (Aljustrel) and S. Domingos massive sulphide deposits, Iberian Pyrite Belt, Portugal. *Ciências da Terra (UNL),* Lisboa, *V,* B56-B59.

Matos, J. X., Petersen, E. U. & Chávez, W. X. (2003b). *SEG-Environmental Geochemistry Field Course, Iberian Pyrite Belt. Portuguese mining sites: S. Domingos, Neves Corvo, Aljustrel, Lousal & Caveira.* IGM, Portugal. 18.

Matos, J. X., Martins, L. & Rosa, C. (2003c). Parque Mineiro da Cova dos Mouros – IGM contribute for the sustainable development of the mining park. *IGME, Pub. Museo Geominero, 2,* 487-494.

Matos, J. X., Pereira, Z., Oliveira, V. & Oliveira, J. T. (2006). The geological setting of the São Domingos pyrite orebody, Iberian Pyrite Belt. In *VII Cong. Nac. Geologia* (283-286). Estremoz, Un. Évora, Portugal.

Matos, J. X., Martins, L. P., Oliveira, J. T., Pereira, Z., Batista, M. J. & Quental, L. (2008a). Rota da pirite no sector português da Faixa Piritosa Ibérica, desafios para um desenvolvimento sustentado do turismo geológico e mineiro. In Paul Carrion (Ed.), *Rutas Minerales en Iberoamérica. Projecto RUMYS, programa CYTED* (136-155). Esc. Sup. Politécnica del Litoral, Guayaquil, Equador.

Matos, J. X., Martins, A., Rego, M., Mateus, A., Pinto, A., Figueiras, J. & Silva, E. (2008b). Roman slag mine wastes distribution in the Portuguese sector of the Iberian Pyrite Belt. *V Cong. Int. SEDPGYM Minería y Metalurgia Históricas SW Europeo* (45). León, Spain.

Matzke, K. (1971). Mina do Lousal. Jazigos Minerais Sul de Portugal, *Liv.-Guia, 4,* 25-32.

McKibben, M. A. & Barnes, H. L. (1986). Oxidation of pyrite in low temperature acidic solutions: Rate laws and surface textures. *Geochimica et Cosmochimica Acta, 50,* 1509-1520.

Montero, I. C. S., Brimhall, G. H., Alpers, C. N. & Swayze, G. A. (2005). Characterization of waste rock associated with acid drainage at the Penn Mine, California, by ground-based visible to short-wave infrared reflectance spectroscopy assisted by digital mapping. *Chemical Geology, 215,* 453-472.

Nieto, J. M., Sarmiento, A. M., Olías, M., Cánovas, C. R., Riba, I., Kalman, J. & Delvalls, T. A. (2007). Acid mine drainage pollution in the Tinto and Odiel rivers (Iberian Pyrite Belt, SW Spain) and bioavailability of the transported metals to the Huelva Estuary. *Environment International, 33,* 445-455.

Nordstrom, D. K. & Alpers, C. N. (1999). Geochemistry of acid mine waters, Part A- Processes, Techniques and Health. In G. S. Plumlee, & M. J. Logson, (Eds), *Environmental Geochemistry of Mineral Deposits,* Reviews in Economic Geology, vol. 6A. Soc. Econ. Geol. Inc., Littleton, 133-160.

Olías, M., Nieto, J. M., Sarmiento, A. M., Cerón, J. C. & Cánovas, C. R. (2004). Seasonal water quality variations in a river affected by acid mine drainage: the Odiel River (South West Spain). *Science of the Total Environment, 333,* 267-281.

Oliveira, J. S. (1997). Algumas reflexões com enfoque na problemática dos riscos ambientais associados à actividade mineira. *Estudos Notas e Trabalhos,* IGM, *39,* 3-26.

Oliveira, J. T. (1990). Stratigraphy and syn-sedimentary tectonism in the South Portuguese Zone. In R. D. Dallmeyer, & E. Martinez Garcia, (Eds.), *Pre-Mesozoic Geology of Iberia,* 334-347.

Oliveira, J. T. & Matos, J. X. (2004). O caminho de ferro da Mina de S. Domingos ao Pomarão: um percurso geo-educacional na Faixa Piritosa Ibérica. *XXIV Encontro Prof. Geociências APG,* 19.

Oliveira, M. L. F. & Matos, J. X. (2002).The exploitation of copper ores and the settlements of Estremadura and south of Portugal during the Calcolithic. *Rev. Ciências Históricas,* Univ. Portucalense, Porto, 123-138.

Oliveira, J. T. (Coord.)., Andrade, A., Antunes, M., Carvalho, D., Coelho, A., Feio, M., Gonçalves, F., Manupella, G., Marques, B., Monteiro, J., Munhá, J., Ramalho, M., Rey, J., Ribeiro, A., Rocha, R. & Zbyszewski, G. (1984). *Carta Geológica de Portugal 1/200000, Notícia Explicativa Fl. 7.* Serviços Geológicos de Portugal.

Oliveira, J. T. (coordenador)., Andrade, A., Antunes, M., Araújo, A., Castro, P., Carvalho, D., Carvalhosa, A., Dias, R., Feio, M., Fonseca, P., Martins, L., Manupella, G., Marques, B., Munhá, J., Oliveira, V., Pais, J., Piçarra, J., Ramalho, M., Rocha, R., Santos, J., Silva, J., Silveira, A. & Zbyszewski, G. (1992). *Carta Geológica de Portugal na escala 1/200.000 Notícia Explicativa da Folha 8.* Serviços Geológicos de Portugal.

Oliveira, J. T., Carvalho, P., Pereira, Z., Pacheco, N., Fernandes, J. P. & Korn, D. (1997). The stratigraphy of the Neves Corvo Mine Region. *Society Economic Geology. Neves Corvo Field Conference,* 86-87.

Oliveira, J. M. S., Farinha, J., Matos, J. X., Ávila, P., Rosa, C., Machado, M. J. C., Daniel, F. S., Martins, L. & Leite, M. R. M. (2002). Diagnóstico Ambiental das Principais Áreas Mineiras Degradadas do País. *Boletim Minas* IGM, *39(2),* 67-85.

Oliveira, J. T., Pereira, Z., Rosa, C. J., Rosa, D. & Matos, J. X. (2005). Recent advances in the study of the stratigraphy and the magmatism of the Iberian Pyrite Belt, Portugal. In Carosi, R., Dias, D. & Iacopini, G. Rosenbaum, (Eds.). *The southern Variscan Belt, Journal of the Virtual Explorer, Electronic Edition, 19/9,* 1441-8142.

Oliveira, J. T., Relvas, J. M. R. S., Pereira, Z., Matos, J., Rosa, C. J., Rosa, D., Munhá, J. M., Jorge, R. C. G. S. & Pinto, A. M. M. (2006). O Complexo Vulcano-Sedimentar da Faixa Piritosa: estratigrafia, vulcanismo, mineralizações associadas e evolução tectonoestratigráfica no contexto da Zona Sul Portugesa. In R., Dias, A., Araújo, P., Terrinha, & J. C. Kulberg, (Eds.), *Geologia de Portugal na Ibéria* (207-244). VII Cong. Nac. Geologia, Univ. Évora, Portugal.

Oliveira, J. T., Pereira, Z., Matos, J. X. & Fernandes, P. (2008). A palinostratigrafia no contexto da interpretação estrutural da Faixa Piritosa. *VIII Reunião Gr. Geologia Estrutural e Tectónica* (107-110). Fac. Ciências Univ. Porto, Portugal.

Pereira, Z., Matos, J. X., Fernandes, P. & Oliveira, J. T. (2007). Devonian and Carboniferous palynostratigraphy of the South Portuguese Zone, Portugal – An overview. *Com. Int. Microflore du Paléozoique* (111-114). INETI, Lisboa, Portugal.

Pereira, Z., Matos, J. X., Fernandes, P. & Oliveira, J. T. (2008). Palynostratigraphy and systematic palynology of the Devonian and Carboniferous sucessions of the South Portuguese Zone, Portugal. *Memórias do INETI*, T. 34, Lisboa, 181.

Pérez-López, R., Álvarez-Valero, A. M., Nieto, J. M., Sáez, R. & Matos, J. X. (2008). Use of sequential extraction procedure for assessing the environmental impact at regional scale of the São Domingos Mine (Iberian Pyrite Belt). *Applied Geochemistry, 23,* 3452-3463.

Pinto, C. (2007). *Caracterização geoquímica das fases aquosas e sólidas associadas aos processos geradores de Drenagem Ácida de Minas (DAM), na envolvente às Minas de Aljustrel*. Msc Thesis to Faculty of Sciences of Lisbon University, Portugal. (Unpublished).

Pinto, A., Mateus, A., Alves, L. C., Matos, J. X., Neng, N. & Figueiras, J. (2007). Detailed slag characterization relevance in environmental and economic assessments., the example of São Domingos (Iberian Pyrite Belt, Portugal). *VI Cong. Geoquímica* (345-348), UTAD Vila Real, Portugal.

Queiroz, N., Pereira, F., Bengala, J., Moreira, J., Freire, J., Viegas, L., Viana, M., Gaspar, O., Pereira, V. & Borralho, V. (1989). *Estudos Notas e Trabalhos SFM*, 50º Aniversário. Porto, Portugal.

Quental, L., Bourguignon, A., Sousa, A. J., Batista, M. J., Brito, M. G., Tavares, T., Abreu, M. M., Vairinho, M. & Cottard, F. (2002). *MINEO Southern Europe environment test site. Contamination impact mapping and modelling* - Final Report. Assessing and Monitoring the environmental impact of mining activities in Europe using advanced Earth Observation Techniques (MINEO) 5yth FP-IST-1999-10337. Mineo project report for European Comission.

Quental, L., Brito, M. G., Sousa, A. J., Abreu, M. M., Batista, M. J., Oliveira, V., Vairinho, M. & Tavares, T. (2003). Utilização de imagens hiperespectrais na avaliação da contaminação mineira em S. Domingos, Faixa Piritosa, Alentejo. *Ciências da Terra* (UNL), Lisboa, nº esp. V. CD-ROM, M33-M36.

Relvas, J. M. R. S., Barriga, F. J. A. S., Pinto, A., Ferreira, A., Pacheco, N., Noiva, P., Barriga, G., Baptista, R., Carvalho, D., Oliveira, V., Munhá, J. & Hutchinson, R. (2002). The Neves-Corvo Deposit, IPB, Portugal: 25 Years after the Discovery. *SEG Special Publication, 9,* 155-176.

Relvas, J. M. R. S., Barriga, F. J. A. S., Ferreira, A., Noiva, P. C., Pacheco, N. & Barriga, G. (2006). Hydrothermal alteration and mineralization in the Neves-Corvo volcanic-hosted massive sulfide deposit, Portugal: I. Geology, Mineralogy, and Geochemistry. *Economic Geology, 101,* 753-790.

Rimstidt, J. D. & Vaughan, D. J. (2003). Pyrite oxidation: a state-of-art assessment of the reaction mechanism. *Geochimica Cosmochimica Acta, 67(5),* 873-880.

Robie, R. A., Hemingway, B. S. & Fisher, J. R. (1978). Thermodynamic properties of minerals and related substances at 298.15 K and 1 Bar (10^5 Pascals) pressure and at higher temperatures. *US Geological Survey Bulletin*, 1452, Washington, 456.

Robinson, R. A. & Stokes, R. H. (1959). *Electrolyte solutions* (2nd ed). London: Butterworths.

Rosado, L., Morais, C., Candeias, A. E., Pinto, A. P., Guimarães, F. & Mirão, J. (2008). Weathering of S. Domingos (Iberian Pyritic Belt) abandoned mine slags. *Mineralogical Magazine*, 72, 489-494.

Rumys Project. Rutas Minerales de Iberoamérica y Ordenación Territorial. www.rumys.espol.edu.ec/

Sáez, R., Pascual, E., Toscano, M. & Almodovar, G. R. (1999). The Iberian type of volcano-sedimentary massive sulphide deposits. *Mineralium Deposita*, 5-6, 549-570.

Salgueiro, A. R., Pereira, H. G., Ferreira da Silva, E., Matos, J. X. & Ávila, P. H. (2007). Principal Components Analysis as a tool for identifying contaminated stream sediment samples in the Aljustrel mining area. *VI Cong. Geoquímica* (266-269) UTAD, Vila Real, Portugal.

Sarmiento, A. M. (2007). *Estúdio de la contaminación por drenajes ácidos de mina de las águas superficiales en la Cuenca del Río Odiel (SO España)*. PhD Thesis, Universidad de Huelva, Huelva, Spain. (Unpublished).

Schermerhorn, L., Zbyzewski, G. & Ferreira, V. (1987). *Carta Geológica 42D Aljustrel e respectiva Notícia Explicativa*. Serviços Geológicos de Portugal.

Silva, E. F., Fonseca, E. C., Matos, J. X., Patinha, C., Reis, P. & Santos Oliveira, J. M. (2005). The effect of unconfined mine tailings on the geochemistry of soils, sediments and surface waters of the Lousal area (Iberian Pyrite Belt, Southern Portugal). *Land Degradation & Development*, 16, 213-228.

Silva, E. F., Patinha, C., Reis, P., Fonseca, E. C., Matos, J. X., Barrosinho, J. & Santos Oliveira, J. M. (2006). Interaction of acid mine drainage with waters and sediments at the Corona stream, Lousal mine (Iberian Pyrite Belt, Southern Portugal). *Environmental Geology*, 50, 1001-1013.

Silva, E. A. F., Bobos, I., Matos, J. X., Patinha, C., Reis, A. P. & Fonseca, E. C. (2009). Mineralogy and geochemistry of trace metals and REE in volcanic massive sulfide host rocks, stream sediments, stream waters and acid mine drainage from the Lousal mine area (Iberian Pyrite Belt, Portugal). *Applied Geochemistry*, 24, 383-401.

Silva, F. (1968). As Minas do Lousal. *Boletim de Minas*, Dir. Geral de Minas e SGP, *5(3)*, 161-181.

Silva, J. B., Oliveira, V., Matos, J. X. & Leitão, J. C. (1997). Aljustrel and the Central Iberian Pyrite Belt. In Barriga, F. J. A. S. & Carvalho, D. (Eds.). *SEG Neves Corvo Field Conference 1997. Society of Economic Geologist, Guidebook Series*, 27, 73-124.

Singer, P. C. & Stumm, W. (1970). Acidic mine drainage: the rate determining step. *Science*, 167, 1121-1123.

Soldevilla, M., Maranón, T. & Cabrera, F. (1992). Heavy metal content in soil and plants from a pyrite mining area in southwest Spain. *Communications Soil Science Plant Analysis*, 23, 1301-1319.

Strauss, G. (1970). Sobre la geologia de la provincia piritífera del SW de la Península Ibérica y de sus yacimientos, en especial sobre la mina de pirita de Lousal (Portugal). *Mem. IMTET*, *77*, 266.

Tailsafe. Sustainable Improvement in Safety of Tailings Facilities – A European Research Project. www.tailsafe.bam.de

Tavares, M. T., Sousa, A. J. & Abreu, M. M. (2008). Ordinary kriging and indicator kriging in the cartography of trace elements contamination in São Domingos mining site (Alentejo, Portugal). *Journal of Geochemical Exploration*, *98*, 43-56.

Thadeu, D. (1989). Portugal. In Dunning, F.W. (Ed), *Mineral Deposits of Europe, Southwest and Eastern Europe with Iceland*. The Institution of Mining and Metallurgy – The Mineralogical Society, 4/5, 197-220.

Tornos, F. (2006). Environment of formation and styles of volcanogenic massive sulfides: The Iberian Pyrite Belt. *Ore Geology Reviews*, *28*, 259-307.

Vara, I. P. (1963). *Piritas de Huelva, su historia, minéria y aprovechamiento*. Madrid, Spain: Summa (Ed.).

Walder, I. F. & Schuster, P. P. (1997). *Acid Rock Drainage. Environmental Geochemistry of Ore Deposits and Mining Activities*. Short Course. SARB Consulting, Inc.

Wagman, D. D., Evans, W. H., Parker, V. B., Shumm, R. H., Halow, I., Bailey, S. M., Churney, K. L. & Nuttall, F. L. (1982). The NBS tables of chemical thermodynamic properties. *Journal of Physical and Chemical Reference Data*, *11* (Suppl *2*), 392.

Webb, J. (1958). Observations on the geology and origin of the San Domingos pyrite deposit, Portugal. *Comunicações dos Serviços Geológicos de Portugal*, *42*, 119-143.

Williams, P. A. (1990). *Oxide Zone Geochemistry*. Chichester, England: Ellis Horwood Limited.

Williamson, M. A. & Rimstidt, J. D. (1994). The kinetics and electrochemical rate-determining step of aqueous pyrite oxidation. *Geochimica et Cosmochimica Acta*, *58*, 5443-5454.

Wolkersdorfer, C. & Bowell, R. (2005). Contemporary Reviews of Mine Water Studies in Europe. *Mine Water and Environment*, *24*, 1-76.

In: Mine Drainage and Related Problems
Editor: Brock C. Robinson, pp. 119-154

ISBN: 978-1-60741-285-4
© 2010 Nova Science Publishers, Inc.

Chapter 3

BIOGEOCHEMISTRY OF ACID MINE DRAINAGE FORMATION: A REVIEW

V. Sheoran[a], A. S. Sheoran[b]* and R. P. Choudhary[b]

[a] Department of Zoology, Faculty of Science, J.N.V. University Jodhpur, India
[b] Department of Mining Engineering, Faculty of Engineering,
J.N.V. University Jodhpur, India

ABSTRACT

Acid Mine Drainage (AMD) and the contaminants associated with it is the most persistent environmental pollution problem, which occurs worldwide in the coal/metal mining regions. It occurs as a result of natural oxidation of sulphide minerals contained in the mining wastes at operating/closed/decommissioned mine sites. Once it develops at a mine, its control can be difficult and expensive. Treatment of AMD usually costs more than control of AMD and may be required for many years after mining activity is ceased. Thus the early diagnosis of the problem would reduce the potential role of AMD formation and controlling the factors responsible for generation. The focus of this paper is on the review of basic biogeochemistry involved in the formation of AMD and presents the various factors, which control the rate and extent of AMD formation.

Keywords: Acid mine drainage; Mining; Mineral processing; Environmental; Pollution.

1. INTRODUCTION

Mining activities generate waste products such as mine overburden and mine tailings (waste soil). The management of these waste materials is an important issue for mining industry worldwide. This becomes challenging especially, if the waste material contains sulphide ores. Dissolution of sulphide ores exposed to oxygen, water and microorganisms

---

* Corresponding author: E-mail: as_sheoran@yahoo.com.

results in acid production and environmentally detrimental Acid mine drainage (AMD), terminology such as Acid Rock Drainage (ARD), are generally interchangeable for the term AMD as used in this review. ARD is often referred as AMD which occur only association with mining activity. (Nordstrom and Alpers, 1999; Bond et al., 2000).

Although this process occurs naturally, mining can promote AMD generation increasing the quantity of the sulphides exposed. Furthermore in metal mining industry, the natural oxidation of the sulphide minerals in the presence of water can generate dilute sulphuric acid which can cause liberation of toxic heavy metals (MEND, 1997; Azapagic, 2004; Ackil and Koldas, 2006; Butler and Seitz, 2007; Butler, 2007). Thus water infiltrating through these mines and mines spoils are therefore often acidic and contains elevated concentrations (relative to unpolluted streams) of sulphate (SO_4^{2-}) and metals such as Fe, Mn, Al, Zn, Cu, Ni, Pb, Cd, As etc. These metals originate from dissolved sulphides, and other associated minerals, and are held in solution because of there greater solubility in acidic liquors (Johnson and Hallberg, 2003). Iron may be present in either ferrous or ferric forms (depending on water pH and Eh) (Langmuir and Whitemore, 1971) is, in most cases, the dominant metal present in AMD, soluble ferric iron imparting the characteristic red colour to extremely acidic mine waters while orange-yellow ferric iron rich sediments ("Ochre" or yellow boy) are common in higher pH streams. This affected water is referred to as AMD which have a serious impact on human health and have ecological implications (Sracek et al., 2004). The generation of AMD is also influenced by the biological activity of microorganisms which assist in breaking down of sulphide minerals (Kuyucak, 2002).

The nature and extent of contamination due to AMD is highly variable depending upon nature of the ore body, mineralogy of local rock material/associated geological strata, climate, mineral engineering constraints and the availability of water and oxygen. Therefore, AMD formation is highly variable from site to site, predicting the potential for AMD treatment can be exceedingly challenging and costly (USEPA, 1994; Akcil and Koldas, 2006).

There are no standard equally applicable mitigation methods to reduce the associate risk of AMD. Every mine is unique in terms of its AMD potential, thus mitigation options also varies from site to site, requires site–specific research to be undertaken. Thus research needs to be carried out to identify the ways in which the generation of AMD can be prevented.

Further the AMD production in mine tailings has been reported extensively by Dubrovsky et al. (1985);Blowes et al. (1991); Morel and Hering (1993); Morin et al. (1988, 1994); Stumm and Morgan (1996); Gunsinger et al. (2006). Briefly the minerals such as pyrite and other metal sulphides are chemically and biologically stable in situations where both oxygen and water are excluded. However following the fracturing and exposure of sulphide containing mineral ores and sulphidic coals the sulphide minerals will oxidize spontaneously to form acidic sulphate rich drainage. AMD, associated with Metal contamination depends on the type and amount of the sulphide minerals oxidised, as well as the type of the gangue minerals present in the rock. The three primary ingredients necessary in its formation are: (1) wastes containing reactive sulphides, including sulphide minerals (S^{2-} or S_2^{2-}), elemental sulphur (S^0) and various sulphur intermediates, Thiosalts $S_2O_3^{2-}$. (2) Water or a high humid atmosphere, and (3) an oxidant, particularly oxygen or air (Kuyucak, 2002).

Without any one of the three essential primary ingredients may likely preclude AMD formation. Like sulphide minerals will not dissolve in water without first oxidizing in air. Conversely, if sulphide minerals occur in a dry environment, and are oxidized in air but there is no contact with water, then the constituents will not be dissolved or mobilized. Similarly if sulphides are submerged beneath surface or ground water that has a low concentration of dissolved oxygen, then little or no oxidation can occur. In majority of cases bacteria play a major role in accelerating the rate of acid generation, the inhibition of these bacterial activities can therefore impede the rate of acid generation (Akcil and Koldas, 2006; Kleinmann, 1980; Kleinmann and Crerar, 1979). Thus AMD if not properly managed, the environmental impacts from AMD can result in serious degradation of water quality.

The most well documented type of water pollution associated with mining is AMD (Kleinmann, 1990), in contrast much less is known about AMD generation from waste mining products. The geochemical environment within the waste rock heaps needs to be well known to prevent the AMD generation itself.

Thus investigation of the geochemistry and biogeochemistry of these environments is a prerequisite for effective management and remediation strategies of AMD. The studies will allow for a more accurate analysis of the mechanism of acidification and neutralization (Klapper and Schultze, 1995; Miller et al., 1996; Blodau, 2006; Seal II et al., 2008).

The objective of this review is to evaluate the recent scientific process in biogeochemistry of AMD formation and the role of various factors including the role of micro organisms which play a major role in accelerating the rate of acid generation; the inhibition of bacterial activity and aforementioned variables can therefore impede the rate of AMD generation.

2. SOURCES OF ACID MINE DRAINAGE AND SOME EXAMPLES FROM INDIA

AMD associated with Surface and Underground mining has long been considered a serious environmental hazard. In such settings, AMD often occurs as low pH water and can therefore be easily identified. More recently, AMD emanating from open pit workings and mining waste heap dumps has become a concern. Little is known of the potential danger posed by these operations, as most are still being worked or maintained. In this method of mining, large volumes of rock are initially subjected to an oxidizing environment. The oxidation of the sulphide minerals and subsequent release of heavy metals and acidity from mine wastes can last for decades and centuries (Blowes and Jambor, 1990; Blowes et al., 1994; Ljungberg and Ohlander, 2001). The sources of AMD includes (Table 1):

Table 1. Sources of Acid Mine Drainage (After Ackil and Koldas, 2006)

Primary sources involves	Secondary sources
Minerals tailings impoundment, Spoil heaps	Mine shaft and Adits
Mine rock dumps	Treatment sludge pounds
Underground and open pit mine workings	Rock cuts
Pumped/nature discharged underground water	Concentrated load-out
Diffuse seeps from replaced overburden in rehabilitated areas	Stockpiles
	Concentrate spills along roads
Construction rock used in roads, dams.	Emergency ponds
Abandoned mines, decommissioned tailing dams, etc.	

Waste rocks, which includes spoil piles, tailings dams and spent heap-leach piles are a significant source of AMD. Acidic overburden exacerbates acidic seep problems as no neutralization of effluent occurs once it has left the ore body (Ackil and Koldas, 2006). One of the areas of concern is acid mine drainage from mine rock piles (waste dumps), produced in high volumes may require extensive control and remediation measures to prevent adverse environmental impact. Thus the proper management of the sources responsible for AMD generation are important from both environmental and on economic perspective.

India is blessed with the occurrence of many minerals, often in large quantities and of strategic importance to it and to other nations. The country has one of the most sophisticated and developed mining industries in the world. The overall goal of managing environmental impacts in India is to design and implement mitigating measures that minimize the residual impact of mining. AMD under normal Indian conditions have been reported from many sites: Handa (1990) reported AMD problem in Jharia coalfield. Shrivastava and Singh (1994) have reported that trace metals: lead, copper, iron, zinc, chromium, manganese and many other metals are responsible for heavy pollution in Jharia coalfields. Handa (1990) also reported that in Shahdol district (Madhya Pradesh) AMD with pH 2-3 and nearby hand pump water had pH of 4.0. Bhole (1994) reported in Assam state the AMD of pH 3.9 in Ledo mine, 3.10 in Tirap and 4.3 pH in Bargolia where as in Madhya Pradesh state pH was reported to be 2.96 in Ravanwara, 6.45 in Chandmeta and 6.64 in Ekalehara. Gupta (1999) reports that the chemical analysis of the three coal mines from the Pench Valley contains trace elements like cobalt, copper, chromium, manganese, nickel and lead and their concentration ranges from 17-36.5 mg/L, 66-105 mg/L, 55-58.5 mg/L, 745-935 mg/L, 59-78 mg/L and 18.5-22 mg/L, respectively. These elements in turn find their way into natural water and chronic exposure to toxic concentration of these elements will be very severe, the cumulative effect of which may produce health hazards in the area. The metals such as arsenic, cobalt, copper, nickel and lead have been categorized into the category of very toxic and relatively accessible from environmental pollution point of view. Sahu (1990) suggested that trace elements such as cobalt, cadmium, manganese and zinc are present in high concentration in respirable coalmine dust. Bandopadhyay (1994); Singh and Sinha (1992) claimed that coal washery effluents showed increase in hardness values of river water, which remained in the range of 178 to 352 mg/L. The iron content of all effluent samples was found to be 200 mg/L, quite higher than the permissible level of 10 mg/L. According to Dubey and Nath (1997); and Handa (1990) at present over one thousand abandoned quarries/subsided areas in the coalmines exist in India. Several of these are giving continuous rise to AMD and several quarries are proposed to be reclaimed, hence planning needs to be carried out while reclaiming so that AMD abatement takes place.

"Panadhro Lignite Project", Gujarat Mineral Development Corporation, produces lignite 15,000-tonnes/day. Lignite is associated with shale containing sulphurous ingredients and traces of pyrite. As the volume of excavation in Panandhro is very high, large pits are created and during rainy season these become virtually ponds of acidic water. In this reference a pit in this area, which incidentally is confluence of two rivers Kali and Korawadi had almost become a small "acid water lake" with storage of 0.5 million m^3 acidic water. The pH of lake water is as low as 2.0 (Purohit et al., 2001).

Currently, AMD is treated by adding lime to neutralize the acid and precipitate heavy metals as hydroxides. These can be flocculated, forming a high density sludge (HDS), which once settled can give a relatively clear overflow, and then discharged into rivers (the HDS process).

3. BASIC CHEMISTRY OF ACID MINE DRAINAGE

The geochemistry of AMD has been the subject of numerous investigations, some common references are Temple and Koehler (1954); Singer and Stumm (1970); Kleinmann et al, (1981); Nordstrom (1982); Williams et al. (1982); Hornberger et al. (1990); Alpers and Blowes (1994); Blowes and Jambor (1994); Evangelou (1995); Nordstrom and Alpers (1999); Plumlea and Logsdon (1999); Komnitsas et al., (2001); Morin et al. (2001); Kuyucak, (2002); Sracek et al., (2004); Blodau (2006); Jambor and Vaughan (2006); Zhao et al. (2007) have reported studies on basic principles, mineralogy, hydrogeology, analytical, remediation techniques and geochemical characterization and generation of acidity in waste rock pile and coal mine lakes. Evangelou and Zang (1995); Belzile et al. (2004) have also reported studies on mechanism of pyrite oxidation and oxidation prevention. Nemati et al. (1998); Bond et al. (2000); Nordstrom (2000); Brake et al. (2001); Kuyucak (2002); Zanker et al. (2002); Johnson (2003); Lee and Chon (2006) have focused on advances in hydro geochemistry and microbiology of acid mine waters. Eary (1999); Shevenell et al. (1999); Bowell (2002) have reviewed the geochemical and equilibrium trends in mine pit lakes.

AMD results when certain sulfide minerals are exposed to oxidizing conditions with oxygen, water, and bacteria. The basic conditions which enable AMD are met when (i) the waste contains the enough sulfide to react chemically and biologically and form acid leachate at a rate faster than it can be neutralized by any alkaline compounds contained in the waste (ii) the physical properties of the waste are such that they allow both water and oxygen to support the chemical and biological reactions, (iii) the climate is the humid enough to cause rainfall infiltration and/or the waste is located such that surface water passes through it and transports the acidic drainage into the environment (Belzile et al., 1997; Hutchison and Ellison, 1992). A series of complex geochemical and micro-biological reactions occurs when water comes in contact with the sulphide minerals in metal mine operation, in coal, refuse or the overburden of a mine operation. The resulting water is usually high in acidity and dissolved metals. Acidity is commonly expressed as milligrams of $CaCO_3$ per liter of solution (mg/l as $CaCO_3$) on the basis of following reaction:

$$2H^+ + CaCO_3 \longrightarrow Ca^{2+} + CO_2 + H_2O \qquad (1)$$

2 moles (2.0 g) of H^+ are neutralized by 1 mole (100.1gm) of $CaCO_3$. On this basis acidity can also be expressed as mill equivalents per liter (meq/l) where 1meq/l equals 50 mg/l as $CaCO_3$ (Rose and Cravotta, 1998). There are many types of sulphide minerals. Two major iron disulfide minerals pyrite, marcasite (FeS_2), both minerals have the same chemical composition but differ crystallographically. The structure of the marcasite is orthorhombic while the pyrite is isomeric (Stumm and Morgan, 1970). Because of the particular structure, marcasite is les stable than pyrite, these are the principal sulfur-bearing minerals in coal but other metal sulphide mineral like Pyrrhotite (FeS), arsenopyrite (FeAsS), chalcopyrite ($CuFeS_2$), $CuFeS_4$, ZnS, PbS, HgS, CoAsS, $Fe_{1-x}S_x$, Fe_3S_4 and other sulfide minerals

containing Sb, Bi, Se and Mo also can produce acidic solutions upon oxidation, Because of its wide distribution in coal and overburden rocks, especially in shales, pyrite (also known as "Fools gold") is recognized as the major source of acidic drainage.

The geochemical process involves a complete oxidation of pyrite as described in the following reactions (Singer and Stumm, 1970; Nordstrom, 1979; Stumm and Morgan, 1981):

In the initial step pyrite reacts with oxygen and water to produce ferrous iron, sulphate and acidity. This reaction generates two moles of acidity for each mole of pyrite oxidized.

$$2\ FeS_2(s) + 7\ O_2 + 2\ H_2O \longrightarrow 2\ Fe^{2+} + 4\ SO_4^{2-} + 4\ H^+$$
Pyrite + Oxygen + Water ⟶ ferrous iron + Sulphate + Acidity (2)

The second reaction involves the conversion of ferrous iron to ferric iron. This reaction has been termed the "rate determining" step for the overall acid generating sequence. This conversion of ferrous iron to ferric iron consumes one mole of acidity. This reaction is pH dependent with the reaction proceeding slowly under acidic conditions (pH 2-3) with no bacteria present and several orders of magnitude faster at pH values near 5. Certain bacteria increase the rate of oxidation from ferrous to ferric iron. Reaction 2 & 3, which involve the oxidation of sulphur and iron, respectively by gaseous or dissolved O_2 can also be mediated by various species of sulphur and iron-oxidising bacteria, notably Acidithiobacillus ferroxidans. The pyrite weathering process along with chemicals has important microbiological components. These bacteria, which require only dissolved CO_2, O_2, a reduce form of Fe or S and minor N and P for their metabolism produce enzymes which catalyze the oxidation reactions, and use the energy released to transform inorganic carbon into cellular matter (Temple and Delchamps, 1953; Kleinmann et al., 1981; Nordstrom, 1982a; Ehrlich, 1990).

$$4\ Fe^{2+} + O_2 + 4\ H^+ \rightarrow 4\ Fe^{3+} + 2\ H_2O$$
Ferrous iron + Oxygen + Acidity → Ferric iron + Water (3)

The third step involves the hydrolysis of ferric iron with water to form the solid ferric hydroxide and the release of additional acidity. Hydrolysis is a reaction which splits the water molecule. Three moles of acidity are generated as a byproduct. Many metals are capable of undergoing hydrolysis. The formation of ferric hydroxide precipitate (solid) is pH dependent. Under very acidic conditions of less than about pH 3.5, the solid mineral does not form and ferric iron remains in solution. At higher pH values, a precipitate forms, commonly referred to as "yellow boy". Figure 1 showing seepage of AMD in an underground copper mine where yellow boy formation is deposited on the side walls of the mine. Most iron released during the initial stages of pyrite oxidation ends up as iron hydroxide due to relatively high pH on pyrite surfaces (Ivanov, 1962; Fornasiero et al., 1992).

$$4Fe^{3+} + 12\ H_2O \rightarrow 4Fe(OH)_3(s) + 12\ H^+$$
Ferric iron + Water → Ferric Hydroxide + Acidity (4)
(Yellow boy)

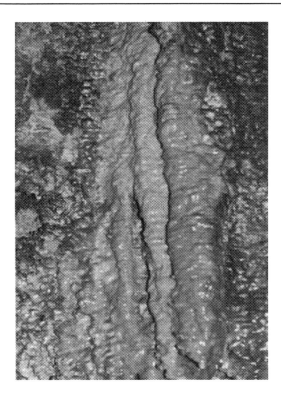

Figure 1. Acid mine drainage seepage in underground Copper Mines

The fourth step involves the oxidation of additional pyrites by ferric iron. The ferric iron is generated by the initial oxidation reactions in steps one and two. This cyclic propagation of acid generation by iron takes place very rapidly and continues until the supply of ferric iron or pyrite is exhausted. Oxygen is not required for the fourth reaction to occur. Note that in this reaction iron is the oxidizing agent, not oxygen (Garrels and Thompson, 1960; McKibben and Barnes, 1986; Moses et al., 1987; Moses and Herman, 1991; Williamson and Rimstidt, 1994).

$$FeS_2 + 14\ Fe^{3+} + 8\ H_2O \rightarrow 15\ Fe^{2+} + 2\ SO_4^{2-} + 16\ H^+$$
Pyrite + Ferric iron + Water → Ferrous iron + Sulphate + Acidiy (5)

In reaction 5, dissolved ferric iron (Fe^{3+}) produced by reaction 4 is the oxidizing agent for pyrite. Kinetic studies show that at acidic pH, rates of pyrite oxidation by Fe^{3+} are much faster than oxidation by O_2 (reaction 3). Therefore, in well-established acid-generating environments, the typical sequence is pyrite oxidation by reaction 5 to produce Fe^{2+}, which is then oxidized to Fe^{3+} by bacteria via reaction 3; the Fe^{3+} is then available for further pyrite oxidation. Although O_2 is not directly consumed in the pyrite-oxidizing step, it is necessary for the regeneration of Fe^{3+} to continue the pyrite-oxidation cycle (Taylor et al., 1984 a,b; Van Everingder and Krouse, 1985; Reedy et al., 1991; Taylor and Wheeler, 1994). As a final step, part or all of the iron may precipitate as Fe (OH)$_3$ or related minerals (reaction 6). Because of the relative insolubility of Fe^{3+}, most dissolved iron in solutions with pH greater than about 3.5 occurs as Fe^{2+}. AMD solutions commonly have pH and Eh (oxidation potential) (Langmuir and Whittemore, 1971; Morin and Hutt, 1998). Near neutral pH iron bearing solutions (pH 5-6) are normally relatively reduced, while more acidic solutions are more oxidizing. At pH less than about 3 (the exact value depending on iron and sulphate content) major amounts of dissolved Fe^{3+} can be present.

The overall chemical reaction describing the oxidation of pyrite is commonly given as:

$$4 FeS_2(s) + 15 O_2 + 14 H_2O \rightarrow 4 Fe(OH)_3(s) + 8 H_2SO_4$$
Pyrite + Oxygen + Water → "Yellow boy" + Sulphuric Acid (6)

In above reaction, solid pyrite, oxygen (O_2) and water (H_2O) are reactants, and solid ferric hydroxide Fe $(OH)_3$ and sulphuric acid are products. The heat energy produced in this reaction for complete conversion of one mole of pyrite to ferric hydroxide amounts to about 1490 kilojoules at 25°C (Naumov et al., 1974; Robie et al., 1978).

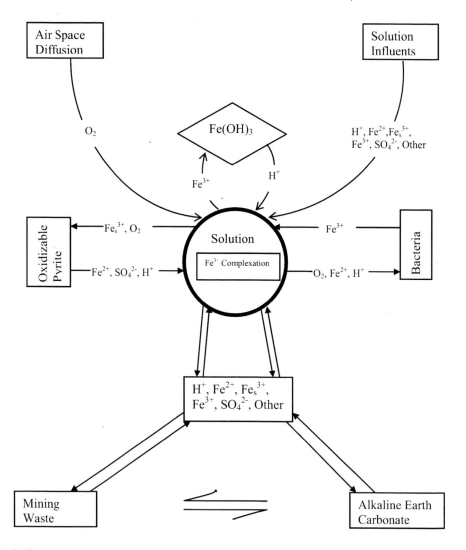

Figure 2. Conceptual schematic diagram of pyrite oxidation in pyrite mine "waste". Ferric iron complexes designated as Fes include all solution species ((Koski et al., 2008)

Hydrolysis and precipitation of iron solids, denoted as Fe(OH)$_3$, generally do not take place until the water is aerated and/or the acid is neutralized. The iron solids, which commonly form a reddish-yellow to yellowish-brown coating on the surfaces, can consist of a variety of amorphous or poorly crystalline ferric oxides, hydroxides or oxyhydroxysulfate minerals including ferrihydrite (variously given as $Fe_3(OH)_4$, $Fe_5HO_8\cdot 4H_2O$, or $Fe_5O_3(OH)_9$), goethite (FeOOH) and schwertmannite ($Fe_8O_8(OH)_6SO_4$) (Chukhrov et al., 1973; Hamilton and Woods, 1981; Goldhaber, 1983; Hood, 1984; Brady et al., 1986; Eggleton and Fitzpatrick, 1988; Fornasiero et al., 1992; Murad et al., 1994; Evangelou, 1995; Bigham et al., 1996). Ferrihydrite and schwertmannite are metastable and may ultimately dehydrate and recrystallize forming hematite (Fe_2O_3) or goethite. In acidic weathering environments, ferric sulphate compounds may form metastable intermediates prior to or in addition to ferric oxyhydroxide compounds (Nordstrom et al., 1979; Nordstrom, 1982a; Rose and Cravotta, 1998). Because the ferric sulphate minerals tend to be soluble and typically form under evaporating conditions, they can be significant sources of acidity and sulphate when later dissolved by runoff or recharge water at surface mines (Nordstrom and Dagenhart, 1978; Olyphant et al. 1991; Cravotta, 1994). Furthermore, other forms of sulphur, such as native S and $S_2O_3^{2-}$, can be intermediate products in the oxidation of pyrite (Nordstrom, 1982b; Goldhaber, 1983; Moses et al., 1987; Kuyucak, 2002), but these tend to oxidize to SO_4^{2-} under surface conditions.

AMD production potential is predicted after understanding the factor, that controlling pyrite oxidation. Such conceptual diagram describe pyrite oxidation, which is shown in the figure 2 in this representation, pyrite oxidation is carried out by O_2 or Fe^{3+}, regeneration of Fe^{3+} is considered to take place by bacteria. Reaction between pyrite oxidation products, carbonates, and mine waste surfaces are also included at the lower portion of the figure 2. This conceptual diagram also adds the clear vision of the pyrite oxidation, which allows one to predict acid production (Evangelou, 1995; Koski et al., 2008).

4. CONTRIBUTING FACTORS OF ACID MINE DRAINAGE

The potential for a mine or its associated waste to generate AMD is a complex geochemical and microbiologically mediated process and is site specific. The acid load ultimately generated from a mine site is primarily a function of the following factors:

4.1. Primary/Generation Factors

4.1.1. Sulphide minerals
Acid mine waste pollution is caused by the physical and chemical weathering of a very common mineral. The main culprit seems to be pyrite or the less abundant marcasite (FeS_2) which may form at lower pH and possibly lower sulphur concentration(Ward, 2002; Blodau, 2006) but AMD can occur also under natural conditions or where sulphides in geologic materials are encountered in metal mining and other deep excavations. The term pyrite is used to collectively refer to all iron disulphide minerals, water, oxygen, ferric iron, bacteria to

catalyse the oxidation reaction, and generate heat. Both water and oxygen are necessary to generate acid drainage.

If metallic minerals such as galena (lead sulphide, PbS), chalcopyrite (iron copper sulphide, $CuFeS_2$,), sphalerite (zinc sulphide, ZnS), molybdenite (MoS_2), millerite (NiS), arseno pyrite (FeAsS). in addition to pyrite and pyrrhotite are present (as is usually the case in the natural oxidation of a mineral deposit and the oxidation of products from the mining of a mineral deposit) then there may be a secondary effect of the oxidation of the iron-sulphur minerals to sulphuric acid and ferric iron. In ores, often pyrrhotite and these metal sulphides occur in appreciable quantities (Evangelou, 1995; Plumlee et al., 1999; Plumlee and Logsdon, 1999; Ljungberg and Ohlander, 2001;). Acidity levels, metal composition and concentration depend on the type and amount of sulphide mineral and the presence or absence of alkaline materials (Stumm and Morgan, 1996). Ferguson and Erickson (1988) have reported that metal ions can be released from the metal sulphide mineral by direct oxidation, bacterial oxidation or acidified ferric sulphate dissolution.

However, ferric iron ion is capable of dissolving many heavy metal sulphide minerals, including those of lead, copper, zinc and cadmium, by the general reaction:-

$$MS + nFe^{3+} \rightarrow Mn^+ + S + nFe^{2+} \qquad (7)$$

Where MS= solid heavy metal sulphide; Fe^{3+}= aqueous ferric iron ion; Mn^+ = aqueous heavy metal ion; S = sulphur, Fe^{2+} = aqueous ferrous iron ion.

It is by this process that significant amount of heavy metals may be solubilised by acid mine drainage.

Chalcopyrite: (Dutrizac and MacDonald, 1974; Murr and Mehta, 1982)

$$CuFeS_2 \rightarrow Cu^{2+} + Fe^{2+} + 2S° + 4e \qquad (8)$$

$$CuFeS_2 + 4Fe^{3+} \rightarrow Cu^{2+} + 5Fe^{2+} + 2S° \qquad (9)$$

$$CuFeS_2 + 4Fe^{3+} + 2H_2O + 3O_2 \rightarrow Cu^{2+} + 5Fe^{2+} + 2H_2SO_4 \qquad (10)$$

$$2CuFeS_2 + 17/2\ O_2 + H_2SO_4 \xrightarrow{bacteria} 2CuSO_4 + Fe_2(SO_4)_3 + H_2O \qquad (11)$$

Formation of copper sulphide has shown in the figure 3 of an underground copper mine where chalcopyrite is being extracted.

Chalcocite: (Dutrizac and MacDonald, 1974; Bruynesteyn and Duncan, 1977)

$$Cu_2S + 4Fe^{3+} \rightarrow 2Cu^{2+} + 4Fe^{2+} + S° \qquad (12)$$

$$Cu_2S + H_2SO_4 + \tfrac{1}{2}O_2 \rightarrow CuS + CuSO_4 + H_2O \qquad (13)$$

$$CuS + 2O_2 \xrightarrow{bacteria} CuSO_4 \qquad (14)$$

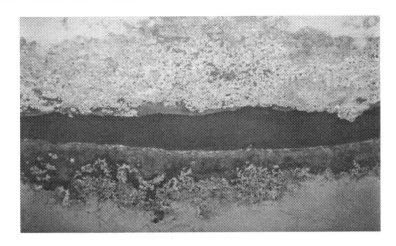

Figure 3. Copper sulphate formation with AMD in an underground copper mine

Sphalerite: (Dutrizac and MacDonald, 1974)

$$ZnS + 2Fe^{3+} \rightarrow Zn^{2+} + 2Fe^{2+} + S^\circ \tag{15}$$

$$ZnS + 2O_2 \xrightarrow{bacteria} ZnSO_4 \tag{16}$$

Galena: (Dutrizac and MacDonald 1974)

$$PbS + Fe_2(SO_4)_3 \rightarrow PbSO_4 + 2FeSO_4 + 2S \tag{17}$$

Millerite: (Bruynesteyn and Hackl 1984)

$$NiS + 2O_2 \xrightarrow{Bacteria} NiSO_4 \tag{18}$$

4.1.2. Water

The volume, depth and fluctuation of water in the mine plays a critical role in balance of acidity and alkalinity in effluent mine drainage. Within unsaturated spoil, water typically fills small pores and occurs as films on particle surfaces. Flow rates of the water vary from relatively rapid movement through interconnected large pores, fractures, and joints to slow movement or nearly stagnant conditions in water films or small pores. Water serves as a reactant, a medium for bacteria, and the transport medium for oxidation products (dissolving and transporting the solutes in the mine drainage). Only minimal amounts of waters are necessary for transporting the iron and aluminum sulfate causing acidity, since these salts are extremely soluble and high concentrations can be readily carried, more than 12 percent (120,000 ppm) total dissolved solids has been observed in mine drainage. The role of water in the formation of AMD is paradoxical, if there is zero relative or absolute humidity no acid producing reaction take place. Even complete flooding will surely retard AMD reactions by limiting availability of oxygen. Any conditions prevailing between these two extremes will lead to formation of AMD (Steger, 1982; Morel and Hering, 1993; Stumm and Morgan,

1996). Water chemistry locally sustains the chemical disequilibrium that drives geochemical and biogeochemical processes. A master variable for almost all processes is the water pH. Rate of pyrite oxidation, ferrous iron oxidation, mineral dissolution, iron precipitation; iron hydroxide transformation and metal and sulphate reduction are strongly water pH dependent.

4.1.3. Oxygen

Atmospheric air containing the oxygen needed for pyrite oxidation is commonly the limiting factor in the rate of acid generation. Both access of air and exposure of pyrite surfaces are promoted by breaking the pyrite bearing rock. The oxygen can gain access either by molecular diffusion through the air filled pore space in the spoil, or by flow of air which is driven through the pore space by temperature or pressure gradients. Oxygen is particularly important to maintain the rapid bacterially catalyzed oxidation at pH values below 3.5. Oxidation of sulphides is significantly reduced when the concentration of oxygen in the pore spaces of mining waste units is less than 1 to 2 percent. A study using XPS (Knipe et al., 1995) confirm that oxygen is the primary oxidant as no evidence of oxidation could be shown when iron sulfide is exposed to deoxygenated water. The oxidation rate is reported to be proportional to the square root of the O_2 partial pressure for pyrite by Mckibben and Barnes (1986) and Williamson and Rimstidt (1994) in the equation (19) & (20) respectively:-

$$r = 10^{-6.77} [O_2]^{0.5} \tag{19}$$

$$r = 10^{-8.19} [O_2]^{0.5} / [H^+]^{0.11} \tag{20}$$

Rates are expressed in mol of pyrite cm^{-2} min^{-1} for equation (19) and mol m^{-2} s^{-1} for equation (20). Diffusion of oxygen in water is a slow process, and solubility of oxygen is a low. The effective exclusion of atmosphere oxygen from pyretic spoil can be achieved by perpetual immersion of the spoil in stagnant ground water (Watzaf, 1992; Guo et al., 1994; Rose and Cravotta, 1998)

4.1.4. Bacteria/ microenvironment

The pyrite weathering process is a series of chemical reactions, where microbiological components catalyze every step in production of AMD. The commission of ferrous to ferric iron in the overall pyrite reaction sequence has been described as the "rate-determining step". This conversion can be greatly accelerated by a species of bacteria, Acidithiobacillus ferrooxidans (acidic tolerant bacteria). Acidophilic ferrooxidans is known for its ability to greatly accelerate the oxidative dissolution of sulphide minerals in mine waste material, thereby increasing acidity (Colmer et al., 1950). These bacteria can thrive under extreme environmental conditions that would prove fatal to most life forms (Whitton and Say, 1975; Allan, 1995; Brake et al., 2001) and several other species thought to be involved in pyrite weathering are widespread in the environment. Bacteria play a catalytic role in AMD generation; temperature and pH appear to be the determining factors for microbial growth in mine waste environments (Majima and Peters, 1966; Nicholson et al., 1988; Konishi, 1990; Nicholson and Scharer, 1994; Kwong, 1995; Janzen, 1996; Johnson and Hallberg, 2003; Belzile et al., 2004). Steger (1982) also reports the important role of temperature in AMD formation. Other microorganisms are also associated with the oxidation of sulphides and the generation of AMD (Kuhnelt, 1955; Bigham et al., 1990; Retallack, 1990; Rossi, 1990; Hackl, 1993; Gould et al., 1994; Ehrlich, 1996; Boult et al., 1997; Nemati et al., 1998; Benner et al., 2000).

The autotrophic bacteria play a catalytic role in the oxidative dissolution of sulphide mineral and *A. ferrooxidans* is recognized as being responsible for the oxidation of iron and inorganic sulphur compounds in areas such as metal mines, mine tailings and coal deposits where these compounds are abundant (Nordstrom and Southam, 1997). *A. ferrooxidans* is acidophilic (acid loving, capable of surviving at low pH's) and has a physiology which is well suited for growth in an inorganic mining environment. It obtains its carbon by fixing atmospheric carbon dioxide and is strictly autotrophic. *A. ferrooxidans* obtains its energy by the oxidation of either metal or sulphur according to the following reactions.

$$Fe^{2+} + 0.25\ O_2 + H^+ \rightarrow Fe^{3+} + 0.5\ H_2O \tag{21}$$

$$H_2S + 2\ O_2 \rightarrow SO_4^{2-} + 2H^+ \tag{22}$$

$$S^O + H_2O + 1.5 O_2 \rightarrow SO_4^{2-} + 2H^+ \tag{23}$$

$$S_2O_3^{2-} + H_2O + 2O_2 \rightarrow 2\ SO_4^{2-} + 2H^+ \tag{24}$$

Pyrite mine tailings leach acid mine drainage originating in large part to the metabolic activity of *A. ferrooxidans*. Reactions 25 and 26 are catalyzed by the bacteria. This increases the rate of chemical weathering of the mine tailings. Reaction products from the bio-catalyzed reactions 25 and 28 contribute to reaction 27 which is abiotic.

$$FeS_2 + 3.5\ O_2 + H_2O \rightarrow Fe^{2+} + 2\ SO_4^{2-} + 2H^+ \tag{25}$$

$$2\ Fe^{2+} + 0.5\ O_2 + 2H^+ \rightarrow 2Fe^{3+} + H_2O \tag{26}$$

$$FeS_2 + 14\ Fe^{3+} + 8\ H_2O \rightarrow 15\ Fe^{2+} + 2\ SO_4^{2-} + 16\ H^+ \tag{27}$$

A. ferrooxidans is generally assumed to an obligate aerobe, but under anaerobic conditions. T. ferrooxidans can be grown on elemental sulphur using ferric iron as an electron acceptor as in reaction 28.

$$S + 6\ Fe^{3+} + 4\ H_2O \rightarrow H_2SO_4 + 6\ Fe^{2+} + 6\ H^+ \tag{28}$$

The above reactions show how *A. ferrooxidans* catalyzes the acid and heavy metal producing reactions that occur in mine tailings resulting in the degradation of water quality downstream of mining operations. These bacteria thrive at optimum temperatures of 25^0C to 35^0C. The rate of reaction becomes 2-11 times faster for 30^0C with ferric iron as oxidant (Janzen, 1996), but become inactive or die as temperatures increase to 53^0C to 60^0C (Tuovinen and Kelly, 1972; Cathles, 1979; Kwong, 1995; Mason and Rice, 2002). Measurements indicate that oxidizing sulphide-rich material can warm internally to temperatures at least as high as 60^0C because of the heat released by the oxidation reactions

(Cathles and Apps, 1975). Some sulphide rich material actually undergoes spontaneous combustion.

4.1.5. pH

pH of the drainage water is an indication of the development of AMD over time, at high pH values, ferrous iron is chemically rapidly oxidized and pyrite oxidation may be maintained without biological mediation and the acidity that is released is quickly neutralized by alkalinity continued in the rock. As this alkalinity is consumed, the pH drops (in stages depending upon the type of alkali minerals) (Singer and Stumm, 1970). At low pH values, microbial mediated oxidation of ferrous iron by acidophilic lithoautotrophic bacteria, such as acidithiobacillus ferrooxidans and leptospirillum ferrooxidans (Pronk and Johnson, 1992; Johnson, 1993) predominates. Generally at this stage, while sulphate acidity and later iron concentrations increase, the concentrations of the other metals (such as copper) are limited by pH solubility controls. At pH values around 4.5, the much more rapid biologically catalyzed oxidation predominates, increasing the rate of acid production. From the onset of pyrite oxidation pH falls (acidity increase) quickly and then stabilizes, typically at values around pH 2.5 to 3. The pH stabilization is normally determined by the optimal habitat requirement of the site specific strain of bacteria. The activity of these bacteria is pH dependent with optimal conditions in the range of pH 2 to 3. Thus, once pyrite oxidation and acid production has begun, conditions are favourable for bacteria to further accelerate the reaction rate. At pH values of about 6 and above, bacterial activity is thought to be insignificant. *A. ferrooxidans* is most active in the pH range 1.0 to 2.5 deriving its energy from redox reactions where reduced sulphur compounds serves as electron donor and oxygen as electron acceptor (Nemati et al., 1998).

At pH (2.5 to 3.0) the products of sulphuric acid and ferric sulphate create conditions where the ferric iron ion itself can act as an oxidant (above about pH 3 the ferric ion is itself hydrolyzed to ferric hydroxide, which precipitates as the familiar rust-coloured stain associated with AMD). In the absence of ferric iron at pH 2.5-3.0, sulphuric acid will dissolve some heavy metal carbonate and oxide minerals, but has little reactive effect on heavy metal sulphides (Dutrizac and McDonald, 1974).

The effect of pH and other potential controls on the oxidation rate of sulfides with ferric iron hydroxides has experimentally not been firmly established in AMD polluted sediments. Currently only incidental observations reports that high pH values discharges sulfide oxidation with ferric iron hydroxides (Kusel et al., 2001; Wendt-Potthoff et al., 2002). It should also be noted that iron sulfide formation may only be controlled by pH and thermodynamic factors in iron rich and carbon limited sediments. In other environments sulfate reduction can be viable process at pH 3 (Christensen et al., 1996; Elliot et al., 1998; Koshorreck et al., 2003; Blodau, 2006).

Kleinmann et al. (1981); Nordstrom (1982a,b); Luther (1987) and Nordstrom and southam (1997) have suggested that the generation of AMD can be understood as three sequential stages. In stage I, while the pH is near-neutral or only slightly acidic, pyrite oxidation by reaction 2 proceeds by a combination of abiotic and bacterial mechanisms, and Fe^{2+} oxidation is primarily abiotic. Any biotic oxidation of pyrite is dominantly by bacteria attached to the surface of pyrite grains. In stage II, pH is generally in the range 3 to 4.5, and Fe^{2+} oxidation is mainly by *T. ferrooxidans*, because abiotic oxidation is so slow. Pyrite oxidation in this transition stage occurs by a combination of reactions 3 and 5, both abiotically and bacterially. In stage III, at pH less than about 3, the concentration of Fe^{3+} becomes high enough that reaction 5 becomes the main mechanism for acid production, with bacterial reoxidation of Fe^{2+} furnishing the Fe^{3+}. In stages I and II, the rate of AMD generation is relatively slow, but in stage III the rate becomes very rapid. This stage is responsible for production of the most acidic AMD. It should be noted that this sequence is based on processes in unsaturated systems with an adequate supply of O_2 and negligible alkaline material, in environments of limited O_2 and/or significant carbonate or other alkaline material.

4.2. Secondary Factors/ Chemical Control Factors: Neutralizing Carbonate Minerals

Regardless of the presence of pyrite and other sulphide minerals, AMD may not be present at a mine site due to the presence of carbonate minerals like calcite ($CaCO_3$), dolomite ($CaMg(CO_3)_2$), magnasite ($MgCO_3$) and Siderite ($FeCO_3$) etc. followed by carbonates that contain various amount of oxidizable cations, such as ankerite (e.g. $Ca_{0.5}Fe(x...y)Mg(y...x)CO_3$), and Magnesian Siderite ($Fe(x...y)Mg(y...x)CO_3$) (Paktunc, 1999; Blodau, 2006) as neutralizing secondary factors (Fripp et al., 2000). These either neutralize the acid produced by oxidation of sulphides or may change the effluent character by adding metal ions mobilized by residual acid. Neutralization of acid by the alkalinity released when acid reacts with is an important means of moderating acid production. The most common neutralizing minerals are calcite and dolomite (Klemow, 2000; Komnitsas et al., 2001; Danielle et al., 2008).

Products from the oxidation reaction (hydrogen ions, metal ions, etc.) may also react with other non-neutralizing constituents. Possible reactions include ion exchange on clay particles, gypsum precipitation, and dissolution of other minerals. Dissolution of other minerals contributes to the contaminant load in the acid drainage. Examples of metals occurring in the dissolved load include aluminum, copper, lead, zinc, and others (Ferguson and Erickson, 1988).

The most common carbonate mineral in coal mining environments is limestone ($CaCO_3$). The factors that affect $CaCO_3$ solubility are pH and pCO_2 (Evangelou 1995)
C = henry constant = 0.0344

$$CaCO_3 \xrightarrow{K_1} Ca^{2+} + CO_3^{2-} \qquad K_1 = 3.98 \times 10^{-9} \qquad (29)$$

$$H_2CO_3 \longrightarrow pCO_2 \qquad pCO_2 = 3.1 \times 10^{-1} \qquad (30)$$

$$H_2CO_3 \xrightarrow{K_2} H^+ + HCO_3^- \qquad K_2 = 4.28 \times 10^{-7} \qquad (31)$$

$$HCO_3^- \longrightarrow H^+ + CO_3^{2-} \qquad K_3 = 4.68 \times 10^{-11} \qquad (32)$$

The solubility of $CaCO_3$ in water depends on two interdependent variables pCO_2 and pH and its solution can only be carried out through successive approximations (Garrels and Christ, 1965).

When pCO_2 changes, pH and Ca also changes. The concentration of Ca reflects the level of alkalinity in water since mass-balance reveals that any calcium quantity (moles) released from $CaCO_3$ must be accompanied by a similar quantity (moles) of CO_3. The important thing reported by Evangelou (1995) is that pCO_2 increases due to lowering pH. Consequently OH^- decreases along with CO_3. Most reactions in natural environment are driven by OH^- not alkalinity. The low pH may not allow heavy metal precipitation even though alkalinity is high. In such situations stronger bases should be used like NaOH, $Ca(OH)_2$ etc.

The initial reaction with an acid solution using calcite is (Lusardi and Erickson, 1985; Cravotta et al., 1990):

$$pH<6.4: CaCO_3 + 2H^+ \rightarrow Ca^{2+} + H_2CO_3(aq), \tag{33}$$

$$: MgCa(CO_3)_2 + 4H^+ \rightarrow Ca^{2+} + Mg^{2+} + 2CO_2 + 2H_2O \tag{34}$$

If gas phase is present, the H_2CO_3 may partly decompose and exsolve into gas phase as follows:

$$H_2CO_3(aq) \rightarrow CO_2(g) + H_2O \tag{35}$$

Upon further neutralization of AMD with carbonate to pH values greater than 6.3, the product is:

$$6.4<pH<10.4: CaCO_3 + H^+ \rightarrow Ca^{2+} + HCO_3^- \tag{36}$$

$$MgCa(CO_3)_2 + 2H^+ \rightarrow Ca^{2+} + Mg^{2+} + 2HCO_3^- \tag{37}$$

If the neutralizing material is lime composed mainly of CaO or $Ca(OH)_2$ than the neutralization reaction is

$$Ca(OH)_2 + 2H^+ \rightarrow Ca^{2+} + 2 H_2O \tag{38}$$

$Ca(OH)_2$ also tends to react with CO_2 from the air to form $CaCO_3$, so that the aged "lime" can contain substantial $CaCO_3$ (Rose et al., 1995).

If the AMD is to be neutralized to pH 6.3 or above and no CO_2 is allowed to exsolve to the gas phase, then the reaction may be written as:

$$FeS_2 + 4 CaCO_3 + 3.75 O_2 + 3.5 H_2O \rightarrow Fe(OH)_3 + 2 SO_4^{2-} + 4 Ca^{2+} + 4 HCO_3 \tag{39}$$

Thus neutralization of products of oxidizing 1 mole of pyrite requires 4 mole of $CaCO_3$ or 400 gm of $CaCO_3$ to 64 gm of pyretic sulphur or 62.5 tonnes of $CaCO_3$ per 1000 tonnes of material with 1% S as pyrite.

In contrast, if all CO_2 escapes to the gas phase and AMD is neutralized to pH 5 then the reaction may be written:

$$FeS_2 + 2\ CaCO_3 + 3.75\ O_2 + 1.5\ H_2O \rightarrow Fe(OH)_3 + 2\ SO_4^{2-} + 2\ Ca^{2+} + 2\ CO_2\ (g) \quad (40)$$

Under these conditions, neutralization of AMD generated by oxidation of 1 mole of pyrite requires 2 moles of $CaCO_3$ or 200 gm of $CaCO_3$ or 31.25 tons of $CaCO_3$ per 1000 tons containing 1% pyritic sulphur. Most favourable situation probably falls between these 2 extremes. Air within most strip mine spoils can contain significant amounts of CO_2 (Lusardi and Erickson, 1985; Cravotta et al., 1994a). so that some CO_2 is exsolving. If O_2 can get into the spoil to drive the pyrite oxidation reaction, then some CO_2 can escape into the open air. On the other hand, in order to provide detectable alkalinity in the effluent some HCO_3^- must be present, thus the theoretical amount of carbonate required to neutralize AMD generally falls intermediate between the two end member cases.

The neutralization potential of silicates varies according to their stoichiometric composition and products of silicate dissolution, because secondary clay minerals, such as montmorillonite and kaolinite, may form (Sherlock et al., 1995)

If the standard tests for neutralization potential are applied (Sobek et al., 1978) then AMD produced by oxidation of 1000 tons of material with 1 % S may be neutralized by interaction with lime at a factor of 31.25 tons of $CaCO_3$ equivalent, equal to 23 tons of $Ca(OH)_2$. Siderite ($FeCO_3$) is quite commonly available in Coal overburdens (Morrison et al., 1990). Dissolution of pure siderite followed by precipitation of $Fe(OH)_3$ generates no net alkalinity, even if CO_2 is exsolved:

$$FeCO_3 + 0.25\ O_2 + 2.5\ H_2O \rightarrow Fe(OH)_3 + CO_2\ (g) + H_2O \quad (41)$$

However, if siderite is exposed to H^+ then neutralization may temporarily take place (Sobek et al, 1978)

$$FeCO_3 + 2H^+ \rightarrow Fe^{2+} + CO_{2\ (g)} + H_2O \quad (42)$$

Here some H^+ has been consumed as a result raising the pH, but dissolved Fe^{2+} represents acidity equivalent to the H^+ consumed so no change in acidity has really occured. When the Fe^{2+} oxidizes and precipitates then H^+ is regenerated. If the CO_2 does not exsolve as gas then the dissolved H_2CO_3 requires additional alkalinity to convert it to HCO_3^- to produce neutralization.

According to Mozley (1989) most of the natural siderite is solid solution containing some Ca, Mg and Mn in addition to Fe. Thus the presence of Ca and Mg in solid solution and its dissolution will contribute some net alkalinity analyses to the reaction of the $CaCO_3$ or $MgCO_3$ component in siderite. At all pH conditions, silicate minerals are buffering the pH (Blowes et al., 1994).

The dissolution of silicate minerals consumes acidity by reactions like:

$$2KAlSi_3O_8\ (K\ Feldspar) + 2\ H^+ + H_2O \rightarrow 2K^+ + Al_2Si_2O_5(OH)_4 + 4\ SiO_2 \quad (43)$$

$$Al_2Si_2O_5(OH)_4\ (Kaolinite) + 6H^+ \rightarrow 3Al^{3+} + 2\ SiO_2 + 5H_2O \quad (44)$$

$$Mg_2Al_2SiO_5(OH)_4\ (Chlorite) + 4\ H^+ + SiO_2 \rightarrow 2Mg^{2+} + Al_2Si_2O_5(OH)_4 + 2H_2O \quad (45)$$

Reactions of these types are responsible for most or all of the dissolved K, Na, Al, Mg, some of Ca, and locally some of the Fe in AMD solutions.

Jaynes et al. (1983) recorded similar high values of O_2 at depth in spoil. Models and experiments by Cathles and Apps (1975) showed that the advection process also operates in pyrite-bearing waste dumps at copper mines.

The advective mode of O_2 transport appears to predominate in spoil that contains a significant proportion of sandstone or other strong rock that leaves appreciable open space between fragments. In contrast, the diffusive mode of transport predominates in less permeable spoil composed of small fragments of weak shale or similar rocks.

AMD generation may also be strongly influenced by the position of alkaline materials relative to pyritic materials. Surface mines with moderate amount of lime stone fragments overlying the coal bearing rocks or with considerable pyrite in the overburden, generally do not produce AMD (Williams et al., 1982; Hornberger, 1985). This lack of AMD generation appears to result from the alkaline quality of water percolating into the spoil after first passing through the carbonate-bearing till.

The reaction of ferric oxides, hydroxides or sulfates with pyrite is a possible means of producing AMD in mine spoil. The ferric ions released by dissolution of these phases could act as oxidant of pyrite. The stoichiometry of the reaction with ferric hydroxide would be as follows:

$$FeS_2 + 14\ Fe(OH)_3 + 26\ H^+ \rightarrow 15\ Fe^{2+} + 2\ SO_4^{2-} + 34\ H_2O \tag{46}$$

Note that this and similar reactions involving ferric minerals consume considerable H^+ in order to occur, but do produce additional acidity as Fe^{2+} in solution and are therefore only significant in acidic environments. Temple and Koehler (1954) proposed that ferric sulfate minerals could be significant sources of acidity, sulfate and ferric ions. The subsequent oxidation of pyrite by Fe^{3+} and/or hydrolysis of Fe^{3+} can produce acidic water, even under water-saturated conditions (Cravotta, 1994).

At a few surface mines, it appears that burial of brush or other organic matter in spoil may generate discharges with the characteristics of AMD. Although the mechanism requires further study, organic matter could, with microbial catalysis, reduce $Fe(OH)_3$ or other ferric compounds to produce ferrous iron in solution:

$$CH_2O(\text{organic matter}) + 4\ Fe(OH)_3 + 8\ H^+ \rightarrow 4\ Fe^{2+} + CO_2 + 11\ H_2O \tag{47}$$

The acidity increases by the formation of CO_2; however, if the CO_2 exsolves, this reaction produces no net additional acidity. Although the reaction produces Fe^{2+}, it consumes considerable H^+ and it does not produce SO_4^{2-}. Some AMD containing low SO_4^{2-} but high Fe may have formed by this mechanism.

Mine drainage containing relatively low SO_4^{2-} concentrations could be caused by dilution or by SO_4^{2-} reduction, which involves the reaction of AMD with organic matter (Rose et al., 1995):

$$2\ CH_2O + SO_4^{2-} \rightarrow H_2S + 2\ HCO_3^- \tag{48}$$

Evidence for sulfate reduction is the fact that H_2S or similar reduced sulfur gases can be smelled at some localities. In addition, this reaction is known to proceed in wetlands and other natural environments where SO_4^{2-} bearing water encounters organic matter. For example, Herlihy et al. (1987) report that about half the SO_4^{2-} in an AMD-affected stream is removed by SO_4^{2-} reduction in the sediments of a shallow lake in Virginia. Reduction of ferric iron by Fe^{3+} reducing bacteria would be required before SO_4^{2-} reducing bacteria would be able to prevail over Fe-reducers (Chappelle and Lovely, 1992). In most AMD, the H_2S will react immediately with Fe to precipitate FeS, because most AMD originates from oxidation of pyrite (FeS_2) and has more S than iron, it is unlikely that SO_4^{2-} reduction will form an effluent with low SO_4^{2-} and high iron. However, any unreacted H_2S can be oxidized to form H_2SO_4 where oxygen is available. Oxidation of organic S in coal could also form an acid SO_4^{2-} bearing solution (Harvey and Dollhopf, 1986).

4.3. Tertiary Factors/Physical Factors

The reactivity differences of various metal sulphides that generate acidic drainage from mine site, tailings and waste heaps have been attributed to differences due to lithologic controls, mineralogy, crystallinity, trace element content, particle size, permeability and mine site hydrology. Though difficult to weigh, each of these factors influences the potential for acid generation and is therefore, an important consideration for long term waste management.

4.3.1. Lithology
Lithology or rock type also influences spoil and drainage quality. Physical characteristics of the rock, such as porosity and accessory minerals can exert various constraints or enhancements to the overall chemical weathering process (Ziemkiewicz, 1991).

4.3.2. Mineralogy and crystallinity
Chemical and microbial pyrite oxidation is directly controlled by mineral and crystal type (Malmstrom et al, 2000) and by particle size, i.e. mass normalized surface area (Stromberg & Banwat, 1990). Among variety of morphologies, such as framboids, octahedral and cubic crystals and conglomerates with irregular surfaces, the small polycrystalline framboids are particularly susceptible to oxidation owing to their high specific surface and high porosity and large surface area (Caruccio et al, 1977; Evangelou and Zang, 1995). Oxidation rates increase with partial pressure of oxygen (William and Rimstidt, 1994; Blodau, 2006).

The rate of acid generation depends on the surface area of pyrite exposed to solution, and on the cystallinity and chemical properties of the pyrite surface (McKibben and Barnes, 1986). This dependence will be most important in initial stages while pH is greater than about 2.5. In general, rock with a high percentage of pyrite will produce acidity faster than rock with low percentage of pyrite. Also, a given mass of pyrite in small particles with high surface area will tend to oxidize more rapidly than the same mass composed of course smoothed surfaced grains. The high surface area of framboidal pyrite atleast partly accounts for its observed high reactivity (Caruccio, 1975; Caruccio et al., 1976). McKibben and Barnes (1986) observed that pyrite surface were pitted after reaction and suggested that the abundance of defects might be crucial in determining the reaction rate, but kinetic experiments on the variety of pytire samples using Fe^{3+} at pH 2 show similar rates (Wiersma and Rimstidt, 1984), so the importance of this effects under strongly acid conditions remains to be demonstrated. Mishra and Osseo-Asare (1988); Kitakaze et al. (1990) and Graham (1991) have shown that pyrite can have vacant positions in its crystal lattice and correlated variations in atomic spacing and physical properties, suggesting another possible cause for different oxidation rates for different pyrite samples. Hammack et al. (1988) also found that sedimentary pyrites were more reactive than could be explained by surface area alone, and suggested that the cause was a difference in crystal structure between sedimentary and hydrothermal pyrite. Another possible variable is the contents in the trace elements in the pyrite, especially elements such as that are clearly related to non-stoichiometry (Wiersma and Rimstidt, 1984).

Different metal sulfide minerals will oxidize at different rates under different conditions, thus accounting for the observed non-uniformity in acid production, also defects in the crystal structure of the metal sulfides may contribute to the observed differences in reactivity. It is likely that the reactivity of the metal sulfides will vary proportionally to the number, size of the crystal defects and also the electrochemical dissolution of the minerals. Pyrite is the most abundant disulfide mineral and is associated in varying amounts with other sulfide/disulfide minerals such as those of nickel, copper, zinc etc. According to Mehta and Murr (1983) and Torma (1988) in the case of two metal sulfide minerals with two different electrical rest potential, the mineral with lower electric rest potential, acting as anode, will be dissolved, while the mineral with higher rest potential, acting as cathode will be galvanically protected. In case of chalcopyrite/pyrite galvanic all ($CuFeS_2/FeS_2$) chalcopyrite with lower rest potential will be dissolved rapidly while pyrite will remain practically intact. Further the bacteria continuously oxidize the elemental sulfur (produced from the anodic oxidation reaction) to sulfate, thus formation of the sulfur barrier on the surface of chalcopyrite is eliminated. The anodic oxidation reaction on chalcopyrite surface can be expressed by:

$$CuFeS_2 \rightarrow Cu^{2+} + Fe^{2+} + 2S^0 + 4e^- \tag{49}$$

While the cathodic oxygen reduction taking place on the FeS_2 surface can be given by

$$O_2 + 4H^+ + 4e^- \rightarrow 2H_2O \tag{50}$$

Thus sum of the reaction can be expressed by :-

$$CuFeS_2 + O_2 + 4H^+ \rightarrow Cu^{2+} + Fe^{2+} + 2S^0 + 2H_2O \tag{51}$$

The elemental sulfur (S^0) and ferrous iron will be oxidized by bacteria according to reactions:

$$4FeSO_4 + 2O_2 + 2H_2SO_4 \rightarrow 2Fe_2(SO_4)_3 + 2H_2O \text{ and} \tag{52}$$

$$2S^0 + 3O_2 + 8H_2O \rightarrow 2H_2SO_4 \tag{53}$$

4.3.3. Particle size and trace element content

Particle size is a fundamental concern because it affects the surface area exposed to weathering and oxidation. Surface area is inversely proportional to particle size. Pyrite in small particles with high surface area will tend to oxidize more rapidly than the same mass composed of coarse smooth–surfaced grains, the high surface area of framboidal pyrite at least partly accounts for its observed high reactivity (Caruccio, 1975; Caruccio et al., 1976). Very coarse grain material, as is found in waste rock dumps, exposes less surface area but may allow air and water to penetrate deeper into the unit, thereby exposing more material to oxidation and ultimately producing more acid. Air circulation in coarse material is aided by wind, changes in barometric pressure and possibly connective gas flow caused by heat generated by the oxidative reaction. In contrast, tailings (fine grain material) may retard air and very fine material may limit water flow, however, finer grains expose more surface area to oxidation. The relationships among particle size, surface area and oxidation play a prominent role in acid prediction methods and in mining waste management units (Stumm and Morgan 1981). Pyrite can occur in grain sizes ranging from invisible to the eye up to several inches. Fine-grained pyrites with a large surface area are much more chemically reactive than the coarser forms. The reactivity of the fine-grained pyrites reflects the fact that acid generating reactions occur at the mineral surface.

The two most common sulfides that generate AMD are pyrite and pyrrohotite. All elements except the eight abundant rock forming elements: oxygen, silicon, aluminium, iron, calcium, sodium, potassium and magnesium are trace elements which vary from site to site in both these minerals by concentration levels, such as ore mineral elements, (Cu in chalcopyrite, Zn in sphalerite, Ni in pentlandite, etc. (Carruccio et al, 1976). Ore mineral trace elements(Cd in sphalerite) or non-ore mineral trace elements (Cu, Co, Ni in pyrite) are often present at concentration below the lower limit of detection of analytical procedures (Hawley and Nichol 1961). Trace heavy elements may have an impact upon the weatherability of sulfides and hence the rate of AMD generation.

4.3.4. Permeability

Physical factors particularly waste rock dump permeability is important. Dumps with high permeability have high oxygen ingress, which contributes to higher chemical reaction rates, hence higher temperature and increased oxygen ingress through convection (Akcil and Koldas, 2006). The physical weathering tendency of a waste controls its permeability in the long term and affect oxygen and water transfer. As the permeability decreases the rate of acid mine drainage production also should slow down (Caruccio and Geidel, 1981a,b). However observation of high sulphide mining wastes indicates that AMD products continue to be released for many decades after mining ceases.

4.3.5. Mine site hydrology

Mine site hydrology plays a critical role in determining drainage quality. The products of pyrite oxidation are free acid and soluble acid salts. If air is excluded and no percolating waters are present, the acid salts generated from the limited available moisture simply reside within the spoil or can be inhibited. When excess moisture is present, the acid weathering products are dissolved and transported with the water moving through the material. Although the basic processes of AMD formation are universal, the importance of any single controlling factor is frequently specific to mine site conditions.

Mine over burden and waste soils (commonly referred to as the mine tailings) are the waste products generated during mining and milling operations. Surface depositions of these waste products results in exposure to the atmosphere precipitation and ground or surface water. Many of these tailings possess the potential to react with oxygen and water generating products that adversely affect the pH and heavy metal composition of near by soils and streams.

Drainage and spoil quality is a product of two competing processes: Acid formation from pyrite oxidation, and generation of alkalinity from dissolution of carbonates and other basic minerals. The acid generation process consists of three phases: initiation, propagation; and termination. The initiation phase can begin as soon as the pyritic materials are exposed to an oxidizing environment; however, the acid load generated is relatively small. In the propagation phase, acid production increases rapidly. In the termination phase, acid production gradually declines. The actual times associated with these phases are at present ill defined, but appear to be on the order of years to decades. Acid drainage due to high pyrite contains high concentration of dissolved metals in it and low base content (Caruccio et al., 1977; Caruccio et al., 1988; Younger and Robins, 2002a,b). The rate of AMD reaction is primarily a function of the hydrogeologic characteristic of the mine and its overburden. Numerous chemical, physical and biological factors interact to control the quality of mine drainage, although the basic process of AMD formation is universal, the importance of any single controlling factor is frequently specific to mine site conditions.

5. AMD Prevention Technology

Remediation technology of large active as well as inactive mine sites has proven to be extraordinarily difficult, complex and expensive, not to mention litigious. The physical and chemical nature of the site makes it difficult to assess the effectiveness of the remediation technology and the relative risk and cost of various alternatives and their contingencies (Nordstrom and Alpers, 1999). There are no easy solutions to these type of environmental problems as mine sites are complex functions of the geology, hydrology, geochemistry, pedology, metrology, microbiology, mining and mineral processing history and their remediation is subjected to considerations of economic limitations, available technology, and potential land use, further the risk of failed remediation.

It is always better to prevent than to cure. Prevention is basically the control at the source level itself at first instance to prevent the AMD generation which is possible when the geochemistry of the AMD generation is known. As both oxygen and water are required to perpetuate the generation of AMD, its exclusion, either of both or one of these can prevent its generation. The objective is to prevent the contact between minerals, water/dissolved oxygen. Various approaches have been evaluated to prevent or minimize the generation of AMD. Such approaches are (Johnson and Hallberg, 2005). (1) Flooding/sealing of underground mines (Donald et al., 1997), (2) Underwater storage of mine tailings (Johnson and Hallberg, 2005), (3) land based storage with sealed covers (MEND, 1997; Peppas et al.,2000; Germain et al., 2003), (4) Blending /alkaline addition (Mehling et al., 1997; Wilson et al., 2000;

Kuyucak, 2000b), (5) Total solidification of tailings(Johnson and Hallberg, 2005), (6) Application of anionic surfactants (Filipek, 1996; Kuyucak, 2002), (7) Microencapsulation (coating) (Evanlagau, 2001; Kuyucak, 2002), (8) Surface Diversion.

One way in which this may be achieved is by flooding and sealing abandoned deep mines. Accessibility of the mine and sealed location are important design considerations. Dry or wet seals placed at easily-reached portal entrance are considerably cheaper than portals with poor access. The dissolved oxygen present in the flooding waters will be consumed by mineral oxidizing (and other) microorganisms present, and replacement of dissolved oxygen by mass transfer and diffusion will be impeded by sealing of the mine. However, this is only effective where the location of the all shafts and adits is known and where influx of oxygen-containing water does not occur. Under water storage has been used for disposing and storing water tailings that are potentially acid producing (Li et al., 1997). To prevent contact between the minerals and dissolved oxygen, shallow water covers may be used, and their effectiveness may be improved by covering the tailings with a layer of sediment and organic material, which has the dual benefit of limiting oxygen ingress and affording some protection against resuspension of the tailings due to actions of wind and waves. Dry covers used for surface storage are constructed from natural or man-made material that retard or divert the movement of water and oxygen into areas containing acid generating rock. Plastic liners are rarely used in mining because covering large volume of waste with a liner is usually too expensive. Covers used for surface storage of reactive minerals spoils may also incorporate an organic layer. The main advantages of organic cover over other types of covers are related with their low hydraulic permeability, high cation exchange capacity and high alkalinity. The sealing layer that covers the spoil is also constructed from clay, although in areas of world that experience acute wet and dry seasons, drying and cracking of the cover can render it less effective than in temporate zones (Swanson et al., 1997, Peppas et al., 2000; Germain et al., 2003).

Another suggested approach for minimizing AMD production is to blend acid generating and acid consuming materials, producing environmentally benign composites (Mehling et al., 1997). A variant on this theme is to add solid phase phosphate, thereby reducing its potential to act as an oxidant of sulphide minerals. However, inhibition of pyrite oxidation using this approach may only be temporary, due to process of armouring of the added phosphate minerals (Belzile, 1997; Evangelou, 1998). Application of the soluble phosphate (together with hydrogen peroxide) is one of the coating technologies. The peroxide oxidized pyrite, producing ferric iron, which react with the phosphate to produce a surface protective coating of ferric phosphate. Other variant of the blending is 'alkaline recharge' employs trenches loaded with alkaline material, usually a combination of soluble sodium carbonate and crushed limestone. The strategy is to charge infiltrating waters with high doses of alkalinity sufficient to overwhelm any acid produced within the backfill. This approach is highly depended on the placement of the alkaline trenches to provide maximum inflow to the acid producing zones. Finally, in as much as lithotrophic (rock eating) iron and sulphar oxidizing bacteria have a pivotal role in generating AMD, various laboratory and field scale test using biocides have been carried out to inhibit their activities in minerals spoils and tailings. This has generally involved the application of anionic surfactants such as sodium dodecyl sulphate (SDS), which are highly toxic to this group of microorganisms. However, the effectiveness of biocide application has been found to be highly variable affording, at best, only short term control of the problem and requiring repeated applications of chemicals.

Microencapsulation is the isolation of minerals by the precipitation of a chemical coating. For this, un-oxidized pyrite is coated to block oxygen transportation to the sulphide mineral surface and prevent the material from reaching with an oxidizing agent to produce ferric ions.

Water management strategy is another option for AMD prevention. Diverting the water, one of the principle reactants in pyrite oxidation, should theoretically stop the production of AMD. Diverting surface water above a mine site to decrease the amount of the water entering the mined area is highly recommended in acid producing areas. This technique can control water volume and direction and minimize the effects of AMD on receiving streams. The diversion is accomplished either by ditching on the uphill side of surface mines or by producing new channels or impervious channels of existing surface stream to convey water across the disturbed area. The study of geochemistry is surely going to help as to implement the prevention program at the source itself.

6. CONCLUSIONS

Acid Mine Drainage (AMD) and the contaminants associated with it is the most persistent environmental pollution problem, which occurs worldwide in the metal as well as coal mining regions. The rate of AMD generation and the chemical character of the AMD vary widely from site to site depending on the mine hydrology and geology relative abundance of acid forming and alkalinity materials and the physical characteristic of spoil (Hornberger, 1985).

Numerous methods for managing sulfide minerals are available to prevent AMD formation and many treatment methods are available. Although a significant amount of research and development have been conducted to better understand the generation of AMD and its control and prevention, it would be mere cost effective if the geochemistry of AMD, its sources are theoretically understood and the control/Preventive methods are implemented to the site itself. The prevention methods include isolating the sulfide minerals, alkaline addition and at some sites by limiting the amount of available oxygen, blending limestone or other neutralizing materials will neutralize AMD as it is formed and minimize the possibility of its mobilization.

The main culprit is pyrite, which in presence of air reacts with water, makes the drainage acidic. More abundant pyrite tends to increase the acidity of drainage, as does decreasing grain size of the pyrite. Iron-oxidizing bacteria and low pH values speed up the acid-forming reaction. Rates of acid formation tend to be slower if limestone or other neutralizers are present. Access of air containing the oxygen needed for pyrite oxidation is commonly the limiting factor in rate of acid generation. Both accesses of air and exposure of pyrite surfaces are promoted by breaking the pyrite-bearing rock. The oxygen can gain access either by molecular diffusion through the air-filled pore space in the spoil, or by flow of air, which is driven through the pore space by temperature or pressure gradients. Because of the complex interactions of all these and other factors, prediction and remediation of AMD is site specific. Once it develops at a mine, its control can be difficult and expensive and treatment of AMD usually cost more than its control and may be for many years after the mining have ceased. Therefore, the prevention of AMD formation would be highly beneficial. Understanding the geochemistry and careful field observation and laboratory analysis combined with an understanding of possible processes is going to be helpful to draw conclusions for the prevention of AMD at a particular location. Much of the efforts to control AMD in the past have been directed at treating the symptoms rather than prevention/Controlling the generation of the problem at the source itself.

Thus early diagnosis of the problem, identification of appropriate prevention/control measures to the source itself would reduce the risk of AMD generation.

REFERENCES

Ackil, A. & Koldas, S. (2006). Acid Mine Drainage (AMD): Causes treatment and case studies. *Journal of Cleaner Production, 14*, 1136-1145.

Allan, J. D. (1995). Stream Ecology: Structure and Function of Running Water. *Chapman & Hall*, New York, 488.

Alpers, C. N. & Blowes, D. W. (1994). Environmental geochemistry of sulfide oxidation. Am. Chem. Soc. Symp. Series 550. *Am Chem Soc., 681.*

Azapagic, A. (2004). Developing a framework for sustainable development indicators for the mining and minerals industry. *Journal of Cleaner Production, 12(6)*, 639-662.

Bandopadhyay, P. (1994). Processing and utilisation of minerals and pollution of water resources. In: S. P Banerjee, (Ed.), Proc. of Nat. Seminar on *"Minerals and Ecology"*, Oxford & IBH Pub., Dhanbad. Jan. 7-8, 267-280.

Belzile, N., Chen, Y. W., Cai, M. & Li, Y. (2004). A review of pyrrhotite oxidation. *Journal of Geochemical Exploration, 84*, 65-76.

Belzile, N., Goldsack, D., Maki, S. & McDonald, A. (1997). Acid mine drainage in the Sudbury area, Ontario. In : Eyles, N(Ed), Environmental Geology of Urban Areas. *Geological Association of Canada*, St. John's Canada, 223-226.

Benner, S. G., Gould, W. D. & Blowes, D. W. (2000). Microbial populations associated with the generation and treatment of acid mine drainage. *Chemical Geology, 169*, 435-448.

Bhole, A. G. (1994). Acid mine drainage and its treatment. Proc. of the Int. Sym. *on the Impact of Mining on the Envir."*, Nagpur,India. Jan. 11-16, 131-141.

Bigham, J. M., Schwertmann, U., Carlson, L. & Murad, E. (1990). A poorly crystallized oxohydroxysulfate of iron formed by bacterial oxidation of Fe(II) in acidic mine waters. *Geochim Cosmochim Acta, 54(10)*, 2743-2758.

Bigham, J. M., Schwertmann, U., Traina, S. J., Winland, R. L. & Wolf, M. (1996). Schwertmannite and the chemical modeling of iron in acid sulfate waters. *Geochimica et Cosmochimica Acta, 60*, 2111-2121.

Blodau, C. (2006). A review of acidity generation and consumption in acidic coal mine lakes and their watersheds. *Science of the Total Environment, 369*, 307-332.

Blowes, D. W. & Jambor, J. L. (1990). The pore water geochemistry and the mineralogy of the vadose zone of sulphide tailings, Waite Amulet, Quebec, Canada. *Appl. Geochem, 5*, 327-346.

Blowes, D. W., Ptacek, C. J., Frind, E. O., Johnson, R. H., Robertson, W. D. & Molson, J. W. (1994). Acid neutralization reactions in inactive mine tailings impoundment and their effect on the transport of dissolved metals. *Proceedings International Land Reclamation and Mine Drainage Conference and Third International Conference on the abatement of Acidic Drainage*, *1*, 429-438.

Blowes, D. W., Reardon, E. J., Jambor, J. L. & Cherry, J. A. (1991). The formation and potential importance of cemented layers in inactive sulfide mine tailings. *Geochim Cosmochim Acta, 55(4)*, 965-978.

D. W., Blowes, & J. L. Jambor, (Eds.). (1994). The environmental geochemistry of sulfide mine wastes. Short Course Handbook 22, Mineralogical association of Canada, *Nepean, Ont., 438*.

Bond, P. L., Druschel, G. K. & Banfield, J. F. (2000). Comparison of Acid Mine Drainage Microbial communities in physically and geochemically distinct ecosystems. *Applied and Environmental Microbiology, 66(11)*, 4962-4971.

Boult, S., Johnson, N. & Curtis, C. (1997). Recognition of a biofilm at the sediment-water interface of an acid mine drainage contaminated stream, and its role in controlling iron flux. *Hydr. Proc., 11*, 391-399.

Bowell, R. J. (2002). The hydrogeochemical dynamics of mine pit lakes. In: Younger PL, Robins NS, editors. *Mine water hydrogeology and geochemistry 198*, Geological Society of London, Geol. Soc. Spec. Publ., 396.

Brady, K. S., Bigham, J. M., Jaynes, W. F. & Logan, T. J. (1986). Influence of sulfate on Fe-oxide formation: Comparison with a stream receiving acid mine drainage. *Clays and Clay Minerals, 34*, 266-274.

Brake, S. S., Dannelly, H. K., Cannors, K. A. & Hasiotis, S. T. (2001). Influence of water chemistry on the distribution of an acidophilic protozoan in an acid mine drainage system at the abandoned Green Valley coal mine, Indiana,USA. *Applied Geochemistry, 16*, 1641-1652.

Bruynesteyn, A. & Duncan, D. W. (1977). The practical aspects of biological leaching studies. Proc 12th int. *mineral processing congr*, sao Paulo, brazil.

Bruynesteyn, A. & Hackl, R. P. (1984). Evolution of acid production potential of waste materials. *Miner Environ, 4*, 5-8.

Butler, T. W. & Seitz, J. C. (2006). Apparent seasonal variations in iron photoreduction in acidic discharge from a former pyrite mine, Oakland, California. *Appl. Geochem, 21*, 1109-1122.

Butler, T. W. (2007). Isotope geochemistry of drainage from an acid mine impaired watershed, Oakland, California, *Applied Geochemistry, 22*, 1416-1426.

Caruccio, F. T. (1975). Estimating the acid potential of coal mine refuse. In: M. J. Chadwick, & G. T. Goodman, (Eds.), The ecology of resource degradation and renewal: London, *Blackwell Science*, 197-203.

Caruccio, F. T., Geidel, G. & Sewell, J. M. (1976). The character of drainage as a function of the occurrence of framboidal pyrite and ground water quality in eastern Kentucky. National Coal Association/Bituminous Coal Research Inc., 6th Symposium on Coal Mine *Drainage Research*, Louisville, Ky., 1-16.

Caruccio, F. T., Ferm, J. C., Horne, J., Geidel, G. & Bagang, B. (1977). Paleoenvironment of coal and its relation to drainage quality. *US Environ Protect Ag Rep*, EPA-600/7-71-067, 108.

Caruccio, F. T. & Geidel, G. (1981a). Estimating acid loads and treatment costs of coal strip mines. Proc Symp. *Surface coal mining and reclamation*, 27-29 Oct 1981, Louisville, Ken.

Caruccio, F. T. & Geidel, G. (1981b). Estimating the minimum acid load that can be expected from a coal strip mine. In: D. H. Grave, (Ed) Proc 1981 *Symp Surface mining hydrology, Sedimentology and reclamation*. University of Kentucky, Lexington,. 437-443.

Caruccio, F. T., Hossner, L. R. & Geidel, G. (1988). Pyritic material: acid drainage, soil acidity & limimg. In Reclammation of surface-mined land, *CRC Press, Boca Roton, 1,* 159-190.

Cathles, L. M. & Apps, J. A. (1975). A model of the dump leaching processing that incorporates oxygen balance, heat balance, and convection. *Metallurgical Transactions, 6(B),* 617-624.

Cathles, L. M. (1979). Predictive capabilities of a finite difference model of copper leaching in low grade industrial sulfide waste dumps. *Mathematical Geology, 1,* 175-186.

Chappelle, F. H. & Lovley, D. R. (1992). Competitive exclusion of sulfate reduction by Fe(III)-reducing bacteria: A mechanism for producing discrete zones of high-iron ground water. *Ground Water, 30,* 29-35.

Christensen, B., Laake, M. & Lien, T. (1996). Treatment of acid mine water by surface reducing bacteria; results from a bench scale experiment. *J. Wat. Res., 30(7),* 1617-1624.

Chukhrov, F. V., Zvyagin, B. B., Ermilova, L. P. & Gorshkov, A. I. (1972). New data on iron oxides in the weathering zone. In: *Proceedings of the International Clay Conference, 1,* Madrid, 397-404.

Colmer, A. R., Temple, K. L. & Hinkle, H. E. (1950). An iron-oxidizing bacterium from the acid drainage of some bituminous coal mines. *J. Bacteriol, 59,* 317-328.

Cravotta, C. A. III. (1994). Secondary iron-sulfate minerals as sources of sulfate and acidity: The geochemical evolution of acidic ground water at a reclaimed surface coal mine in Pennsylvania, In: Alpers, C.N. and Blowes, D.W., Environmental geochemistry of sulfide oxidation: Washington, D.C., *American Chemical Society Symposium Series, 550,* 345-364.

Cravotta, C. A. III, Brady, K. B. C., Gustafson-Minnich, L. C. & DiMatteo, M. R. (1994a) .Geochemical and geohydrological characteristics of bedrock and mine spoil from two methods of mining at a reclaimed surface coal mine in Clarion County, PA, USA. U.S. *Bureau of Mines Special Publication,* SP-06B, 242-249.

Cravotta, C. A. III., Brady, K. B. C., Smith, M. W. & Beam, R. L. (1990). Effectiveness of the addition of alkaline materials at surface coal mines in preventing or abating acid mine drainage - Part I, Geochemical considerations. In: *Proceedings of the 1990 Mining and Reclamation Conference,* Charleston, W. V., April 23-26, , Morgantown, WV, West Virginia Univ., *1,* 221-223.

Danielle, M. C., Huminicki, J. & Rimstidt, D. (2008). Neutralization of sulfuric acid solutions by calcite dissolution and the application to anoxic limestone drain design, *Applied Geochemistry, 23,* 148-165.

Dincer, T., Al-Mugrin, A. & Zimmermann, U. (1974). Study of the infiltration and recharge through the sand dunes in arid zones with special reference to the stable isotopes and thermonuclear tritium. *J. Hydrol, 23,* 79-100.

Donald, S. B., Holl, N., Landine, P. & Welch, D. (1997). Recent hydrological developments in uranium mine tailings management in Canada: the key Lake experience, *In IAEG Symposium on Engineering Geology and the Environment,* 233-243.

Dubey, A. & Nath, R. (1997). Environmental consequences of acid mine drainage. *J. Mine Tech., 18(1 & 2),* 60-62.

Dubrovsky, J. M., Cherry, J. A., Reardon, J. E. & Vivyurka, A. J. (1985). Geochemical evolution of inactive pyretic tailings in the Eliot Lake uranium district: 1. The Groundwater zone. *Can. Geotech. J*, *22*, 110-128.

Dutrizac, J. E. & McDonald, R. J. (1974). Ferric ions as a leaching mediums. *Miner Science Engg.*, *6*, 59-100.

Eary, L. E. (1999). Geochemical and equilibrium trends in mine pit lakes. *Appl. Geochem*, *14(8)*, 963-987.

Eggleton, R. A. & Fitzpatrick, R. W. (1988). New data and a revised structural model for ferrihydrite. *Clays and Clay Minerals*, *36*, 111-124.

Ehrlich, H. L. (1990). Geomicrobiology (2nd): New York, *Marcel Dekker, Inc.*, 646.

Ehrlich, H. L. (1996). *Geomicrobiology*. Marcel Dekker, New York..

Elliot, P., Ragusa, S. & Catcheside, D. (1998). Growth of sulfate-reducing bacteria under acidic conditions in an upflow anaerobic bioreactor as a treatment system for acid mine drainage. *Water Resources*, *32(12)*, 3724-3730.

Evangelou, V. P. & Zhang, Y. L. (1995). A review: pyrite oxidation mechanisms and acid mine drainage prevention. *Crit Rev Environ Sci Technol.*, *25(2)*, 141-199.

Evangelou, V. P. (1995). Pyrite oxidation and its control: CRC Press, *293*.

Evangelou, V. P. (2001). Pyrite microencapsulation technologies: principles and Potential field application, *Ecological Engineering*, *17*, 165-178.

Evans, D. R. & Rose, A. W. (1995). Experiments on alkaline addition to coal mine spoil. In: Proceedings of Sudbury '95, *Mining and the Environment*, 49-58.

Ferguson, K. D. & Erickson, P. M. (1988). Pre-Mine Prediction of Acid Mine Drainage. Environmental Protection Kapilano 100, Park Royal, *West Vancouver*, B.C. V71 1A2, Canada, 25-43.

Filipek, J. (1996). Acid rock drainage control. *Journal of Mining Environment Management*, 16-21.

Fornasiero, D., Ejit V. & Ralston, J. (1992). An electro kinetic study of pyrite oxidation. *Colloids and Surfaces*, *62*, 63-73.

Fripp, J., Ziemkiewicz, P. F. & Charkavorki, H. (2000). Acid mine drainage treatment Tech. Notes Collection ERDC TN-EMRRP-SR, May, 14. U.S. *Army Engineer Research and Development Center*, Cicksburg, MS. 1-7. www.wes.army.mil/el/emrrp.

Garrels, R. M. & Thompson, M. E. (1960). Oxidation of pyrite in ferric sulfate solution. *American Journal of Science*, *258*, 57-67.

Garrels, R. M. & Christ, C. L. (1965). Solutions, Minerals and Equilibria, Freeman, *Cooper and Company*, San Francisco, CA.

Germain, D., Tasse, N. & Cyr, J. (2003). Rehabilitation of mine tailings by simultaneous prevention of oxidation and treatment of acid effluents using a wood-waste cover, 6th *ICARD*, 2003, Cairns, QLD, 263-271.

Goldhaber, M. B. (1983). Experimental study of metastable sulfur oxyanion formation during pyrite oxidation at pH 6-9 and 30^{oC}. *Amer. J. of Science*, *28*, 193-217.

Gould, W. D., Bechard, G. & Lortie, L. (1994). The nature and role of microganisms in the taoling environment. In: J. L. Jambor, D. W. Blowes, (Eds.), *The Environmental Geochemistry of Sulfide Mine Wastes*. Mineralogical Association of Canada, Ottawa, Canada, 185-199.

Graham, U. M. (1991). Relationships between formational mechanisms and variations in properties in pyrite crystals synthesized between temperatures of 150 and 350 degrees Celsius. Ph.D. thesis, *Pennsylvania State University*, 166.

Gunsinger, M. R., Ptacek, C. J., Blowes, D. W. & Jambor, J. L. (2006). Evaluation of long-term sulfide oxidation processes within pyrrhotite-rich tailings, Lynn Lake, Manitoba, *Journal of Contaminant Hydrology, 83*, 149-170.

Guo, W. (1993). *Numerical simulation of coupled heat transfer and gas flow in porous media with applications to acid mine drainage*: University Park, Pennsylvania State University, Ph.D. thesis..

Guo, W., Parizekm R. R. & Rose, A. W. (1994). The role of thermal convection in resupplying O_2 to strip coal-mine spoil. *Soil Science, 158*, 47-55.

Gupta, D. C. (1999). Environmental aspects of selected trace elements associated with coal and natural waters of Pench Valley coalfield of India and their impact on human health. *Int. J. of Coal Geol, 40*, 133-149.

Hackl, R. (1993). Elements of microbiological systems. In: R., Hackl, E. Peters, & D. Dreisinger, (Eds.), *Introduction to Biohydrometallurgy*, Short Course, Univ. of British Columbia,(Chapter 1), 1-25.

Hamilton, I. C. & Woods, R. (1981). An investigation of surface oxidation of pyrite and pyrrhotite by linear potential sweep voltametry, *Journal of Electroanal. Chem., 118*, 327-343.

Hammack, R. W., Lau, R. W. & Diehl, J. R. (1988). Methods for determining fundamental chemical differences between iron disulfides from different geologic provenances. U.S. Bur. *of Mines Inf. Circ*. IC-9183, 136-146.

Handa, B. K. (1990). Abatement of water pollution in mining industry. "*Impact of Mining on Envir.*", R. K. Trivedy, & M.P. Sinha, (Ed.). Ashish Pub.House, Pub., New Delhi, 227-252.

Harries, J. R. & Ritchie, A. I. M. (1985). Pore gas composition in waste rock dump undergoing pyritic oxidation. *Soil Sci., 140*, 143-152.

Harvey, K. C. & Dollhopf, D. J. (1986). A*cid production from organic sulfur, in Acid mine soil reclamation advancements in the northern plains*: Montana State University, Reclamation Research Publication, *86(01)*, 54-60.

Herlihy, A. T., Mills, A. L., Hornberger, G. M. & Bruckner, A. E. (1987). The importance of sediment sulfate reduction to the sulfate budget of an impoundment receiving acid mine drainage. *Water Resources Research, 23*, 287-292.

Hood, W. C. & Oertal, A. O. (1984). A leaching column method for predicting effluent quality from surface mines, *in Proc. Sym. Surf. Min. Hydr. Sedim. Recl*., University of Kentucky, Lexington, KY, 271.

Hornberger, R. J. (1985). Delineation of acid mine drainage potential of coal-bearing strata of the Pottsville and Allegheny Groups in western Pennsylvania. University Park, The Pennsylvania State University, M.S. thesis, 558.

Hornberger, R. J., Smith, M. W., Friedrich, A. E. & Lovell, H. L. (1990). Acid mine drainage from active and abandoned coal mines in Pennsylvania. In: Majumdar, S.K., Miller, E.W. and Parizek, R.R., *Water resources in Pennsylvania-Availability, quality, and management*. The Pennsylvania Academy of Science, 432-451.

Hutchison, I. P. G. & Ellison, R. D. (1992). Mine waste management—A resource for mining industry professionals, *regulators and consulting engineers*, Lewis Publishers, 654.

Jambor, J. L. & Vaughan, D. J. (2006). Mineralogy and geochemistry of acid mine drainage and metalliferous minewastes-Preface. *Applied Geochemistry, 21*, 1249-1250.

Janzen, M. P. (1996). *Role of ferric iron, trace metal content*, and crystal structure on pyrrhotite oxidation. MSc thesis. University of Waterloo. Canada, 175.

Jaynes, D. B., Rogowski, A. S., Pionke, H. B. & Jacoby, E. L. (1983). Atmosphere and temperature changes within a reclaimed coal strip mine. *Soil Science, 136*, 164-177.

Johnson, D. B. (1993). Biogeochemical cycling of iron and sulfur in leaching environments. *FEMS Microbial Rev., 11*, 63-70.

Johnson, D. B. (2003). Chemical and microbiological characteristics of mineral spoils and drainage waters at abandoned coal and metal mines, *Water Air Soil Pollut. Focus, 3*, 47-66.

Johnson, D. B. & Hallberg, K. B. (2003). The Microbiology of Acidic mine waters, *Research in Microbiology, 154*, 466-473.

Johnson, D. B. & Hallberg, K. B. (2005). Acid mine drainage remediation options: a review, *Science of Total Environment, 338*, 3-14.

Karathanasis, A. D. & Thompson, Y. L. (1995). Mineralogy of iron precipitates in a constructed acid mine drainage wetland. *Soil Sci. Soc. Am. J, 59*, 1773-1781.

Kelly, D. P. & Wood, A. P. (2000). Reclassification of Some species of thiobacillus to the newly designated genera acidthiobacillus gen. Nov., Halothiobacillus gen. Nov. and thermithiobacillus gen. nov. *Int. J. Sysst. Evl. Microbial, 50*, 511-516.

Kelly, D. P. & Jones, C. A. (1978). Factors affecting metabolism and ferrous iron oxidation in suspensions and batch culture of Thiobacillus ferrooxidans: relevance to ferric iron leach solution regeneration, in: L. E., Murr, A. E. Torma, & J. A. Brierly, (Eds.), *MetallurgicalApplications of Bacterial Leaching and Related Microbiological Phenomena*, Academic press, New York, 19-43.

Kitakaze, A., Graham, U. M. & Ohmoto, H. (1990). Anisotropy, crystal structure, density, and S/Fe ratios of hydrothermal pyrites (abs.). *Geological Society of America Abstracts with Program, 22(7)*, A362.

Klapper, H. & Schultze, M. (1995). Geogenically acidified mining lakes—living conditions and possibilities of restoration. *Int Rev Hydrobiol, 80(4)*, 639-653.

Kleinmann, R. L. P. (1980). Bactericidal control of acid mine drainage in surface mines and coal refuse. In : Proce. Of Nat. Symp. on Surface mine Hydrology,Sedimentology and Reclamation , *Univ. of Kentucky*, Lexington,KY, 333-337.

Kleinmann, R. L. P. & Crerar, D. A. (1979). *Thiobacillus Ferrooxidance* and the formation of acidity in simulated coal mine environment .*Geomicrobiology, 1(4)*, 373-388.

Kleinmann, R. L. P., Crerar, D. A. & Pacelli, R. R. (1981). Biogeochemistry of acid mine drainage and a method to control acid formation. *Mining Engineering, 33*, 300-303.

Kleinman, R. L. P. (1990). Acid mine drainage, U.S. bureau of mines researches and develops: control methods for both coal and metal mines. *J. Environ. Sci. Technol., 24 (9)*, 1278-1285.

Klemow, K. M. (2000). Environmental effect of mining in the anthracite regions, problems and possible solutions. Proc.of "Resources Oversight Hearing on the Abandoned Mine Reclamation", *Needs of the Pennsylvania Anthracite Fields*. Jan. 24, 1-10.

Knipe, S. W., Mycroft, J. R., Pratt, A. R., Nesbitt, H. W. & Bancroft, G. M. (1995). *X-ray photo-electron spectroscopic study of water adsorption on iron sulphide minerals. Geochem*, Cosmochim. Acta 59, 1079-1090.

Komnitsas, K., Paspaliaris, I., Ziberchmidt, M. & Groudev, S. (2001). Environment impacts at coal waste disposal sites-efficiency of desulfurization technologies. *Global Nest. Int. J*, *3(2)*, 109-116.

Kuhnelt, W. (1955). An introduction to the study of soil animals. In: Kevan, McE. (Ed.), Soil Zoology. *Butterworths Scientific Publications*, London, 2001, 3-22.

Konishi, Y. (1990). Bacteria dissolution of pyrite by Thiobacillus Ferrooxidans. *Bioprocess Eng.*, *5*, 231.

Koschorreck, M., Wendt-potthoff, K. & Geller, W. (2003). Microbial sulfate reduction at low pH in sediments of an acidic lake in Argentina. *Environ Sci Technol.*, *37(6)*, 1159-1162.

Koski, R. A., Munk, L., Foster, A. L., Shanks III, W. C. & Stillings, L. L. (2008). Sulfide oxidation and distribution of metals near abandoned copper mines in coastal environments, Prince William Sound, Alaska, USA, *Applied Geochemistry*, *23*, 227-254.

Kuo, E. Y. & Ritchie, A. I. M. (1999). The impact of convection on the overall oxidation rates in sulfidic waste rock dumps. Proceedings of Conference: *Sudbury '99-Mining and the Environment*, Paper AD21, 9-18.

Kusel, K., Roth, U., Trinkwalter, T. & Peiffer, S. (2001). Effect of pH on the anaerobic microbial cycling of sulfur in mining-impacted fresher water lake sediments, *Environmental and Experimental Botany*, *46*, 213-223.

Kuyucak, N. (2002). Role of microorganisms in mining: generation of acid rock drainage and its mitigation and treatment. *The European Journal of Mineral Processing and Environmental Protection*, *2(3)*, 179-196.

Kwong, E. C. M. (1995). Abiotic and biotic pyrrhotite dissolution. MSc thesis. *University of Waterloo*, Waterloo, Ontario.

Langmuir, D. (1997). *Aqueous Environmental Geochemistry*. Prentice Hall, Upper Saddle River, NJ.

Langmuir, D. & Whittemore, D. O. (1971). Variations in the stability of precipitated ferric oxyhydroxides, in Non-equilibrium Systems in Natural Water Chemistry: *American Chemical Society Advances in Chemistry Series, No. 106*, 209-234.

Lee, J. S. & Chon, H. T. (2006). Hydrogeochemical characteristics of acid mine drainage in the vicinity of an abandoned mine, Daduk Creek, Korea, *Journal of Geochemical Exploration*, *88*, 37-40.

Lefebvre, R. (1994). *Characterization and numerical modeling of acid mine drainage in waste rock dump*, PhD thesis, Universite' Laval., *375*.

Lefebvre, R., Ge'linas, P. & Isabel, D. (1993). Heat transfer during acid mine drainage production in a waste rock dump, La Mine Doyon, Que'bec, *Report GREGI 93-03 submitted to CANMET*, 46.

Lefebvre, R., Hockley, D., Smolensky, J. & Lamontagne, A. (2001). Multiphase transfer processes in a waste rock piles producing acid mine drainage: 2. Applications of numerical simulation. *J. Contam. Hydrol*, *52*, 165-186.

Li, M. G., Aube, B. C. & St-Arnaud, L. C. (1997). Considerations in the use of shallow water covers for decommissioning reactive tailings. *Proceedings of the Fourth International Conference on Acid Rock Drainage*, May 30–June 6, 1997, Vancouver, BC, *Vol. I*, 115-130.

Ljungberg, J. & Öhlander, B. (2001). The geochemical dynamics of oxidising mine tailings at Laver, northern Sweden. *Journal Geochemical Exploration*, *74(1-3)*, 57-72.

Lusardi, P. J. & Erickson, P. M. (1985). Assessment and reclamation of an abandoned acid-producing strip mine in northern Clarion County, Pennsylvania. In: *Symposium on Surface Mining, Hydrology, Sedimentation, and Reclamation: Lexington*, Ky., University of Kentucky, 313-321.

Ivanov, V. I., (1962). Effects of some factors on iron oxidation by cultures of Thiobacillus Ferrooxidans, *Microbiology*, (Engl. Transl.) *31*, 645.

Luthar, G. W. III. (1987). Pyrite oxidation and reduction: Molecular orbital theory consideration, Geochem. et. *Cosmochem. Acta, 4*, 2665.

Majima, H. & Peters, E. (1966). Oxidation rates of sulfide minerals by aqueous oxidation at elevated temperatures. Trans. Metall. Soc. *AIME, 236*, 1409-1413.

Mason, L. J. & Rice, N. M. (2002). The adaptation of Thiobacillus ferrooxidans for the treatment of nickel– iron sulphide concentrates. *Min. Eng., 15*, 795- 808.

McKay, L. D., Gillham, R. W. & Cherry, J. A. (1993). Field experiments in a fractured clay till: 2. Solute and colloid transport. Water Resour. *Res., 29*, 3879- 3890.

McKibben, M. A. & Barnes, H. L. (1986). Oxidation of pyrite in low temperature acidic solutions--Rate laws and surface textures. *Geochimica et Cosmochimica Acta, 50*, 1509-1520.

Mehling, P. E., Day, S. J. & Sexsmith, K. S. (1997). Blending and layering waste rock to delay, mitigate or prevent acid generation: a case review study. *Proceedings of the Fourth International Conference on Acid Rock Drainage*, May 30-June 6, 1997, Vancouver, BC, Vol. II, 953-970.

Mehta, A. P. & Murr, L. E. (1983). Fundamental studies of the contribution of galvanic interaction to acid-bacterial leaching of mixed metal sulfides, *Hydrometallurgy, 9*, 235.

MEND. (1997). Proceedings of the Fourth International Conference on Acid Rock Drainage. Vancouver, B.C., May 31-June 6, 1997, Vol. 1-4.

Miller, G. C., Lyons, W. B. & Davis, A. (1996). Understanding the water quality of pit lakes. *Environ Sci Technol., 30(3)*, 118A-123A.

Mishra, K. K. & Osseo-Asare, K. (1988). Aspects of the interfacial electrochemistry of semiconductor pyrite. *Journal of Electrochemical Society, 135*, 2502-2509.

Morel, F. M. M. & Hering, J. G. (1993). Principles and applications of aquatic chemistry. Wiley-Interscience, New York.

Morin, K. A. & Hutt, N. M. (1998). Kinetic tests and risk assessment for ARD. 5th Annual BC metal leaching and ARD workshop, December 9-10, *Vancouver*, Canada, 1-10.

Morin, K. A. (2001). Prediction of minesite-drainage chemistry through closure using operational monitoring data. *J Geochem Explor, 73*, 123-130.

Morin, K. A., Cherry, J. A., Dave, N. K., Lim, T. P. & Vivyurka, A. J. (1988). Migration of acidic groundwater seepage from uranium-tailings impoundments: 1. Field study and conceptual hydrogeochemical model. *J. Contam. Hydrol, 2*, 271-303.

Morin, K. A., Home, I. A. & Riehm, D. (1994). High frequency geochemical monitoring of the seepage from minerock dumps, BHP mineral's copper mine, *British Columbia. Proceedings of Pittsburg Conference*, U.S. Bureau Mines Special Publications SP 06A-94, 355-364.

Morrison, J. L., Atkinson, S. D. & Sheetz, B. E. (1990). Delineation of potential manganese sources in the coal overburdens of western Pennsylvania. In: Proceedings of the (1990) Mining and Reclamation Conference and Exhibition, Charleston, West Virginia, April 23-26, 1990, *Morgantown*, W.V., West Virginia University *Vol-1*, 249-256.

Moses, C. O. & Herman, J. S. (1991). Pyrite oxidation at circumneutral pH. *Geochimica et Cosmochimica Acta, 55*, 471-482.

Moses, C. O., Nordstrom, D. K., Herman, J. S. & Mills, A. L. (1987). Aqueous pyrite oxidation by dissolved oxygen and by ferric iron. *Geochimica et Cosmochimica Acta, 51*, 1561-1571.

Mozley, P. S. (1989). Relationship between deposition environment and the elemental composition of early diamagnetic siderite. *Geology, 17*, 704-706.

Murad, E., Schwertmann, U., Bigham, J. M. & Carlson, L. (1994). Mineralogy and characteristics of poorly crystallized precipitates formed by oxidation of Fe in acid sulfate waters. In: Alpers, C.N. and Blowes, D.W., Environmental geochemistry of sulfide oxidation. Washington, D.C., *American Chemical Society Symposium Series, v-550*, 190-200.

Murr, L. E. & Mehta, A. P. (1982). Characteization of leaching reactions involving metal sulfides in wastes and concentrates utilizing electron microscopy and microanalysis techniques. *Resource Conserv, 9*, 45-57.

Naumov, G. B., Ryzhenko, B. N. & Kodakovsky, I. L. (1974). Handbook of thermodynamic data: Publication PB226 722, U.S. Dept of Commerce, Natl. *Tech. Inf. Serv., 328*.

Nemati, M., Harrisonm, S. T. L., Hansford, G. S. & Webb, C. (1998). Biological oxidation of ferrous sulphate by thiobacillus ferrooxidans: a review on the kinetic aspects, *Biochemical Engineering Journal, 1*, 171-190.

Nicholson, R. V., Gillham, R. W. & Reardon, E. J. (1988). Pyrite oxidation in carbonate buffered solution: 1. Experimental kinetics: *Geochimica et Cosmochimica Acta, 52*, 1077-1085.

Nicholson, R. V. & Scharer, J. M. (1994). Laboratory studies of pyrrhotite oxidation kinetics. In: C. N. Alpers, & D. W. Blowes, (Eds.), Environmental Geochemistry of Sulfide Oxidation. *ACS Symposium Series, vol. 550*, Washington, DC, 14-30.

Nordstorm, D. K. & Southam, G. (1997). Geomicrobiology of sulfide mineral oxidation; In J. F. Banfield, & K. H. Nealson, (Ed.), *Geomicrobiology; interaction between microbes and minerals*, Mineralogical Society of America 35, Washigtone D.C., 361-390.

Nordstrom, D. K. & Alpers, C. N. (1999). Negative pH, effluorescent mineralogy and consequences for environmental restoration at the iron mountain superfund site, California, *Proc. Natl. Acad. sci., 96*, USA, , 3455-3462.

Nordstrom, D. K. & Dagenhart, T. V. (1978). Hydrated iron sulfate minerals associated with pyrite oxidation: Field relations and thermodynamic properties (abst.). Geol. Soc. Amer. *Abstracts with Program, 10(7)*, 464.

Nordstrom, D. K. (1982a). Aqueous pyrite oxidation and the consequent formation of secondary iron minerals, J. A., Kittrick, D. S. Fanning, & L. R. Hossner, (Eds). Acid sulfate weathering. *Soil Science Society of America*, 37-63.

Nordstrom, D. K. (1982b). The effect of sulfate in aluminium concentrations in natural waters: some stability relations in the system Al_2O_3–SO_3–H_2O at 298^{oK}. Geochim. Cosmochim. *Acta, 46*, 681-692.

Nordstrom, D. K. (2000). Advances in hydrochemistry and Microbiology of Acid mine waters, *Ins. Geol. Rev., 42*, 499-515.

Nordstrom, D. K. & Ball, J. W. (1986). The geochemical behavior of aluminium in acidified surface streams. *Science, 232*, 54-56.

Nordstrom, D. K., Plummer, L. N., Wigley, T. M. L., Wolery, T. J. & Ball, J. W. (1979). A comparison of computerized chemical models for equilibrium calculations in aqueous systems. In: Jenne, E.A. (Ed.), Chemical Modeling in Aqueous Systems. Am. Chem. Soc. Symp. Ser., vol. 93. *Am. Chem. Soc.*, Washington, DC, 857- 892.

Olyphant, G. A., Bayless, E. R. & Harper, D. (1991). Seasonal and weather-related controls on solute concentrations and acid drainage from a pyritic coal-refuse deposit in southwestern Indiana, U.S.A. *Journal of Contaminant Hydrology*, 7, 219-236.

Paktunc, A. D. (1999). Minerological constraints on the determination of neutralization potential and prediction of acid mine drainage. *Environ Geol, 39(2)*, 103-112.

Peppas, A., Komnitsas, K. & Halikia, I. (2000). Use of organic covers for acid mine drainage control, *Minerals Enginerring, 13(5)*, 563-574.

Plumee, G. S. & Logsdon, M. J. (1999). The environmental geochemistry of mineral deposits, *Rev. Econ. Geol.*, 6, 133-160.

Plumlee, G. S., Gray, J. E., Roeber, M. M., Coolbaugh, M., Flohr, M. & Whitney, G. (1995). The importance of geology in understanding and remediating environmental problems at Summitville. In Proceedings: Summitville Forum '95, *Colorado Geological Survey Special Publication*, 38, 13-22.

Plumee, G. S., Smith, K. S., Montour, M. R., Ficklin, W. H. & Mosier, E. L. (1999). Geologic controls on the composition of natural waters and mine waters draining diverse mineral-deposite types. In: G. S. Plumee, & M. J. Logsdon, Editors. The environmental geochemistry of mineral deposits, *Rev Econ Geol*, 6, 373-407.

Pronk, J. T. & Johnson, D. B. (1992). Oxidation and reduction of iron by acidophilic bacteria. *Geomicrobiology*, 10, 153-171.

Purohit, A. K., Pancholi, D. A., Mathur, S. N. & Chakradhar, B. (2001). Environmental impact - assessment and management plan at lignite mine Panandhro. Proc. of Nat. Seminar on "*Mining Envir. and Society*", Udaipur 13-14 Oct., 67-74.

Reedy, B. J., Beattie, J. K. & Lowson, R. T. (1991). A vibratiob spectroscopic ^{18}O study of pyrite oxidation. *Geochimica et Cosmochimica Acta*, 55, 1609-1614.

Retallack, G. J. (1990). Soils of the Past. *Unwin Hyman*, Boston, MA.

Robbins, E. I. (1998). Historical overview and future directions of microbial role in the acidic coal mine drainage system. In: Proceedings, *15th Annual American Society for Surface Mining and Reclamation Meeting*, St. Louis, MO, 174-191.

Robie, R. A., Hemingway, B. S. & Fisher, J. R. (1978). Thermodynamic properties of minerals and related substances at 298.15K and 1 bar (105 Pascals) pressure and at higher temperatures. U.S. *Geological Survey, 1452*, 456.

Rose, A. W. & Cravotta, C. A. (1998). Geochemistry of coal mine drainage, *in* B. C., Brady, T., Kania, W. M. Smith, & R. J. Hornberger, (Eds)., Coal Mine Drainage Prediction and Pollution Prevention in Pennsylvania: *Pennsylvania Department of Environmental Protection*, 1.1-1.22p. http://www.ott.wrcc.osmre.gov/library/pub/cmdpppp.htm.

Rose, A. W., Phelps, L. B., Parizek, R. R. & Evans, D. R. (1995). Effectiveness of lime kiln flue dust in preventing acid mine drainage at the Kauffman surface coal mine, Clearfield County, Pennsylvania. In: *Proceedings of the 1995 Meeting of the American Society for Surface Mining and Reclamation*, ed. by G.E. Schuman and G.F. Vance, 159-171.

Rossi, G. (1990). *Biohydrometallurgy*. McGraw-Hill, Hamburg, Germany.

Sahu, K. C. (1990). Heavy metal pollution in mining. "*Impact of Mining on Envir.*" Eds. R. K. Trivedy, & M. P. Sinha, Ashish Pub.House, Pub., New Delhi, 63-90.

Schindler, D. W., Turner, M. A., Stainton, M. P. & Linsey, G. A. (1986). Natural sources of acid neutralizing capacity in low alkalinity lakes of the precambrian shield. *Science, 232(4752)*, 844-847.

Seal II, R. R., Hammarstrom, J. M., Johnson, A. N., Piatak, N. M. & Wandless, G. A. (2008). Environmental geochemistry of a Kuroko-type massive sulfide deposit at the abandoned Valzinco mine, Virginia, USA, *Applied Geochemistry, 23*, 320-342.

Sherlock, E. J., Lawrence, R. W. & Poulin, R. (1995). On the neutralization of acid rock drainage by carbonate and silicate minerals. *Environ Geol., 25*, 43-54.

Shevenell, L., Connors, K. A. & Henry, C. D. (1999). Controls on pit lake water quality at sixteen open-pit mines in Nevada. *Appl Geochem, 14*, 669-687.

Shrivastava, B. K. & Singh, G. (1994). Trace metal contamination of soils in coal mine environment – a case study of a part of Jharia coalfield. Proc. of Second Nat. Seminar on *"Mineral and Ecology"*, Ed. Banerjee S.P., Oxford & IBH Pub., Dhanbad, Jan. 7-8, 379-388.

Singer, P. C. & Stumm, W. (1970). Acid mine drainage: the rate limiting step. *Science, 167*, 1121-1123.

Singh, G. & Sinha, D. K. (1992). The problem of acid mine drainage its occurrence and effects. Proc.of "Envir. Management of Mining Operations", *Dept. of Envir. and Forest, Govt. of India*, 156-167.

Sobek, A. A., Schuller, W. A., Freeman, J. R. & Smith, R. M. (1978). Field and laboratory methods applicable to overburdens and minesoils. U.S. *Environmental Protection Agency Environmental Protection Technology*, EPA-600/2-78-054: 203.

Sracek, O., Choquette, M., Gelinas, P., Lefebvre, R. & Nicholson, R. V. (2004). Geochemical characterization of acid mine drainage from a waste rock pile, mine Doyon, Quebec, Canada, *Journal of Contaminant Hydrology, 69*, 45-71.

Steger, H. F. (1982). Oxidation of sulfide minerals: VII. Effect of temperature and relative humidity on the oxidation of pyrrhotite. *Chem. Geol., 35*, 281- 295.

Stumm, W. & Morgan, J. J. (1996). Aquatic chemistry — chemical equilibria and rates in natural waters. *Wiley-Interscience*, New York.

Stumm, W. & Morgan, J. J. (1981). *Aquatic Chemistry*, 2nd ed. Wiley, New York, 780.

Swanson, D. A., Barbour, S. L. & Wilson, G W. (1997). Dry-site versus wet-site cover design. *Proceedings of the Fourth International Conference on Acid Rock Drainage*, May 30-June 6, 1997, Vancouver, BC, Vol. IV, 1595-1610.

Temple, K. L. & Delchamps, E. W. (1953). Autotrophic bacteria and the formation of acid in bituminous coal mines. *Applied Microbiology, 1*, 255-258.

Taylor, B. E. & Wheeler, M. C. (1994). Sulfur- and oxygen-isotope geochemistry of acid mine drainage in the western United States, in C. N. Alpers, & D. W. Blowes, (Eds)., Environmental geochemistry of sulfide oxidation: Washington, D.C. *American Chemical Society Symposium Series, 550*, 481-514.

Taylor, B. E., Wheeler, M. C. & Nordstrom, D. K. (1984a). Oxygen and sulfur compositions of sulfate in acid mine drainage: Evidence for oxidation mechanisms, *Nature, 308*, 538.

Taylor, B. E., Wheeler, M. C. & Nordstrom, D. K. (1984b). Stable isotope geochemistry of acid mine drainage: Experimental oxidation of pyrite. Geochim. et Cosmochim. *Acta, 48*, 2669.

Temple, K. L. & Koehler, W. A. (1954). Drainage from bituminous coal mines. West Virginia University Bulletin (Series 54, no. 4-1). Engineering Experiment *Station Research Bulletin*, *25*, 35.

Torma, A. E. (1988). Leaching of metals, in Rehm, H.J. and Reed, G. (Ed.), Biotechnology6B, *VCH Verlagsgesellschaft*, Weinheim, Germany, 367-399.

Tuovinen, O. H. & Kelly, D. P. (1972). Biology of Thiobacillus ferrooxidans in relation to the microbiological leaching of sulphide ores. Z. Allg. *Mikrobiol*, *12*, 311-346.

USEPA, (1994). *Technical document of acid mine drainage prediction*. Office of SolidWasteWashington, USA, *48*.

Van Everingden, R. O. & Krouse, H. R. (1985). Isotope composition of sulphates generated by bacterial and abiological oxidation. *Nature*, *315*, 395-396.

Ward, C. R. (2002(. Analysis and significance of mineral matter in coal seams, *International Journal of Coal Geology*, *50*, 135-168.

Watzlaf, G. R. (1992). Pyrite oxidation in coal waste under water saturated and unsaturated conditions. In: Proceedings of the 1992 National Meeting of the American Society for Surface Mining and Reclamation: Princeton, W.V., *American Society for Surface Mining and Reclamation*, 191-205.

Wendt-Potthoff, K., Frommichen, R., Herzspurng, P. & Koschorreck, M. (2002). Microbial Fe(III) reduction in acidic mining lake sediments after adition of an organic substrate and lime. *Water Air Soil Pollut Focus*, *2(3)*, 81-96.

Whitton, B. A. & Say, P. J. (1975). Heavy metals. In: B. A. Whitton, (Ed.), *River ecology*. Blackwell, Oxford, 286-311.

Wiersma, C. L. & Rimstidt, J. D. (1984). Rates of reaction of pyrite and marcasite with ferric iron at pH 2. *Geochimica et Cosmichemica Acta*, *48*, 85-92.

Williams, E. G., Rose, A. W., Parizek, R. R. & Waters, S. A. (1982). Factors controlling the generation of acid mine drainage. University Park, Pennsylvania State University, Final Report on U.S. *Bureau of Mines Research Grant*, G5105086: 256.

Williamson, M. A. & Rimstidt, J. D. (1994). The kinetics and electrochemical rate determining step of aqueous pyrite oxidation. *Geochimica et Cosmochimica Acta*, *58*, 5443-5454.

Wilson, G. W., Newman, L. L. & Ferguson, K. D. (2000). The co-disposal of waste rock and tailings. ICARD 2000, Society of Mining, *Metallurgy and Exploration*, *2*, 789.

P. L. ounger, & N. S. Robins, (ed). 2002a. Mine water hydrogeology and geochemistry, Vol. 198,Geological Society of London, *Geol. Soc. Spec. Publ.*, *396*.

Younger, P. L. & Robins, N. S. (2002b). Challenges in the characterization and prediction of the hydrology and geochemistry of mined ground. In: P. L. Younger, & N. S. Robins, editors. *Mine water hydrogeology and geochemistry*, *Vol. 198*, Geological Society of London, Geol. Soc. Spec. Publ., *396*.

Zanker, H., Moll, H., Richter, W., Brendler, V., Hennig, C., Reich, T., Kluge, A. & Huttig, G. (2002). The colloid chemistry of acid rock drainage solution from an abandoned Zn-Pb-Ag mine. *Applied Geochemistry*, *17*, 633-648.

Zhao, F., Cong, Z., Sun, H. & Ren, D. (2007). The geochemistry of rare earth elements (REE) in acid mine drainage from the Sitai coal mine, Shanxi Province, North China. *International Journal of Coal Geology*, *70*, 184-192.

In: Mine Drainage and Related Problems
Editor: Brock C. Robinson, pp.155-188

ISBN: 978-1-60741-285-4
© 2010 Nova Science Publishers, Inc.

Chapter 4

STATUS OF ACID MINE DRAINAGE AND METAL LEACHING STUDIES IN GHANA

Thomas M. Akabzaa[1], Bernard Kortatsi[2] and Justine Seyire[3]
[1]Department of Earth Sciences, University of Ghana,
[2]Water Research Institute of the CSIR, Ghana,
[3]Environmental Protection Agency (EPA), Ghana

ABSTRACT

This chapter reviews the status of acid mine drainage studies and management in Ghana. Acid drainage, and the contaminants associated with it, has been acknowledged as the single largest environmental problem facing the mining industry worldwide. Metals mobilised from acid rock/mine drainage conditions invade the environment with disastrous consequences. Fortunately, knowledge of both local and regional geology and geochemical processes can lead to a better understanding of the phenomenon and its environmental impacts. This knowledge has constituted the bases of management strategies, defining practices in acid drainage prediction, prevention, control, treatment and monitoring, developed to reduce risk of formation of acid drainage from mine wastes in most developed countries. These management approaches rely on the correct identification of mine rock, both in the pre-mining phase and as an ongoing part of mine planning. The Birimian and Tarkwaian Rocks Systems are the main source of gold in Ghana. Gold mineralisation in the Birimian occur in the form of quartz-veined gold in mainly carbonaceous phyllites while mineralisation in disseminated sulphides is in the metavolcanics. Base metal sulphides and sulphosalts including arsenopyrite, pyrite, sphalerite, pyrrohorite, and terahedrites dominate ore mineralogy. Although mineralisation in the Tarkwaian is not directly associated with sulphide-bearing rocks, dolerite intrusions among the Tarkwaian country rocks have been reported to contain considerable sulphide minerals. These sulphide-bearing lithologies present potential acid generating sites. Various studies of stream drainage emanating from some mine sites have recorded pH values as low as 2.4 with associated high doses of As (23. mg/L), Mn (15 mg/L) and Fe (103mg/L). Low pH (<3.4) and high trace metals have been reported from some boreholes in mining areas. Areas outside mining centres with similar geology have water bodies with similar low pH and high metal content. Unfortunately, acid mine drainage and metal leaching management practices are still a very gray area in Ghana. Mining environmental management in itself is a novelty in Ghana. The guidelines for mining environmental management and Environmental Regulations are two key tools for mining environmental management in Ghana. However, these do not have definite benchmarks, protocols and methodologies to guide companies in the evaluation of rocks for acid generating potential in Environmental Impact Assessment (EIA) processes. Limited studies on the subject matter has been carried out by mining companies on the mine scale and largely restricted to the mine site, as part of the project's EIA.

INTRODUCTION

Mining can influence the quality of water in a variety of major ways. These include generation of acid mine drainage that can contaminate water sources; the release of trace metals in unacceptable proportions into the water system thereby rendering them potentially dangerous for faunal and floral consumption; groundwater salinisation facilitated by deep mining encountering stagnant and connate or fossil waters that may be of very high salinity, that may contaminate fresh water; contamination from chemical agents used in recovery of minerals such as gold.

Among all the aspects of mining that impact negatively on the environment, the phenomenon of acid rock and/or mine drainage (ARD/AMD) and the heavy metal contamination associated with it is widely regarded as the most intense and persistent mining related environmental problem (Da Rosa and Lyon, 1997). Acid rock/mine drainage (ARD/AMD) is a naturally occurring process produced from the oxidation of sulphur to sulphates and leading to the production of acidity, sulphate, and iron whenever water containing oxygen comes into contact with sulphur present as sulphides in mineralized rock and mine wastes (Nriagu, 1978; Smith & Skema, 2001). This process is usually accelerated when mining exposes metal sulphides in rocks to oxygen and water, allowing rapid oxidation of the sulphides and dissolution of their products. The transport of such oxidation products culminates in drainage waters with anomalous loading of dissolved elements that reflect the mineral composition of the drained rocks. Jones (1998) pointed out that the pH of such drainage and the concentrations of metals are a function of the balance between acid generating and acid neutralizing minerals and their elemental composition and are site specific.

The phenomenon is particularly a source of surface and ground water pollution, and where it has occurred it has resulted in serious ecological disasters. For example, Schreck (1998) reports that 35 acidified mining lakes exist in Central Germany alone. These acidified lakes emanate from rising groundwater that has filled abandoned lignite mines resulting in AMD. Taylor (1998) reports of death of aquatic ecosystems, soil contamination, degradation of food chain and contaminated groundwater as a result of the drainage of contaminated, low pH water from acid generating abandoned mines at Rum Jungle and other locations into waterways in Australia. Abandoned coal mines have been identified as the largest source of pollution in Pennsylvania, with an estimated 4,000 km of streams impaired due to AMD (Smith and Skena, 2001). David et al. (1997) have pointed out that the persisting historical nature of mining-related water contamination is amply reflected in the names of some rivers, such as the north Cornish Red River, the Ochre Dykes of Derbyshire and South Yorkshire, the Yellow Stream of Lancashire, the Norwegian Raubekken and the Spanish Rio Tinto.

There is growing unanimity of opinion that acid drainage, and the contaminants associated with it, is the single largest environmental problem facing the mining industry worldwide (Ferguson and Erickson, 1998; U.S. Forest Service, 1993; USEPA, 1994; Kuyucak, 2001; Smith and Skena 2001). It has also been identified as the largest liability facing the mining industry in Canada, the United States, Australia and Europe, resulting in an estimated $US22 billion, $C2–5 billion, $A60 million, $US300 million and DM13 billion committed to its treatment, abatement and other liabilities cost in the USA, Canada, Australia, Sweden and the former East Germany respectively (Taylor 1998; MEND, 2001)

It is now well understood that a thorough knowledge of both local and regional geochemical processes can lead to a better understanding of the environmental impacts of acid drainage and the associated metal leaching. Such an understanding is critical for the development of acid drainage prediction and treatment strategies, and also for the evolution of environmental guidelines for its prevention and treatment. Consequently, in the last two decades, efforts have been made globally to understand the phenomenon in major mining centres throughout the developed world. These efforts have included the development of innovative technologies for prediction, prevention and treatment and the evolution of comprehensive regulatory frameworks for the management of ARD by governments and mining companies. These efforts over the years have also resulted in a large body of literature on the subject. In fact, metal leaching and acid rock drainage policy and guidelines are fast becoming standard practice for all government environmental agencies in the developed world. This is the result of the growing realisation that the cost of treating such problematic waste can be very high, but if the sources can be identified, then their selective treatment may be more cost effective than the treatment of much larger quantities of non-problematic material with which problematic material is arbitrarily classified (Downing et al., 2000).

THEORETICAL OVERVIEW OF ACID DRAINAGE PREDICTION METHODS

An essential part of controlling AMD production is the ability to predict it. There are several methods that can be used to predict AMD production. However the theoretical basis for the evolution of these methods to predict acid drainage is universal. It is premised on the fact that the propensity of a given rock type to generate ARD is a function of variables such as the presence of acid generating material—sulphur content of sulphides; the content of acid consuming materials, such as carbonates and the amount and type of potential contaminants present (Sobek et al., 1978; Cladridge & Downing, 1993; USEPA, 1994; Evangelou, 1995).

The pre-requisite for the onset of AMD is the generation of acid at a rate faster that it can be neutralised by any alkaline material in the waste during the weathering of sulphide deposits. The sulphide oxidation process is usually illustrated using pyrite (FeS_2), because it is the commonest mineral causing AMD and is found in coal and metal mines. However, other sulphide minerals such as pyrrhotite, marcasite, galena, sphalerite, chalcopyrite can cause AMD. The oxidation of pyrite is in four (4) basic steps represented by the stated equations 1 to 4 below (Salomons, 1995). Salomons (1995) identified three stages in the process.

In Stage 1: Conditions are at a pH above 4.5, high sulphate and low Fe concentrations with little or no acidity. In this, equation 1, AMD proceeds both biotically and abiotically and the disulphide ion (-1) is oxidised to sulphate (+6). The sulphate of the Fe^{2+} is soluble in water. Some amount of acid is generated in the process. Equation 2 is abiotic and slows down with decreasing pH.

$$2FeS_2 + 7O_2 + 2H_2O \longrightarrow 2Fe^{2+} + 4SO_4^{2-} + 4H^+ \quad \text{............} \quad \text{Equation 1}$$

$$4Fe^{2+} + 10H_2O + O_2 \longrightarrow 4Fe(OH)_3(s) + 8H^+ \quad \text{............} \quad \text{Equation 2}$$

$$2Fe^{2+} + O_2 + 2H^+ \longrightarrow 2Fe^{3+} + H_2O \quad \text{............} \quad \text{Equation 3}$$

$$FeS_2 + 14Fe^{3+} + 8H_2O \longrightarrow 15Fe^{2+} + 2SO_4^{2-} + 16H^+ \quad \text{............} \quad \text{Equation 4}$$

Stage 2: The pH is between 2.5-4.5 and it produces high sulphate levels, acidity, as well as an increase in total Fe. The Fe^{3+}/Fe^{2+} ratio is still low. Equation 2 is biotic, especially determined by the action of *Thiobacillus ferro-oxidans*.

Stage 3: Occurs at pH values well below 2.5 and is characterised by high sulphate as well as Fe level and high Fe^{3+}/Fe^{2+} ratios. Equation 3 is wholly determined by bacterial oxidation. Equation 4 is determined by the rate of equation 3. Fe^{2+} is also oxidised to Fe^{3+} with the production of a large amount of acid (H^+ product), especially by considering the combined equations.

Equation 5 is the overall equation for the oxidation of both the Iron and Sulphur.

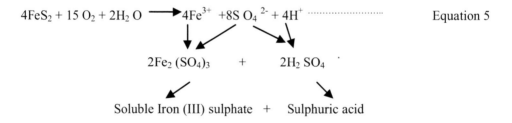

From the oxidation reactions, it is clear that acidic conditions result from the ultimate formation of sulphuric acid. The pH of waters reach levels as low as 2.3 and occasionally 1.8 (Maidment, 1995) and there have even been instances where sub-zero pHs have been measured (Da Rosa & Lyon, 1997). This in turn causes other minerals to dissolve because the solubilisation of heavy metals increases with increasing acidity. Metals such as iron, copper, aluminum and lead tend to dissolve and mobilise easily in the acidic waters associated with AMD. Metal leaching in many rock types, typically occur as acid levels drop to below 5.5 or 6 on the pH scale. Some elements can, however, remain soluble at neutral or even alkaline pHs e.g. zinc, arsenic, manganese, selenium, cadmium, molybdenum (Eppinger et al, 1999).

By considering equations 1-4, the intensity of acid generation is said to be dependent, to a large extent, on the rate and extent of oxidation. The intensity of acid generation is governed by chemical parameters such as pH, temperature, oxygen concentration in the aqueous and gaseous phases, chemical activity of Iron - III (Fe^{3+}), sulphide mineral content and the surface area of the exposed metal sulphides. Biological parameters include biological activation energy, population density of the bacteria, rate of bacterial growth and supply of nutrients. Apart from the catalytic activity of sulphur - and iron - oxidising bacteria, *T. ferro-oxidans*, there are also *T. thio-oxidans* and even fungi and yeasts such as *Pullularia pollulans, Spicaria divaricata*, filamentous iron bacteria of the genus *Metallogenium*. which can also speed up the reaction. These bacteria are indigenous to environments that contain metal sulphides, oxygen (atmospheric or aqueous) and acid conditions (Salomons, 1995; Kuyucak, 2001).

There are secondary factors that control acid consumption or the alteration of the products from acid generation reactions. Neutralisation of the AMD is found to occur when carbonate minerals are found in the substrata. Examples of these are the presence of calcite and dolomite rocks. In addition, some residual CaO usually left over from the metal extractive process will neutralise AMD to a degree, only dependent on the quantity of carbonate present with respect to the AMD. The relation between acid generation by pyrite oxidation and the neutralisation of the acid by $CaCO_3$ is represented by equation 6.

$$4FeS + 8CaCO_3 + 15 O_2 + 6H_2O \longrightarrow 4Fe(OH)_3 + 8SO_4^{2-} + 8Ca^{2+} + 8CO_2 \quad ....\text{Equation 6}$$

$$CaSO_4$$

Neutralisation of the generated acid is, further reduced by the products of the reaction in Equation 6. In actual fact the $CaSO_4$ and Iron hydroxide generated coat particles, thus preventing further neutralisation. Tertiary factors producing and controlling AMD are the physical aspects of the waste material and hydrological factors, which affect acid production, its migration and consumption. Coarse-grained wastes promote oxygen advection and diffusion resulting in higher acid production and vice-versa, while contaminants concentrations in sediment often vary according to the organic matter content as well as the grain size of the matter, with fine grained, organic rich sediments exhibiting relatively high levels of pollutants (Salomons, 1995; Kuyucak, 2001, Seyire, 2000, Mogollon et al., 1995; Miller et al., 1995).

THE STATUS OF ACID DRAINAGE STUDIES IN GHANA

The growing efforts to understand the phenomenon of acid drainage and its menace and the growing literature on it have been disproportionately confined to the developed countries. The subject is a grey area in most mining centres in Africa in particular. For example, Ghana, despite been a major player in the mining industry in Africa, with a long history of mining, does not have an elaborate framework for acid drainage management. Mechanised commercial mining dates back to the 18th century while artisanal mining by indigenous people is said to have been around for over five hundred years (Junner, 1935). The scale of mining activity, however, particularly increased tremendously after policy changes in 1986 by the Provisional National Defence Council government The mining sector, one of the main sources of foreign exchange for the country, was considered a priority sector for these reforms (Jonah, 1987; Akabzaa, 2000). Since this period, sixteen new mines have come on stream while about a dozen others are at mine development stage. In total, there are more than twenty medium sized to large-scale gold mining projects fully developed or being developed, along with one mine each of manganese and bauxite under extensive expansion and a diamond mine under divesture. There are also, in addition, over 100 historic gold mine sites scattered all over the country.

Many of these mines are situated within lithological units with sulphur bearing rocks. It is the oxidation of sulphidic mine waste, exposed country rocks and excavations that normally leads to generation of acid rock/mine drainage. Water discharges from such an environment, such as mine pits, adits and other tunnels and tailings piles will be laced with a variety of pollutants that poses a threat to not only aquatic life in streams and rivers but also to downstream water users. Such an environment may supply various chemical pollutants to ground and surface water systems and encourage heavy metal mobilisation. These pollutants also contaminate soils with serious consequences to local ecosystems.

The prevention of these potential impacts of acid drainge is usually achieved through technical evaluations of the likelihood and extent of ARD in proposed projects based on elaborate guidelines provided by the relevant environmental protection authorities. Unfortunately, the environmental regulatory framework for the mining industry in Ghana was not promoted with the same zeal and timing as mining investment. Environmental regulations for the management of the mining industry were introduced in 1999 as Legislative Instrument (LI1652), and these were preceded by guidelines that were introduced in 1994. These regulations give broad guidelines for discharge of industrial wastewater (EPA, 1994; 1999). The guidelines on acid mine drainage management are part of the general guidelines for the preparation of Environmental Impact Assessment of mining projects and are very scanty.

Under these guidelines, mining companies are normally required to investigate the rocks in their project domain for their acid generation potential. This is to enable them categorise mine waste for safe disposal and to design mitigating measures to combat acid mine drainage, the effect of which usually lingers much longer after the closure of the mine. This requirement would normally form part of the companies' Environmental Impact Assessment (EIA) study presented to the Environmental protection Agency (EPA) before mining leases are granted. However, weakness in the EIA guidelines and processes in Ghana have been highlighted by various studies (Salami, 1999; Salami, 1999b; Akabzaa, 2000; Kuma et al., 2002; Seyire, 2001). Kuma et al. (2002), while acknowledging that existing regulations clearly require mining EIA studies to include investigations of surface and groundwater systems, highlighted the need for more prescriptive guidelines to enrich the hydrogeological content.

Studies of various Environmental Impact Assessment Statements of mining projects in the country in recent times indicate that acid mine drainage and metal leaching evaluation is not normally taken seriously by some companies (Salami, 1999b; Seyire, 2001). Yet there are several reported cases of heavy metal pollution of community water sources in some mining centres in the country (Amonoo-Neizer, & Busari, 1980; Amuzu, 1991; Smedley, et al., 1993; Armah, et al., 1998; Acquah, 1996). Some of these cases of water resource contamination have been at the centre of tension between mining companies and local communities, sometimes turning violent and leading to temporary stoppage of mine production.

In Ghana, apart from diamond, which is produced from alluvial sources, all other major minerals - gold, bauxite and manganese - won in commercial quantities- are derived from metasedimentary and metavolcanic rocks of the Birimian system and the paleo-sedimentary rocks of the Tarkwaian System. They are dominated by greenstone and slate sequences that have metamorphosed to greenschist or less commonly amphibolite or granulite facies in the case of the Birimian and paleo-conglomeratic units in the case of the Tarkwian (Ensenloiir and Hirdes, 1992; Milesi et al., 1992). Elsewhere, these rock units have been reported to have considerable acid mine drainage and metal leaching potential (Ashley, 2002; Cox and Bagby, 1986). However, in Ghana very few companies mining these type-rocks acknowledge any possibility of acid mine drainage or significant metal leaching in their operational areas as reflected in the content of the various Environmental Assessment Statements (Anon, 1990; 1994; 1996; 1998).

In particular, mines exploiting gold from the Banket Formation of the Tarkwaian rock system do not consider acid rock/mine drainage and metal leaching possibilities in their EIAs studies (Anon, 1990; 1995). The Banket Formation is the major gold bearing lithological unit of theTarkwaian rock system. The generally held opinion is that gold in ores here is free milling and not directly associated with sulphides (Anon, 1990, Smith 2001). Therefore the rocks have no potential for acid mine drainage. Consequently, the issue of acid mine drainage is usually given very cursory study and, in some cases, even rarely mentioned in EIA studies of projects in the area.

However, pyrite and other sulphide- bearing rocks are not uncommon in the Tarkwaian. Recent excavations in the Tarkwaian at Damang have exposed considerable amount of sulphides in the rocks. In addition many communities within the area have complained of contamination of their domestic water sources. However, because of the general belief that the rocks have no potential for acid mine drainage, the subject of acid rock/mine drainage is rarely revisited in environmental audits and other review studies. In some cases the communities' complaints have often been considered vexatious, on grounds that the ores have no acid generation potential.

While there is general acknowledgement that sulphide ores associated with Birimian rocks have possibilities for acid mine drainage, very few companies give the acid mine drainage evaluation serious attention in their impact studies. The excuse is that these ores contain carbonate alterations whose buffering effect is more than the acid generating capacity of any available sulphides (Anon, 1994; 1996; 1998). In many sulphide ore deposits, trace metals such as Cadmium, Mercury and metalloids, e.g. Antimony and Arsenic are usually nearly always present in the ores and leached when the sulphides are exposed to air and water. These elements, many of which are mobile even at neutral pH are highly toxic even in small quantities, particularly in soluble form, which can be absorbed by living organisms (Walker et al., 2004; Lee et al., 2002)

Empirical medical records from the study area suggest that reported cases of mental disorders are on the increase. Manganese and mercury have long been known to cause neurological disorders (Smedley and Kinniburgh, 2002; Jamieson and Pryzbylowicz, 1997; Songsori et al., 1994). Similarly, medical research in the area showed that arsenic dermatitis have been observed in the Bogoso area where Bogoso Gold limited operates an open pit mine (Akabzaa, 2000). Songsori et al., (1994) also reported that research has showed abnormal mercury levels in hairs taken from small-scale miners and their families in the area. These findings underscore the need for further investigation into the distribution and concentrations and sources of these pollutants in the area.

There are limited studies of some mining companies carried out on the mine scale and largely restricted to the mine site mostly as part of the project's Environmental Impact

Studies. Companies such as Ashanti Goldfields Company (AGC), now Anglogold Ashanti, Bogosso Gold Ltd, and Newmont (Ghana) Limited, have conducted appreciable amount of work on AMD, the rest have given the problem a very cursory attention. Even for these few exceptional cases for which ARD evaluation has be done in Ghana, the companies have been criticised for using unconventional in-house non conventional Acid Base Accounting procedures without complementary tests (Anon, 2005). Some of these mine scale studies have been reviewed here.

ACID MINE DRAINAGE STATUS OF AGC OBUASI MINE SITE

Ashanti Goldfields Company (Obuasi) Ltd has had over hundred years of underground mining and about ten years of open pit mining. The ore is primarily free gold within quartz veins and lenses; disseminated refractory gold within sulphide minerals (mostly pyrite and arsenopyrite) formed at the fringes of the shear zones, and oxidised ore derived from sulphides (Amanor & Gyapong, 1991; Minerals Commission, 1991). While the high sulphide content of the ore is suggestive that AMD is likely to occur in underground workings, data in EIA baseline showed that analyses of mine water pumped from three shafts had high hardness and pH values of 7.1-8.0. These values, according to the report, are indicative of alkaline conditions and a surplus of acid neutralising capacity. Similarly, at the Sansu open pit, it concluded that since the waste rock contained large amounts of alkali, it is unlikely that acid effluent would be generated by rainwater percolating through the waste dumps even though no geochemical analyses were carried out to support these assertions. It was further assumed that no AMD was expected given the oxidised nature of the ore and would only create a problem, if and, when the underlying sulphide deposits occurring at depth were eventually mined (Minerals Commission, 1991).

However, Smedley et al. (1993) in a study carried out in Obuasi reported highly acidic waters (pH 3.9- 6.9) for streams, wells and boreholes sampled. In 1998, AGC Ltd. commissioned a report into the back- filing of Dokyiwa Pits at Obuasi. A more comprehensive analysis for predicting AMD generation was carried out. This showed the possibilities and conditions under which AMD is likely to occur. The solid phase tailings were found to contain varying degrees of Au, S and As for the Oxide Treatment Plant (OTP) and Sulphide Treatment Plant (STP) respectively. For instance, mean values of Gold (Au), Sulphur (S) and Arsenic (As) for the Oxide Treatment Plant (OTP) tailings for 1996-1998 were 0.54 g/t, 0.10% and 0.08% respectively. Mean values of Au, S and As for the Sulphide Treatment Plant (STP) tailings for 1996-1997 were 1.6 g/t, 0.39% and 0.87% respectively. It can be inferred that tailings contain some amount of As and S (in sulphides) which, when allowed to oxidise, may produce AMD.

Even though the tailings were found to contain very low levels of sulphur (0.10-0.39%) these could produce AMD whenever Neutralisation Potential Ration (NPR) <3 and no AMD if NPR$\geq$3. In effect, the balance between the neutralisation potential (NP) and the acid production (AP) give a quantitative measure (NPR= NP÷AP) for predicting AMD generation. From Table 1, it is obvious AMD may occur for OTP but not STP tailings. However, STP also had the possibility of being reactive if the calcareous constituents are reduced through uncontrolled neutralisation. These conclusion could only be said to be very conservative since the application of more rigorous criteria as such as those required by the United States Forestry Services could have probably shown more acid generating units. This scheme classifies a material as non- acid producing when NPR > 5 and not 3 as in this case (USAFS, 1994).

The Bogoso Mine Site

At Bogosu Gold Ltd. (BGL) previously known as Canadian Bogosu Resources Ltd. (CBR), ore reserves are approximately, 9 Mt of sulphide ore grading 3.9 g/t and 1 Mt of oxide ore grading 3.2 g/t. The gold in this mine occurs in two forms. The first is primary sulphide ore, similar to the Obuasi ores and contains, on the average, 5% pyrite, 3% arsenopyrite with about 80% of all the ore in this mine being this type. The high sulphide content implies once these are oxidised, the ore could be acid generating. Second is ore in quartz – carbonate rich rock that has a similar sulphide composition to the first one, but a carbonate content of up to 10% of the ore. As a result, this could provide sufficient buffering capacity to neutralise the acid formed from the sulphides. No mention of the potential for AMD was made in the EIA of the mine (Minerals Commission, 1991).

However, latter studies indicate the potential for acid mine drainage in the mine. Acid water of pH 3 was reportedly pumped out from the old Chujah pit and the abandoned Marlu shaft implying the overall capacity of mine rocks to form AMD. The problem of acid mine drainage generation from the operations of the Bogoso Gold Mine Limited has now been identified as a major environmental concern for the company. The problem was first identified in 1992 during routine environmental audit of the mine when widespread pools of ARD were found in several mining areas associated with sulphide bearing material deposited on the surface (EAU, 1994). In 1994 a recommendation was made that the source of ARD be characterized to differentiate acid forming from non-acid forming material. Initial characterization work was performed in 1994 (Miller, 1994) using Net Acid Generation (NAG) test. The consultants considered material that returned NAGpH values of less than 3 as possible acid generating and those with NAGpH greater than 5 as non-acid generating (Table 2). The findings confirmed that transition and primary water rock is a potential ARD concern and that 50% of transition and primary waste rock is potentially acid generating. The findings showed that sulphide and carbonate content of primary rock varied widely with zones of potentially acid forming, non-acid forming and acid consuming material in the primary zone. Transition zone waste rock was found to generate acid more rapidly upon exposure, presumably due to selective leaching of carbonates from this zone.

Table 1. The AMD potential of STP and OTP Tailings

Parameter	STP Tailings	OTP Tailings
S%	0.32	0.11
Acid Production, AP	10.02	3.43
Neutralisation Potential, NP	46.01	0.2
NPR = NP/AP	**4.6**	**0.06**

Source: EIA Report, 1998. AGC Ltd.

Table 2. Summary of NAG Tests

Borehole No.	No. of samples	Lithology	NAGpH Range	Average
Chujah Pit Hanging Wall and Chujah Stage 3 Hanging Wall				
4	7	Graphitic phyllites	2.24 – 2.90	2.53
6	31	Phyllites/greywacke	3.28 – 7.94	6.04
7	2	Phyllites/greywacke	4.2 – 4.24	4.22
10	29	Phyllite	2.48 – 8.65	5.25
11	4	Phyllite	2.30 – 7.89	3.87
13	22	Phyllite	2.54 – 7.85	4.22
14	10	Phyllites/graphitic phyllite	2.54 – 7.73	6.51
15	6	Phyllites	4.99 – 6.30	5.91
16	26	Phyllites	5.36 – 8.69	7.12
18	9	Graphitic Phyllites	3.98 – 7.65	5.95
Bogoso North Pit Hanging Wall				
5	2	Graphitic Phyllites	3.16 – 4.38	-
8	2	Graphitic Phyllites	2.84 – 3.49	4.89
9	2	Graphitic Phyllites	2.82 – 4.0	5.38
11	2	Greywacke	2.48 – 3.81	9.7
15	6	Tuff/graphitic phyliite	5.37 – 6.71	-

Source (Anon, 1994)

The consultants recommended that a characterization system be developed to distinguish potentially acid generating, non-acid generating, and acid consuming waster rock types to effectively control ARD. They recommended the NAG test as a rapid screening tool that should be used to classify waste rock for disposal. Also in the Annual Environmental Report of the Environmental Protection Agency (EPA) of 1997, the mine was queried for having back-filled some pits that were generating AMD. The report also indicated that transitional ores, which were also generating AMD, had not been properly attended to in certain places. The mine managers were advised to take acid mine drainage management seriously (EPA 1997). Following these recommendations a major ARD assessment was concluded in 1999 (Anon, 1999) with strong acknowledgement that acid mine drainage is a major problem facing the mine. The ARD/AMD assessment identified areas of potential ARD and proposed management protocols that could help mitigate the problem and improve the management of ARD, particularly in the tailings dam and pits, sub-grade ore stockpiles, and dumps. The mine operators have now recognized that proper assessment of ARD potential and development of a proper site mitigation, control and management strategy for ARD should be a critical component of environmental management.

Newmont Ghana Gold (Ahafo Project) Mine Site

Newmont (Ghana) Limited has two major projects in Ghana, the Ahafo and Akyem projects in the Brong Ahafo and Eastern Regions of Ghana, respectively. The Ahafo project is an open cast operation that started in late 2005 while the Akyem project is still at mine development stage. The Ahafo Project exploits gold hosted in shear structures that extend over the length of the mining lease area, covered by Birimian metasediments and metavolcanics with granitoid intrusions. Primary quartz vein and sulphide hosted ore, transition and oxide ore is mined here. The gold occurs in association with pyrite and less commonly with Arsenopyrite. Total estimated ore tonnage is 140 million grading 2.29g/t, but experts say this estimate is conservative and could double (Anon, 2003).

Acid mine drainage assessment was conducted as part of the project's EIA process in 2002. Acid Base Accounting tests were conducted on 125 samples covering nine proposed mining sites. According to the conclusions of this study, all the samples, except ten, showed low potential for acid generation to totally non-acid generating. That all the samples, including tailings had very high acid neutralisation capacity with positive NNP. That further calculation of the NPR indicated that all the samples are non-acid forming since they returned NPR > 2.5. In 2003, additional tests were conducted to ascertain the acid generating potential of waste rock. Acid neutralisation potential was determined by Net Carbonate Value (NCV) method by titrations with acidity corrections to evaluate the effect of iron carbonates on acid neutralisation potential for waste rock characterisation. According to the NCV results, most of the material, including the sulphide zones could be classified as neutral to basic. The overall conclusions were that AMD is not an issue in the mine, however, routine monitoring during the life of the mine was recommended.

The AMD procedures used in the EIA for this project have been evaluated by the United States Environmental Protection Agency (USEPA) at the request of some civil society groups. The USEPA has criticised the methods used by the company in its AMD assessment as been less rigorous and unconventional. They proposed that the company employ conventional static AMD assessment methods and complement them with kinetic tests. The company has duly complied and has a kinetic programme in place at the mine site (Anon, 2006).

The Other Mines

In most of the other mines, although a complete geochemical assessment of the mine rock types was not done, most reports have tended to suggest that the occurrence of AMD is unlikely (Minerals Commission 1991; NSR 1992). They have most often justified this conclusion with pH of drainage in these mine sites which they describe as generally neutral to basic. This predominant use of near neutral to slightly alkaline pH values in determining AMD is highly and grossly misleading. Other factors proposed for the non-occurrence/ formation of AMD are not adequately representative of the situation on the ground. For instance, the fact that oxidised ores is being mined and hence the exclusion of the formation of AMD is lame. It is common knowledge that the purely oxidised layer is only a few tens of hundreds of metres thick and is often underlain by sequences of intermediate transition zone of mixed oxide and sulphide ores, which finally grade into the metallurgically richer sulphide zone. Therefore, in any normal operation, oxide ores would be depleted within a few months to years of mining, and the fact that these companies have been mining for a long time implies they have possibly moved into the sulphide mineralisation zone.

Using the case of Ashanti (Obuasi, specifically) as an example, AMD was not a threat since mining was largely restricted to the oxidised ore when Environmental Consultants carried out the study, in April 1991. There is ample evidence that the company has moved from mining the oxide and unpredictably high transition ore into sulphide ores. ARD assessment of the Bogoso mine with similar geology as the Obuasi mine suggest that the transition ores are more susceptible to acid generation due to early selective leaching of carbonates in this material.

In a few cases where these earlier studies had identified instances of acid generation in some mines, especially underground mines in Prestea, Obuasi and Konongo, the acid generation potential of these mines has been downplayed. For instance in a report in 1991, the Minerals Commission concluded that AMD identified in some mine shafts was expected to be confined entirely to the underground and hence restricted to acidic mine water pumped from the shafts (Minerals Commission, 1991). This conclusion cannot be entirely true as aquifers are re-charged by the downward movement of water through the unsaturated zone. Lateral and downward percolation of AMD, may therefore contaminate an aquifer if present and especially in the Ashanti belt where ore bodies are closely associated with the deep seated and laterally persistent graphitic shear zones(Amanor & Gyapong (1991).

METAL LEACHING STUDIES

In comparison to acid mine drainage studies, considerable work has been done with respect to assessment of metals in water bodies, however many of these studies stop short of identifying the sources of these metals. Recent metal leaching studies have been conducted by the authors but limited the Tarkwa mining district of Ghana (Akabzaa, 2004; Kortatsi, 2000)

These studies assessed the combined impact of several mines on the quality of surface and groundwater in the Tarkwa mining district. The area is drainaged by mine-impacted water from the Tarkwa, Teberebie, Iduapriem, Bogoso, Presetea, Damang, Wassa and Sankofa large-scale gold mines, the Nsua manganese mine, and scattered artisanal gold mining sites within the area. These studies analysed water samples from streams, boreholes, hand dug wells, and mine spoil. Scatter plots of trends among measured parameters were used to assess drainage quality and differential impacts. Rock and mine spoil samples were analysed for their mineralogical composition and to identify possible acid generating and consuming components. Static geochemical acid rock drainage prediction methods were employed to assessment the acid generating possibilities of these samples.

Drainage quality exhibits wide seasonal and spatial variations with strong influence of geology on the water chemistry. High metal load regimes are associated with low and high pH regimes and coincident with acid-generating and buffering zones, respectively. The most degraded waters emanate from the Prestea and Iduapriem mine sites, while the Tarkwa mine showing minimal metal loading (Akabzaa 2004).

Areas with low pH (< 5.5), and high sulphate ions and trace ions were identified and are suggestive of acid mine drainage while sites with high pH (< 7.5), HCO_3^-, subdued SO_4^{2-}, and high trace ions are suggestive of sites where acid neutralization is effective.

High metal sources are largely confined to mining operations in the Birimian formation with ores containing more than 2% sulphides. However, restricted high metal regimes are observed in drainage within the Tarkwaian Formation associated with scatted sulphide bearing dolerite dykes in the operational areas of Tarkwa and Damang mines. Earlier studies disputed sulphides in the Tarkwaian formation until recently when acid generating dykes discovered in operating pits.

Thirty-one samples of rock units likely to be encountered in mining operations, waste rock and tailings dumps were tested for their acid generating potential using Acid Base Accounting (ABA) and Net Acid Generation pH (NApH) tests. For the ABA tests, variables including Paste pH, Sulphide-sulphur (%), Maximum Acid Potential (MAP), Neutralization Potential (NP) and the Net Neutralization Potential (NNP), all expressed in $CaCO_3$ equivalent tons/1000 tons material were determined, according to the method of Smith et (1974) and modified by Sobek et al. (1978). The NAGpH of the reaction solution was determined by titration to pH 7. This gives a value for the net acidity generated (NAG) by the acid generation and neutralization reactions occurring in the sample as outlined in Lawrence and Sheske (1997) and Miller (2000).

The mineralogy of the ores and associated rock types, particularly the presence and distribution of sulphides and carbonates dictate the acid generating possibilities of the rock units. Sulphide minerals such as pyrite, arsenopyrite, pyrrhotite, chalcopyrite and sphalerite were found to be common in many of these mine sites and may contribute to the maximum potential acidity. In a similar vain, various carbonate minerals such as rhodochrosite, kutnahorite, calcite, dolomite and ankerite, were identified as the dominant carbonate mineralogy, and likely to contribute neutralizing potential. Siderite, although a ubiquitous carbonates in these rocks, particularly rocks from Prestea is not likely contribute to net neutralization potential Seal and Foleyl (2002).

The results of acid base accounting and NAGpH tests are presented in table 3. Identified potentially acid generating samples are listed in table 4 based on a variety of methodologies. Established ARD studies have defined various criteria for characterizing the results of ABA. Similar criteria exist for classifying samples according to their NAGpH. Classification methodologies for the cauterization of results of acid base accounting developed variously by Price et al., (1998) Ferguson et al (1998), MEND (2000) and USEPA Forest Service 1994) were used to determine acid generating and non-acid generating samples from the ABA results. The classification of the Net Acid Generation (NAG) tests results were based on the criteria outlined by the British Columbia Acid Mine Drainage Task Force and a similar one by the Australian Mineral Industries Research Association (MEND, 1989; Greenhill, 2000.

Both ABA and NAGpH testsidentified various possible acid generating lithologies, likely to be encountered during mining in all the three mines investigated. At the Prestea mine site, phyllite, greywacke, mineralised quartz reefs and old tailings dumps at Bondaye exhibit potential for acid generation. At the Nsuta mine site, phyllite, encapsulating the carbonate ore, and old manganese oxide tailings being reworked are potentially acid generating. At the Tarkwa mine, scattered basic dykes intruding the paleo conglomeratic Tarkwaian formation hosting the mine were found to be acid generating. This is contrary to popular belief that the mine has no acid generating possibilities because it is situated in the non-sulphide bearing paleo-conglomeratic rocks. Table 4 is a summary of the possible acid generating samples using different classification schemes.

Table 3. Results of Acid Base Accounting

Location	Sample Id	Sample Type	Paste pH	Sulphur (%)	MAP	NP	NNP	NPR=NP/MAP	NAGpH
Tarkwa Mine	GFLMTS	Tailings from old dump	7.8	0.0001	0.003	44.3	44.30	14063.49	6.4
	GFRC01	Felsic dyke	8.2	0.0279	0.879	45.57	44.69	51.85	6.0
	GFRC02	Quartz vein cutting dyke	8.2	0.0029	0.091	53.16	53.07	581.94	6.0
	GFRS01	Weathered basic dyke	7.6	0.0457	1.440	20.25	**18.81**	14.07	6.0
	GFRS02	Fine grained quartzite	8.0	0.0001	0.003	29.11	29.11	9241.27	5.6
	GFRS03	Fine grained Conglomerate	7.8	0.0464	1.462	32.91	31.45	22.52	6.2
	GFRS05	Fine grained Quartzite	8.0	0.0556	1.751	63.29	61.54	36.14	6.6
	GFRS07	Fresh basic Dyke	7.8	0.0371	1.169	11.39	**10.22**	9.75	6.7
Nsuta Mine	GMCTS01	Reworked Manganese Oxide tailings	4.8	0.1584	4.990	21.85	**16.86**	**4.38**	**4.6**
	GMRS01	Phyllite above ore zone, Hill D, South Crest	6.3	0.3979	12.534	61.91	49.376	**4.94**	**4.8**
	GMRS02C	Greenstone from Hill D South Crest	8.0	0.0554	1.745	170.89	169.14	97.93	7.0
	GMRS03C	Transition ore, Hill D South Crest	8.0	0.0369	1.162	126.58	125.42	108.90	7.1
	GMRS04	Carbonate Ore from Hill South Crest	8.1	0.0288	0.90	110.13	109.22	121.40	6.1
	GMRS04C	Reworked Manganese Oxide tailings	7.5	0.0461	1.452	139.24	137.79	95.89	7.0
	GMRS05	**Greenstone Hill D North Crest**	7.3	0.0297	0.936	21.99	21.05	20.30	6.1
	GMRS06	Transition ore	8.3	0.0371	1.169	120.25	119.08	102.90	6.1
	GMRS07	Carbonate ore from Hill D North Crest	8.3	0.0297	0.936	31.65	30.71	33.83	6.5
	GMRS09	**Phyllite below ore zone**	7.2	0.5743	18.092	207.59	189.50	11.47	**5.0**
Prestea Mine	PGRS01	Quartz vein	8.1	0.1386	4.366	221.52	217.15	50.74	6.6
	PGRS02	Dolerite dyke	8.0	0.0279	0.879	30.38	29.50	34.57	6.8
	PGRS03C	Phyllite (foot wall)	6.8	0.7861	24.762	103.13	78.37	4.16	4.7
	PGRS04	Ankerite alteration	8.2	0.0371	1.169	284.81	283.64	243.71	7.2
	PGRS06	Meta greywacke	6.8	0.4954	15.605	71.27	55.66	**4.57**	6.4
	PGRS07	Mineralized Quartz Reef	6.5	0.4346	13.690	67.09	53.40	**4.90**	**4.8**
	PGRS09	Unmineralized Quartz Reef	8	0.0371	1.169	39.24	38.07	33.58	5.8
	PGRS10	Siderite alteration	8	0.1201	3.783	292.41	288.63	77.29	8.6
	PGRTSB01	Tailings dump from Bondaye	3.7	0.2961	3.027	13.92	**10.89**	**4.60**	**4.5**
	PGRTSM	Old tailings from Tuapim	6.9	0.1016	3.200	60.76	57.56	18.99	6.1

Table 4. Identified possible acid generating lithologies

Method	Acid Base Accounting Tests			NAGpH Tests
	Price et al. ('1998)	Ferguson et al. (1998)	USEPAFS (2000)	MEND(2000) and Greenhill (2000)
Criteria	NP < 1	NNP < 20CaCO3/ton	NPR of 5:1	NAGpH less than 5
Acid Generating samples	None	GMCTS01, GRTDB01, GFRS01 and GFRS07	GMCTS01, GMCRS01 PGRTSB01, PGRS03C, PGRS06 and PGRS07	GMCTS01, GMCRS01, GMCRS09, PGRS03C, PGRS07 and PGRTSB01

Table 5. Alkalinity data for the few boreholes for which the original data exist

Location	Borehole Number	Total alkalinity		Alk_{1983} - Alk_{2001}
		Alk_o (1983) (mg l^{-1} $CaCO_3$)	Alk_t (2001) (mg l^{-1} $CaCO_3$)	
Benso	21/E/13-3	40.0	32.0	8.0
Benso	21/E/13-4	40.0	36.0	4.0
AdumBanso	21/H/23-4	115.0	100.0	15.0
AdumDominase	21/G/70-3 21/G/70-3	60.0 65.0	40.0 52.0	20.0 13.0
Nsuaem	18/I/98-3	85.0	80.0	5.0

Acidification Status of Groundwater in Tarkwa-Prestea Area

Several natural processes (oxidation of sulphides and nitrogen compounds, ion exchange reactions and metal ions hydrolysis) can cause groundwater acidification (pH increase). Weathering processes usually expose fresh sulphide rocks to the air and water (eg rain) and therefore chemical reactions that produce acid mine drainage take place even where there is no mining. Acid mine drainage is therefore not exclusive to mining alone. It can also take place once sulphide rocks are exposed to the atmosphere by weathering. However, mining dramatically accelerates the process of acid generation because it breaks open the subsurface and exposes large volumes of fresh sulphide rocks to atmosphere in a relatively short time. Furthermore, by crushing rocks for the extraction of gold the surface area of the rock is being increased and therefore the potential for acid generation is increased. Additionally, finely ground sulphide bearing tailings left over from gold ore beneficiation processes can also be a major source of acid mine drainage.

According to Grimvall et al. (1986), however, several natural neutralizing reactions will consume large quantities of hydrogen ions before any changes in pH can be observed. Carbonate weathering would normally give rise pH >6; while silicate and alumino-silicate buffering would give rise to $5.0 \leq pH \leq 6.0$. Similarly, cation exchange (exchange involving H^+, Na^+, Ca^{2+} etc.) reactions would give $4.0 \leq pH \leq 5.0$ while metal buffering (dissolution of metal hydroxides) would give pH >4.0 (Grimvall et al., 1986). Except for a few samples in which the pH ≤ 4.0, the groundwater pH is predominantly $5 \leq pH \leq 6.0$ indicating mainly silicate or alumino-silicate buffering.

Water composition and its change with time as a result of acidification can often be qualitatively understood but detailed quantitative descriptions are scarce in the literature (Grimvall et al., 1986). Edmunds and Kinniburgh (1986) in a regional survey of groundwater in the United Kingdom observed that low alkalinity groundwaters were most susceptible to acidification. Similarly, Hultberg and Wenbald (1980) indicated that alkalinity decreases over time are indeed a good warning against groundwater acidification in future. Acidification (Ac) is thus often defined as loss of alkalinity (Henriksen, 1979). Grimvall et al. (1986) hypothesised that the best quantitative indicator for groundwater acidification is the hardness to alkalinity ratio. Thus hardness to alkalinity ratio or hardness-alkalinity plot is commonly used to evaluate the acidification status of surface or groundwaters (Grimvall et al., 1986; Jacks et al., 1984; Caritat, 1995).

Figure 1 is a plot of total hardness against alkalinity for representative groundwater samples taken from the Tarkwa -Prestea area. According to Caritat et al. (1998) a slope of 1:1 on this kind of diagram epitomises waters that are in equilibrium with atmospheric CO_2 and minerals such as carbonates in the soil, regolith and bedrock. On the other hand, a slope of 2:1 or greater (alkalinity to hardness ratio of 0.5 or less) signifies the influence of strong acids, usually interpreted to be of anthropogenic origin. In Figure 1 most of the groundwater samples fall along or close to the line with slope 1:1 suggesting equilibrium with CO_2 and rock minerals or natural acidification and buffering. In this case influence of anthropogenic acidification if any, may be very minimal. The four samples that plot close to or to the left side or above the line with slope 2:1 line point to the fact that though acidification of the groundwater samples within the area could predominantly be due to natural weathering processes, acidification due to anthropogenic factors (mainly mining activities) does occur.

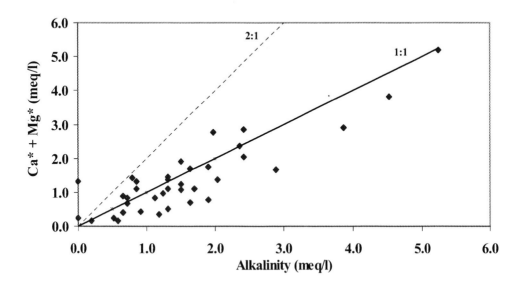

Figure 1. Scatter plot of total hardness versus alkalinity of representative groundwater samples from the Tarkwa-Prestea area

Since acidification (Ac) has been defined as loss of alkalinity (Henriksen, 1979), it implies that

$$Ac = Alk_o - Alk_t \qquad \text{(Equation 6)}$$

where Alk_o = the original or pre- acidification alkalinity,
Alk_t = the present-day alkalinity (Henriksen and Kirkhusmo, 1982).

An attempt has been made to use equation 6 to compute the acidification capacity of the groundwater. The present alkalinity and original alkalinity data for a few of the boreholes are presented in Table 5.

The data in Table 5 shows that the average acidification capacity in 6 boreholes over a period of 18 years is 10.8 mg l^{-1} of $CaCO_3$. However the 1983 data, which is the data at the time of borehole or well drilling (pre- acidification alkalinity) is very limited. Furthermore, the accuracy of the 1983 data could not be guaranteed because it was not stated how the measurements were carried out; whether they were carried out in the field or in the laboratory. Therefore, this computed acidity may not be a reliable result and consequently has not been used in acidification status assessment. Nonetheless, according to Henriksen (1980) and Henriksen and Kirkhusmo (1982), the original alkalinity could be estimated from the sum of its non-marine calcium and magnesium concentrations (Carbonates derived from fossil shell materials). Thus the present day data can be used to estimate acidification according to equation 7. Consequently a modified empirical formula for the computation of acidification based on the same definition of acidification but depending only on the present data is given as follows

$$Ac = 0.93 \, (Ca^* + Mg^*) - 14 - Alk + Al \qquad \text{Equation 7}$$

where all Ca^* ad Mg^* are non-marine calcium and magnesium respectively, Alk is alkalinity, Al is aluminium concentration and concentrations are expressed in meq l^{-1} (Henriksen, 1980; Henriksen and Kirkhusmo, 1986). Approximately 100 km from the sea, the Tarkwa-Prestea area is apparently too remote for its hardness to be derived from marine sources more so when the underlying rocks display crystalline characteristics. Additionally, there is no petrographic evidence to suggest fossiled derived corbonate in the study area. Thus the calcium and magnesium concentrations of groundwater in the Tarkwa-Prestea area are essentially non-marine (non-fossiled). Consequently, the formula in equation 8 could be applicable to groundwater from the Tarkwa-Prestea area. The above formula has, therefore, been used to compute the acidification capacity or status of groundwater in the Tarkwa-Prestea area (Table 6). Similarly, the acid neutralising capacity (ANC) of water, calculated based on the formula

$$ANC = (Ca^{2+} + Mg^{2+} + Na^+ + K^+) - (NO_3^- + SO_4^{2-} + Cl^-) \qquad \text{Equation 8.}$$

where all concentrations are expressed in meq l^{-1} (Stumm, 1992), is also presented in Table 7. The relationships between ANC, NANC (net acid neutralising capacity) and Ac are presented in Figure 7-3. The groundwater is low in acid neutralising capacity, varying from -1.7 meq l^{-1} to 4.26 meq l^{-1} with a mean value of 1.15 meq l^{-1} confirming buffering agents other than carbonates. Nonetheless, Ac of the groundwater is also very low varying between -15.1 meq l^{-1} and -10.7 meq l^{-1} with a mean value of -14.1 meq l^{-1}. It is obvious from Figure 7-3 that though the ANC is low it remains positive for most samples while Ac remains negative. This gives a positive net acid neutralising capacity. The connotation of this is that in spite of the low acid neutralising capacity, the groundwater still has the potential to neutralise acids probably due to the presence of alumino-silicates and some mafic rocks. Even for those samples with negative ANC the net acid neutralising capacity is positive suggesting, at least, a low degree of acid neutralising potential.

For those samples that show negative acid neutralising capacity suggesting strong acids from anthropogenic origin, oxidation of sulphides predominantly pyrite and arsenopyrite due to mining would be expected to play a major role in groundwater acidification. Oxidation of pyrite and arsenopyrite were described by equation 6 and (7.6), respectively, in the earlier part of this chapter. Assuming a complete and congruent reaction, the stoichiometric oxidation of pyrite to generate acid gives molar Fe^{2+}/SO_4^{2-} ratios of 0.5 and 1 for equations 6 and 7, respectively.. Nonetheless, sulphate concentration in shallow groundwater in the study area is remarkably low (1.0-53.0 mg l^{-1}; mean = 13.0 mg l^{-1}; median =7.0 mg l^{-1}, and most of the Fe^{2+}/SO_4^{2-} molar ratios (median value of 0.02) are also generally either much lower or higher than the stoichiometric values for pyrite and arsenopyrite oxidation.

Additionally, the complete absence of Fe^{2+} from some of the boreholes with the lowest pH values (e.g. Tamso 20-I-89-1 pH=3.9) giving Fe^{2+}/SO_4^{2-} molar ratio = 0.0 suggest that either iron co-precipitated with other trace metals as ferric oxyhydroxide complex or oxidation of pyrite or arsenopyrite is not responsible for the acidity associated with the wells with negative acid neutralising potential notably the Tamso boreholes. However, the Eh-pH conditions do not favour the precipitation of iron as ferric oxyhydroxide in these wells. Thus oxidation of pyrite or arsenopyrite may not be responsible for the strong acidity associated with the Tamso boreholes.

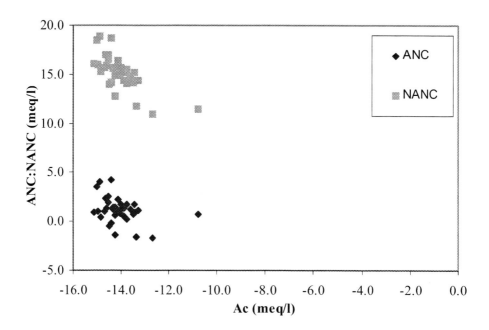

Figure 2. Scatter plot of acid neutralising capacity and net acid neutralising capacity versus acidification capacity for groundwater in the Tarkwa–Prestea area

Table 6. Computed net acidification potential of groundwater in the Tarkwa-Prestea area

Town	BH No./ type	Ca	Mg	K	Na	NO3	Cl	SO4	Cl	Al	(Ca*+Mg*)	Alk	ANC	Ac	ANC-Ac
Prestea	Gwcc (13)	0.5	0.6	0.0	0.9	0.1	0.2	0.0	0.7	0.0	1.1	1.7	1.3	-14.6	15.9
Prestea	Gwcc (12)	1.0	1.0	0.0	0.8	0.0	0.0	0.0	0.3	0.0	2.0	2.4	2.5	-14.5	17.0
Hemang	20-13- 65-4	0.2	1.1	0.1	1.0	0.3	0.2	0.1	0.8	0.0	1.3	0.9	1.2	-13.6	14.8
Ankobra	/B/065-2	0.4	0.5	0.0	0.6	0.0	0.5	0.0	0.2	0.0	1.0	1.3	1.4	-14.3	15.7
Ankobra	B/065-1	0.1	0.7	0.0	0.4	0.0	0.0	0.1	0.2	0.0	0.8	1.9	0.9	-15.1	16.0
K Nirmpa	20-C-01-2	1.2	0.6	0.0	1.0	0.0	0.9	0.4	0.9	0.0	1.7	1.9	1.5	-14.2	15.7
Odumase	44-1-45-1	0.2	0.5	0.0	0.6	0.0	0.0	0.0	0.3	0.0	0.7	1.6	1.0	-15.0	16.0
Odumase	44-1-45-4	0.4	0.9	0.6	2.0	0.4	0.0	0.0	2.2	0.0	1.3	0.0	0.7	-10.7	11.4
Odumase	300m, sch	0.0	0.4	0.0	0.4	0.0	0.2	0.6	0.1	0.0	0.4	0.7	0.6	-14.2	14.9
Beposo	44-E-73-1	0.6	0.3	0.0	0.8	0.0	0.0	0.0	0.4	0.0	0.8	1.1	1.2	-14.3	15.6
Beposo	44-E-73-2	0.4	0.7	0.0	1.2	0.0	0.0	0.2	0.9	0.0	1.1	1.5	1.3	-14.3	15.5
Bawdie	44-E-49-2	1.1	0.6	0.0	1.1	0.0	0.2	0.0	0.5	0.1	1.7	2.9	2.3	-14.6	16.9
Dawurampong	44-C-32-2	0.2	0.4	0.0	0.5	0.0	0.0	0.0	0.1	0.0	0.5	1.3	1.0	-14.7	15.7
Hiawia	44-E61-1	0.1	0.6	0.0	0.8	0.0	0.0	0.0	0.2	0.0	0.7	0.7	1.3	-14.0	15.3
Hiawia	44-E61-2	0.1	0.8	0.0	1.0	0.0	0.0	0.1	1.2	0.0	0.8	0.7	0.5	-13.9	14.4
Bogoso clinic	44-I-28-1	0.5	0.7	0.0	0.3	0.0	0.1	0.0	0.2	0.0	1.2	1.5	1.4	-14.3	15.7
Insu	47-0-98-1	0.8	0.7	0.1	1.0	0.3	0.0	0.1	1.1	0.0	1.4	0.8	0.9	-13.4	14.3
Insu	47-0-98-2	1.0	0.9	0.0	1.1	0.1	0.0	0.0	1.2	0.0	1.9	1.5	1.7	-13.5	15.2
Insu	47-0-98-3	0.8	0.6	0.0	1.0	0.0	0.5	0.0	0.7	0.0	1.4	1.3	1.7	-14.0	15.7
Huni Valley	W 121	1.0	0.4	0.0	0.7	0.0	0.8	0.0	0.1	0.0	1.4	2.0	2.0	-14.5	16.5
Aboso Nsuem	Borehole	3.7	1.5	0.2	1.0	0.2	0.4	0.4	1.5	0.0	5.2	5.2	4.3	-14.4	18.7
Atwereboanda	Borehole 1	1.4	1.5	0.1	0.8	0.1	0.0	0.6	1.4	0.0	2.9	2.4	1.7	-13.7	15.5
Atwereboanda	Borehole 2	0.6	0.5	0.1	0.5	0.0	0.1	0.3	0.3	0.0	1.1	1.3	1.2	-14.3	15.4
Tarkwa	Sch. of Mines	0.2	0.2	0.0	0.5	0.3	0.0	0.0	0.1	0.0	0.3	1.2	0.5	-14.8	15.3
Tarkwa	Sec. Sch	2.6	1.2	0.0	0.7	0.0	0.0	0.1	0.3	0.0	3.8	4.5	4.1	-14.9	18.9
Tamso	20-1-89-2	0.1	0.1	0.0	0.4	1.8	0.0	0.0	0.4	0.1	0.2	0.2	-1.6	-13.3	11.8
Tamso	20-1-89 -1	0.1	0.2	0.1	1.0	2.1	0.0	0.0	0.9	0.1	0.2	0.0	-1.7	-12.7	11.0
Benso	21-E-13-4	0.1	0.1	0.0	0.3	0.6	0.4	0.0	0.2	0.0	0.2	0.6	-0.2	-14.4	14.2
Benso	21-E-13-3	0.1	0.1	0.1	0.7	2.1	0.1	0.0	0.3	0.0	0.2	0.5	-1.4	-14.2	12.8

Table 6. (Continued)

Town	BH No./ type	Ca	Mg	K	Na	NO3	Cl	SO4	Cl	Al	(Ca*+ Mg*)	Alk	ANC	Ac	ANC-Ac
Adum Banso	21-H-23-4	1.1	0.6	0.0	0.9	0.1	0.0	0.3	1.3	0.0	1.7	1.6	0.8	-14.0	14.8
Adum Banso	21-H-23-3	1.4	0.9	0.1	0.8	0.0	0.0	0.1	0.9	0.0	2.4	2.4	2.3	-14.1	16.4
AdumDoninase	21-G-70-4	0.6	0.5	0.1	1.0	0.3	0.0	0.0	1.2	0.0	1.1	0.9	0.7	-13.5	14.2
AdumDoninase	21-G-70-3	0.5	0.4	0.0	1.0	0.5	0.0	0.0	1.2	0.0	0.9	0.7	0.3	-13.8	14.0
Nsuaem		2.2	0.7	0.0	0.8	0.0	0.5	0.0	0.2	0.0	2.9	3.9	3.6	-15.0	18.5
Nsuaem	18-I-58-3	0.8	0.7	0.0	0.8	0.1	0.0	0.0	0.9	0.0	1.5	1.3	1.3	-13.9	15.2
Simpa	18-E-77-3	2.1	0.7	0.1	1.8	0.0	0.0	0.4	3.1	0.0	2.8	2.0	1.2	-13.3	14.4

** Ca^* = non-marine calcium, Mg^* = non-marine magnesium, Alk = alkalinity, ANC = acid neutralisation capacity, Ac = acidification

Acidification Status of Surface Waters in Tarkwa-Prestea Area

Table 7 presents a summary of the chemical characteristics of some surface water bodies in Birimian rock areas within the concession of the Bogoso Gold Limited. These data were taken from the Bogoso Gold Limited's mine records. Table 7 indicatesthat the electrical conductivity values (mean = 476µS cm^{-1}) and sulphate concentrations (mean = 391 mg l^{-1}) are relatively high compared with the values obtained for the domestic wells and boreholes that have sulphate concentrations between 1 mg l^{-1} and 53 mg l^{-1} with a mean value of 14 mg l^{-1}). Additionally, these surface waters have relatively heavier suite of trace metals. Furtheermore, pH values are variable with a mean value of 5.4 and a range 2.4-7.3.

A review of the chemical data indicates mean sulphate concentrations of 391 mg l^{-1} and a maximum value of 1840 mg l^{-1} with the mean and the maximum calcium concentrations of 22 mg l^{-1} and 81 mg l^{-1} respectively. These do not support gypsum or anhydrite dissolution to produce SO_4^{2-}. The Fe/SO_4 molar ratio of less that 0.1 of the surface waters is too low to be supported by the stoichiometry of pyrite or arsenopyrite oxidation and so also the near neutral pH. Nonetheless, the fact that water influenced by sulphide oxidation is either mostly low in pH and high in metals or, in some cases, has neutral pH ($\approx$ 7.0) but high level of sulphates and dissolved trace metals (Iribar et al., 2000), suggests that the near neutral pH, the observed high sulphate and relatively high heavy metal concentrations that most of the surface waters samples have shown clearly indicate that the surface waters are influenced by sulphide oxidation. The pH of surface waters is near neutral probably as a result of sulphide oxidation followed by neutralization by alkalinity provided by dissolution of silicates, mainly ferromagnesian silicates. The low Fe/SO_4 molar ratio is probably the result of iron co-precipitating with other trace metals particularly arsenic as Fe-oxyhydroxides.

Aluminium concentrations are very high accompanied with very low pH in a few of the surface waters notably Chujah stage 3 Pit (Al = 3.4mg l^{-1}; pH = 3.4); Chujah 3 Spring (Al = 42.5mg l^{-1}; pH = 2.4); Bogoso North Pit (Al = 7.4mg l^{-1}; pH = 2.9) and a few groundwaters as at Odumase (44-I-45-4) (Al = 2.3mg l^{-1}; pH = 3.9) Tamso (20-I89-1) (Al = 1.3 mg l^{-1}; pH = 3.9). This observation suggests that due to the extreme acid condition, additional acid neutralisation capacity is derived from dissolution of sulphate minerals particularly alunite (Langmuir, 1997) and clay minerals such as kaolinite and gibbsite. Typical alunite dissolution to produce Al^{3+} is as follows

$$KAl_3(SO_4)_2(OH)_6 + 6H^+ = K^+ + 3Al^{3+} + 2SO_4^{2-} \qquad \text{Equation 9}$$

Speciation using Phreeqc for Windows (Parkhurst and Appelo, 1999) confirmed the production of Al^{3+} in some boreholes in the Tarkwa – Prestea area through alunite or kaolinite dissolution. For instance, in the borehole (44-I-45-4) at Odumase the saturation indices for alunite and kaolinite were 0.29 and 0.11 respectively suggesting alunite and kaolinite dissolution proceeded until the groundwater had reached equilibrium with both minerals and no longer able to dissolve more. Alunite and kaolinite thus probably become the major contributors of aluminium to the groundwater. Similar reactions occur in some of the surface waters.

TRACE METAL POLLUTION IN GROUNDWATER

A second kind of impact of mining activities on groundwater quality in a mining area as the Tarkwa–Prestea area is contamination of water resources as a result of pollution from toxic trace metals. Trace metals are dissolved metallic elements that occur or can be present in groundwater, often at concentrations well below 0.1 mg l^{-1}, but occasionally much higher (Freeze and Cherry, 1979). Prior to mining, trace metals are locked in rocks underground and are mostly in the reduced state (Carlos et al., 1997). During mining, the rocks are unearthed and the metals set free as wind blown soil that return to earth in rain or as dry fallout. Furthermore, the metals can be set free through the dissolution of rocks by CO_2 charged water or acid mine drainage and can spread through the environment particularly water resources in dangerous concentrations that can have adverse impact on human health.

The trace metal content of representative groundwater samples taken from the Tarkwa-Prestea have not revealed the trace metal loadings initially anticipated of a mining area. The generally low trace metal content of the groundwaters may have resulted from the fact that the rocks of the Tarkwaian formation, especially the rocks containing the gold ores, are devoid of most trace metals (Marston et al., 1992). Furthermore, if the analyses of metals in the waste rock from the Bogoso Goldmine in the Birimian area would be accepted as representative of average metal contents of the Birimian rocks, then although sulphide ores that form significant portion of the Birimian rocks are known to be associated with high, heavy metal concentration, it could be said with certainty that the Birimian rocks ores are only medium in heavy metal contents. Trace metals that occur in relatively high concentration in the rocks include aluminium, arsenic, barium, iron and manganese. As a result, the trace metals concentrations in the boreholes and wells also follow the same trend and show aluminium, arsenic, barium, iron, manganese, mercury and nickel as the trace metals having concentrations significantly above the background values in a few of the boreholes and wells in the Tarkwa-Prestea area.

Most trace metals may produce kidney damage. Mercury, chromium, uranium, lead and cadmium are listed in textbooks as potent nephrotoxins (Doul et al., 1980). Renal injury has also been demonstrated after exposure to arsenic, gold, iron, antimony, platinum and thallium (Maher, 1976). The few trace metals that occur in significant quantities in a few of the boreholes and wells and therefore could pose quality problem in the Tarkwa-Prestea area are discussed below.

TRACE METAL POLLUTION IN SURFACE WATER

Summary of selected descriptive statistics of measured parameters from various mine-impacted drainage are presented in Table 8. Individual point sources of metal ion loading, anomalous pH, and elevated conductivity and alkalinity values is discernable. Spatial discrimination of the data from mine impacted areas show that the Prestea mine site has had more impact on the immediate drainage area than any of the areas under study. This is followed by the mines at Iduapriem and Teberibie, Nsuta, Damang, and Tarkwa in that order. The Tarkwa mine area exhibits very low metal loading, modest specific conductivity and alkalinity values, and near-neutral pH values.

Table 7. Summary of chemical characteristics of surface water samples from the Bogoso mine site

	Location	pH	EC. μS cm^{-1}	Al μg/l	As μg/l	Ca mg/l	Co μg/l	Cu μg/l	Fe mg/l	Mg mg/l	Na mg/l	Ni μg/l	Zn μg/l	SO$_4$ mg/l
T2	Stream draining tailings spill S dam	5.0	1310	35	13	13	168	5	37	16	189	9	32	622
C1	Chujah pit	7.0	750	28	8	63	15	3	0	66	11	67	59	412
C2	Seep into Chujah pit	5.6	150	12	18	3	64	4	19	7	7	134	116	44
C3	Chujah stage 3 Pit	3.4	220	3410	5	4	51	133	0	4	4	112	206	74
C4	Odumase Creek	6.5	90	68	13	7	1	8	1	5	11	4	41	-
C5	Nankafa East Pit	7.1	400	76	5	65	0	3	0	7	6	1	9	214
C6	Nankafa Creek	6.8	230	22	7	18	26	2	0	14	11	27	17	-
C7	Chujah 3 Pit	3.9	410	800	14	26	120	209	1	20	5	275	125	190
C8	Chujah 3 'Spring'	2.4	2900	42500	2730	81	1160	2720	277	75	9	1860	1080	1840
C9	Odumase h/w Pit	4.7	20	713	30	2	168	43	4	2	4	27	27	-
C10	Mansahaie pit	5.5	30	15	57	1	2	2	0	1	7	5	8	-
C11	Bogo Creek	6.4	160	39	175	8	1	15	6	4	12	4	35	-
C12	Mansi River	6.9	50	44	5	5	1	3	2	3	7	2	7	-
N1	Pond north of Marlu Dump (West)	7.3	430	24	104	39	0	36	0	29	4	3	4	173
N2	Pond north of Bogoso north dump	3.6	490	774	14	20	172	125	1	27	6	183	106	247
N3	Bogoso North Pit	2.9	730	7430	119	11	268	4320	7	10	5	441	111	236
N4	Marlu Pit	6.7	560	67	16	48	5	23	0	31	10	37	7	250
S1	Abliifa pit(small pond)	4.5	20	34	13	1	1	2	0	0	6	2	8	-
S2	Subri Creek	6.6	100	34	10	9	1	2	0	5	8	2	5	-

Table 8. Summary of measured parameters for dry and wet seasons
(except pH and otherwise stated)

Parameter	Dry Season				Wet Season			
	Mean	StD	Min	Max	Mean	StD	Min	Max
PH	6.3	0.9	3.1	8.6	8.6	0.9	3.4	9.8
Temp°C	26.7	2.1	24.2	32.6	32.6	1.8	23.3	30.5
Cond (µS/cm)	231	296	33	1264	1264	643	33	4320
Alka	46	36	0	162	162	216	0	1552
HCO_3^-	56	44	0	198	131	264	0	1893
SO_4	54	82	5	407	6	47	1	254
Ca	15	16	0	0	198	18	0	112
Cl	18	19	2	2	72	20	0	124
Mg	14	23	0	131	120	1020	0	4534
Na	-	-	-		407	94	3	825
K	-	-	-	-	180	4	0	17
PO_4	-	-	-	-	645	0	0	1
NO_3	-	-	-	-	15	1	0	6
SiO_2	-	-	-	-	103	6	0	26
Zn	0.656	1.095	0.030	6.250	6.250	0.275	0.001	1.444
Cu	0.121	0.710	0.009	5.840	1.090	2.858	0.001	25.570
Mn	0.848	1.856	0.030	15.000	21.380	4.273	0.029	26.270
Ni	-	-	-	-	8.000	1.239	0.001	10.980
Fe	5.677	16.344	0.030	103.000	1.817	8.172	0.174	59.170
Pb	0.209	0.220	0.029	1.090	0.001	0.067	0.001	0.493
Cd	-	-	-	-	0.001	0.001	0.001	0.008
As	0.558	2.732	0.001	21.380	0.056	3.589	0.001	28.950
Hg (ppb)	1.276	1.125	0.700	8.000	-	-	-	-

The geology and geochemistry of the ore-bearing rocks and, to some extent, the ore processing method employed by these mines have considerable control on the quality of drainage water. The Prestea, Iduapriem, and parts of Teberibie are located within meta-volcanic and meta-sedimentary sequences with sulphide and altered carbonate mineralisation. At Prestea, the ore has traditionally been processed by roasting, with reported release of substantial sulphide oxidation products, including arsenic, into the atmosphere (Amono-Naiser 1984).The Prestea had the highest concentration of trace ions and sulphate in drainage water. Peak values were observed in streams taking drainage from old tailings dumps at Bondai near Prestea, with pH values near 4. However, in other areas, the pH of the drainage was alkaline, though water degradation was evident, as indicated by elevated sulphate, Mn, Fe, Zn, Cu, and As concentrations. The high alkalinity values of drainage in some areas demonstrated the dominance of acid-buffering carbonates.

Although the Iduapriem mine site appears less impacted than Prestea, the single most impacted point source was identified here. A buried dyke of stripped mine overburden with considerable sulphide mineralogy sends acid drainage to one of the streams draining the area. Two small streams carrying leachate from this source join the Bediebawuo stream further downstream. Relatively low pH (3.5 – 4.5) values, accompanied by very high metal concentrations and low alkalinity values, were recorded at sampling points along these two streams. A strong negative correlation between pH and trace and major ions and the strong positive correlation among trace and major ions in the Iduapriem area suggest that these ions are likely derived locally from the water-rock interaction. The waste site is currently been managed as potentially acid forming by the mine owners

At Nsuta, manganese ore is comprised of manganese carbonates and oxides with subordinate sulphides as accessory minerals in the enclosing country rocks. The processing of manganese ore entails washing the excavated oxide or carbonate to remove manganese impurities; the resulting waste slurry, containing Mn and associated Fe, is usually discharged into the drainage system. Not surprisingly, the Nsuta area records the highest Mn concentrations. Black slurry of manganese oxide is very common in streams draining the area. The level of degradation increases towards the centre of active mining and processing activities. However, the concentrations of other measured metal ions are generally lower here than at Prestea and Iduapriem. Although cadmium concentrations were generally below detection in the entire study area, detectable concentrations were observed in the Nsuta area with a maximum of 8µg/L in drainage waters from the manganese mine. Ramdohr (1980) has intimated that Cd in leachates may normally be controlled by sphalerite, because in sphalerite, a small part of the Zn is often isomorphously substituted by Cd. An energy dispersive spectra (EDS) probe of some sphalerite grains in samples from Nsuta indicated that Cd was present but generally in very low concentrations, not exceeding 0.02 atomic weight percent.

The Tarkwa and Damang mines are located within the paleosedimentary Tarkwaian sequences. The ore at Damang is dominated by conglomerates, with subordinate mineralized basic intrusives. The conglomerates are largely quartz and devoid of sulphides and carbonates; however, the mineralized intrusives have considerable sulphides. In contrast, mineralisation in the Tarkwa mine is confined to only the conglomeritic horizons.

SOURCES OF MAJOR AND TRACE METALS IN DRAINAGE

The spatial distribution of zones of high trace and major ion concentrations and the preponderance of some ions in specific places clearly shows that these zones of high trace and major ion concentrations represent either point or diffuse sources of pollutants in the drainage. The spatial distribution of high trace element regimes, in particular strongly reflect the influence of geology on the water chemistry. Principal Component Analysis (PCA) was applied to the data from the entire study area, data sets from Iduapriem, Nsuta, Damang and Prestea mines sites, respectively, in order to extract principal factors corresponding to the different sources of variation in the water chemistry. The Kaiser criterion (Kaiser, 1960) was used to extract principal components from the data and rotated using varimax normalization for each location. The water chemistry appears mainly controlled by weathering and dissolution of carbonates, silicates and sulphides, with strong incorporation of a variety of trace elements. The strong influence of geology on the hydrochemistry is well illustrated by the PCA results (Table9). The hydrochemistry of drainage from Prestea, Nsuta and Iduapriem mine sites underlain mainly by metavolcanics and metasedimetns of the Birimian formation reflects the dissolution of carbonates, silicates and the simultaneous oxidation of sulphides that constitute the dominant mineralogy at these sites (Akabzaa, 2004, Nyame et al. 2002). The influence of a variety of dissolution of manganese carbonates on the hydrochemistry of drainage from the Nsuta manganese mine site is well highlighted by the PCA results. Similarly the results show limited oxidation of sulphides and more pronounced buffering activity from the dissolution of carbonates, major constituent minerals of scattered dolerite dykes in the Tarkwaian formation that underlie the Damang mine site, also possibly account for the drainage chemistry.

The distributions of manganese points to both diffuse and point sources. High manganese loading is within mines sites in the Birimian formation and the highest recorded values of 24.40mg/L and 26.27 mg/L for the dry and wet seasons, respectively are recorded from the Nsuta mine site (Table 8.) where manganese is being exploited from manganeferous formations within the the Birimian metavocanics. Similarly areas marking the contact between Birimian metavolcanic and metasedimentary rocks, generally dominated by manganese-bearing horizons (gondite), exhibit high manganese values. These rocks are common throughout areas underlain by Birimian rocks, such as Prestea and Iduapriem mine sites and can be described as diffuse sources. High manganese loading of 25.42 mg/L and 10.26 mg/L have been recorded in the Iduapriem and Prestea mine sites (Table 10), respectively, underlain by Birimian rocks. However, the manganese carbonate and oxide ore that are exposed at Nsuta represent point sources of manganese.

The distribution of arsenic, copper, and nickel exhibits similar patterns. Higher values are located in drainage associated with Prestea, Iduapriem and Nsuta, underlain by Birimian rocks. The concentration of these trace elements are also high in drainage from the Damang mine site in the Tarkwaian formation and are likely to be associated with sulphide bearing dolerite dykes intruding Tarkwaian rocks which are mineralised. Price and Errington (1998) observed that such high zones of concentration of metals occur even in neutral pH drainage, and often result from localised acid-generating sources.

Sulphate concentrations were highest in streams taking leachate from the Iduapriem mine (8 to 407 mg/L), followed by drainage within the Prestea mine site (10 – 262 mg/L) and three other streams draining the Damang mine (7 to 101 mg/L) (Table 10) . Sulphate can be used as a measure of the rate of sulphide oxidation, though it is also occasionally dominant in areas where significant acid-buffering reactions are taking place, such as in the mine pits at Prestea and Teberibie. In such situations, depending on the reaction kinetics and the preponderance of carbonate material over acid-generating species, high sulphate concentrations may be correlated with moderately alkaline pH and high Ca^{2+} concentrations. Bicarbonate was also observed to be very high in some non-mining environments, likely resulting from decay of organic matter, and cannot be entirely attributed to rock weathering.

The highest recorded mercury concentrations in the drainage were coincident with areas of very active artisanal mining activities. Augmented Hg levels were also identified with water draining historic tailings dumps at Prestea and Tarkwa. In these areas, Hg concentrations are several times above the maximum allowable limits in drinking water established by Ghana's Environmental Protection Agency (EPA) and the World Health Organization (WHO). These observations are not surprising as Hg is the chemical used in gold extraction by traditional small miners and, before the introduction of cyanide leaching technology, Hg amalgamation was the technology of choice for gold extraction in large-scale mines, which were then mining only free milling gold in quartz reefs (Quashie et al., 1981).

Table 9. Rotation Principal Component Analysis Loading for Mine-impacted Drainage

	Entire Study Area			Iduapriem			Nsuta			Damang			Prestea		
	\multicolumn{15}{c}{Components}
	1	2	3	1	2	3	1	2	3	1	2	3	1	2	3
pH	0.74	0.03	-0.49	0.30	0.86	0.10	-0.29	-0.47	0.65	0.76	-0.09	0.01	0.90	0.01	-0.43
Cond.	0.89	0.02	0.28	0.39	0.83	0.26	-0.18	0.55	0.76	0.52	0.16	0.39	0.99	0.11	-0.05
Alka	0.89	0.02	0.02	0.46	0.71	0.41	-0.81	-0.06	0.15	0.70	0.63	0.15	0.98	0.97	-0.13
HCO$_3$	0.89	0.89	0.29	0.46	0.71	0.41	-0.81	-0.06	0.15	0.70	0.63	0.15	0.98	0.10	-0.13
Ca	0.80	-0.17	-0.32	0.83	0.27	-0.12	-0.72	0.34	0.51	0.84	0.02	-0.12	0.93	-0.09	-0.01
Mg	0.69	0.15	0.88	0.44	0.03	0.44	0.25	0.88	0.19	0.01	0.22	-0.61	0.75	0.31	-0.52
Cl	0.16	-0.07	0.12	0.36	0.27	0.82	-0.22	0.04	0.85	0.09	0.53	0.41	0.55	0.83	0.03
SO$_4$	0.27	0.58	0.54	0.87	0.03	0.16	0.05	0.81	0.19	-0.04	0.27	0.77	-0.07	0.84	-0.38
Mn	0.22	0.55	0.03	-0.19	-0.06	0.76	0.75	-0.01	-0.50	0.54	0.20	0.33	0.33	0.94	-0.00
Fe	-0.37	0.69	-0.16	0.38	0.82	0.18	0.38	-0.90	0.11	-0.58	-0.10	0.14	-0.44	0.88	0.14
Cu	0.13	0.75	-0.49	0.71	-0.19	0.06	0.79	-0.41	-0.30	0.22	0.78	-0.26	0.11	0.98	0.17
Zn	-0.14	0.31	-0.30	-0.24	0.02	-0.88	0.70	-0.52	0.30	-0.13	-0.57	0.29	0.12	0.46	0.85
Pb	-0.11	0.68	-0.33	-0.09	-0.49	-0.76	0.41	-0.80	-0.25	-0.02	0.84	-0.22	-0.04	0.90	0.13
As	-0.14	0.42	0.47	0.17	0.02	-0.46	0.80	0.08	0.35	-0.10	0.84	0.25	-0.47	0.21	0.86
Hg	-0.27	-0.04	0.47	0.52	-0.72	0.16	-0.06	-0.05	-0.76	-0.38	0.19	-0.50	-0.34	-0.41	0.74
Total Variance Explained (Rotation Sum of Squared Loadings)															
Eigenvalues	5.9	2.5	2.3	4.9	4.2	3.5	5.2	5.0	3.7	4.1	3.7	2.5	7.7	5.4	2.9
% of variance	34.8	14.8	13.4	28.9	24.6	20.8	30.6	29.2	21.6	23.8	21.7	14.5	45.3	31.9	17.0
Cumulative(%)	34.8	49.6	62.9	28.9	53.5	74.4	30.6	59.8	81.5	23.8	45.5	60.	45.3	77.2	94.2

Table 10. Descriptive water quality parameters statistics for various mine sites from dry season sampling

Iduapriem	pH	Temp	Cond	Alkal	HCO3-	Ca	Cl	SO4	Mg	Ca Hard	T Hard	Mn	Fe	Cu	Zn	Pb	As	Hg(ppb)
Mean	5.68	27.6	237.1	32.6	39.8	16.2	31.9	72.4	5.2	40.5	59.8	0.768	3.226	0.054	1.040	0.183	0.013	1.6
Median	5.98	27.2	102.6	27.5	33.6	9.6	20.0	20.9	2.7	24.0	40.0	0.735	1.310	0.020	0.500	0.140	0.001	1.0
StD	1.19	1.8	338.2	23.4	28.6	16.2	27.0	127.7	6.6	40.6	62.7	0.306	3.972	0.141	1.285	0.148	0.028	0.8
Min	3.09	25.3	41.6	0.0	0.0	3.2	10.0	8.0	1.0	8.0	0.0	0.240	0.030	0.009	0.029	0.029	0.001	0.7
Max	6.82	32.1	1211.0	85.0	104.0	56.1	120.0	407.0	25.2	140.3	240.0	1.230	14.310	0.610	4.160	0.640	0.093	3.0
Tarkwa																		
Mean	6.66	25.6	62.2	24.0	29.3	5.6	4.5	13.8	10.2	14.0	56.0	0.460	0.305	0.020	0.380	0.275	0.003	0.7
Median	6.66	25.6	62.2	24.0	29.3	5.6	4.5	13.8	10.2	14.0	56.0	0.460	0.305	0.020	0.380	0.275	0.003	0.7
StD	0.05	0.3	19.0	8.5	10.3	3.4	0.7	5.1	0.7	8.5	11.3	0.453	0.290	0.000	0.042	0.021	0.003	0.0
Min	6.62	25.4	48.8	18.0	22.0	3.2	4.0	10.2	9.7	8.0	48.0	0.140	0.100	0.020	0.350	0.260	0.001	0.7
Max	6.69	25.8	75.6	30.0	36.6	8.0	5.0	17.4	10.7	20.0	64.0	0.780	0.510	0.020	0.410	0.290	0.005	0.7
N suta																		
Mean	6.59	28.4	181.6	56.6	69.1	15.2	28.8	23.7	7.9	38.1	70.6	0.960	4.231	0.012	0.278	0.291	0.005	1.4
Median	6.42	28.5	197.0	52.5	64.1	15.2	25.0	25.1	9.9	38.1	75.0	0.835	1.560	0.009	0.260	0.210	0.003	0.9
StD	0.84	1.7	57.6	16.6	20.4	5.7	14.1	12.1	5.0	14.4	25.7	0.289	6.934	0.008	0.228	0.260	0.007	1.0
Min	5.85	26.4	54.6	30.0	36.6	6.4	15.0	6.3	1.7	16.0	30.0	0.660	0.021	0.007	0.021	0.060	0.001	0.7
Max	8.57	30.6	243.0	85.0	104.0	22.4	60.0	45.1	13.3	56.1	100.0	1.400	20.520	0.030	0.750	0.860	0.021	3.0
Damang																		
Mean	6.36	26.1	177.2	31.5	38.4	6.8	3.9	35.1	9.9	17.0	57.9	0.693	14.103	0.509	0.758	0.378	0.002	1.0
Median	6.46	25.4	89.3	31.0	37.8	7.2	3.0	19.5	6.8	18.0	45.0	0.555	3.225	0.025	0.645	0.280	0.001	0.7
StD	0.26	2.2	311.2	10.6	12.9	2.9	1.9	29.1	8.6	7.2	32.9	0.600	28.623	1.679	0.507	0.295	0.002	0.5
Min	5.82	24.2	62.5	14.0	17.1	0.0	2.0	6.8	2.9	0.0	28.0	0.120	1.000	0.010	0.040	0.080	0.001	0.7
Max	6.69	32.6	1164.0	46.0	56.1	12.0	8.9	101.1	32.1	30.0	146.0	2.140	103.000	5.840	2.160	1.090	0.009	2.0
Prestea																		
Mean	7.02	25.2	464.6	76.0	92.8	25.5	7.8	83.7	48.5	63.8	263.2	2.670	20.164	0.103	1.882	0.285	0.011	1.0
Median	7.14	25.1	271.9	52.0	63.4	16.9	6.0	11.8	20.9	42.1	179.0	0.195	3.915	0.025	0.830	0.185	0.001	0.7
StD	0.78	0.7	513.2	68.3	83.5	28.5	6.0	114.4	55.0	71.1	260.5	6.047	32.615	0.200	2.418	0.328	0.015	0.5
Min	5.76	24.4	37.7	12.0	14.6	2.4	4.0	10.2	0.5	6.0	10.0	0.060	0.021	0.007	0.110	0.030	0.001	0.7
Max	7.89	26.3	1264.0	162.0	198.0	72.1	19.9	261.9	131.0	180.3	645.0	15.010	82.330	0.510	6.250	0.940	0.035	2.0
Wells																		
Mean	5.63	25.1	210.2	40.0	48.8	16.9	24.4	19.4	9.9	42.1	83.0	2.205	0.021	0.007	0.071	0.075	0.001	0.7
Median	5.63	25.1	210.2	40.0	48.8	16.9	24.4	19.4	9.9	42.1	83.0	2.205	0.021	0.007	0.071	0.075	0.001	0.7
StD	0.47	0.4	243.0	42.4	51.8	18.2	30.2	14.8	11.3	45.4	91.9	3.077	0.000	0.000	0.070	0.035	0.000	0.0
Min	5.29	24.8	38.3	10.0	12.2	4.0	3.0	8.9	1.9	10.0	18.0	0.029	0.021	0.007	0.021	0.050	0.001	0.7
Max	5.96	25.3	382.0	70.0	85.4	29.7	45.7	29.8	17.9	74.3	148.0	4.380	0.021	0.007	0.120	0.100	0.001	0.7

Analysis of tailings samples from the historic tailings dumps at Bondaye and Tarkwa where the mines processed gold with mercury showed augmented levels of mercury in these tailings.

Most mine-impacted stream waters in the Prestea and Nsuta area in particular have Ni, Fe, Cu, As, and Mn values above both Ghana's EPA and WHO drinking water standards. Average concentration of Mn in drainage in the entire area is several times above both sets of allowable guide values. Similarly, As and Fe values in drainage from the Tarkwa mine area are several times above the allowable limits. The average concentration of Ni and Cu in Prestea also exceed these guideline values.

Many of the communities in these areas have few alternative sources of domestic water supply and largely rely on these streams. High concentration of these hazardous elements in these streams therefore has some serious health implications. Communities in this area have high incidence of skin lesions, which are associated with metal toxicity. However, there has not been conclusive medical investigation of the exact causes of the lesions in these mining impacted communities, although arsenic dermatitis has been reported in some communities near Bogoso (Songsore et al. 1994).

METAL LOADING IN MINE PITS AND WATER BODIES WITHIN MINING AREAS

Metal loading, as expected, is observed to much higher in mine pits and water bodies within active and dormant mine sites. Monitoring data from the Bogoso mine site with acid mine drainage problems show metal loading in drainage with the mine site being several fold water exist in surface water from the entire Tarkwa mining district (table7). These largely indicate that the metal leaching has been facilitated by acid mine drainage in the area.

CONCLUSIONS

Acid mine drainage study is a grey but growing area in Ghana and has many challenges. The few studies have so employed mainly static geochemical acid prediction methods such as acid base accounting and NAG test. In view of the inherent shortcomings of these methods, there is need for complementary use of kinetic acid rock drainage predictive methods.

One of the reasons for the limited studies in acid prediction in the country is the absence of clear regulation on acid drainage prediction, mitigation and management. The Ghana Environmental Protection Agency and Geoscience Institutions in the country have a lot to learn from mining countries like Canada, the U.S.A. and Australia where cutting edge acid mine drainage management regulation and research exist.

Although there is no strong evidence that metal leaching associated with acid mine drainage is a serious threat to groundwater resources in general, high metal loads have been identified in a few boreholes in the Bogoso area were acid mine drainage has been identified as a problem. High metal loading is however pervasive in most surface water bodies in the mining areas and is more intense within the periphery of some mines.

REFERENCES

Acquah, P. C. (1996). Natural Resources Management and Sustainable Development – The Case for Gold. In: *Mining and Environment Research Network (MERN) Bulletin, No.9*, 39-44.

Akabzaa, T. M. (2004). *Evaluation of Evaluation of Metal Leaching and Acid Rock Drainage Risks in the Nsuta-Tarkwa-Damang-Bogoso-Prestea Mining Axis*, South western Ghana. PhD thesis, University of Ghana, unpublished, 50-85.

Akabzaa, T. M. (2000). Boom and Dislocation. *The Environmental and Social Impacts of Mining in the Wassa West District of Ghana*. Third World Network (TWN) - Africa.

Amanor, J. A. & Gyapong, W. A. (1991). The geology of Ashanti Goldfield. In: Kesse, G. O. (Ed.) Proceedings of the International Conference on the geology of Ghana with special emphasis on gold, 9- 16 October, Accra. *Geological Society of Ghana*, Accra. Vol. 1, C1-18.

Amanor, J. A. & Gyapong, W. A. (1991). The geology of Ashanti Goldfield. In: Kesse, G. O. (Ed.) Proceedings of the International Conference on the geology of Ghana with special emphasis on gold, 9- 16 October, Accra. *Geological Society of Ghana*, Accra. Vol. 1, C1-18.

Amuzu, A. T. (1991). Water pollution and the gold mining industry in Ghana. In: G. O. Kesse, (Ed.) Proceedings of the International Conference on the geology of Ghana with special emphasis on gold, 9- 16 October, Accra. *Geological Society of Ghana*, Accra Vol. 4.

Appelo, C A. & Postma, D. (1999). *Geochemistry*, Groundwater and Pollution. AA Balkema, Rotterdam, The Netherlands, 10-69

Armah, A. K., Darpaah, G. A. & Carboo, D. (1998). Heavy metal levels and physical parameters of drainage ways and wells in three mining areas in Ghana. In: *Journal of the Ghana Science Association*, Vol. 1 No. 1, July, 113- 117.

Amonoo-Neizer, E. H. & Busari, G. L. (1980). Arsenic status of Ghana soils- Contamination of soils near gold smelters. Ghana *J. Sci.*, 20 (1and2): 57-62 Economy. *Resource Policy, Vol. 27*, 61-75.

Anonymous, (1990). Environmental Baseline Statement and Impact assessment Draft Report on Ghana Australia Gold Limited. *Environmental Advisory Unit*. Liverpool, UK, 89.

Anonymous, (1994). Environmental Baseline Study, 1994. Obenemase Gold Mines Ltd. *Konongo Concession*, Ghana, CR 725. 112.

Anonymous, (1996). *Environmental Impact Statement*, EIS, 1996. Bibiani Goldfields Ltd. 95.

Anonymous, (1996). *Environmental Impact Statement*, EIS, 1995. Goldfield Tarkwa Ltd. 195.

Anonymous, (1998). *Environmental Impact Assessment*, EIA, Report for backfilling on Dokyiwa Pits, Obuasi, AGC Ltd. Ashanti Region, Ghana. Contract No. C14/9. 103.

Anonymous, (2003). *Environmental Impact Assessment*, EIA, Report for the Newmont Ghana Gold Limited (Ahafo Project) 222-223.

Anonymous, (2006). USEPA communication to Ghana EPA, unpublished., 5.

Ashley, E. Y., R. (2002). Geoenvironmental Model for Low-Sulphide Gold-Quartz Veins deposits, in R. R. Seal II, N. K. Foley, (Eds.), *Progress on Geoenvironmental Models for Selected Mineral Deposit Types*. U.S. Geological Survey Open-File Report 02-195, 176-195.

Balistrieri, L S., Box, S. E. & Bookstrom, A. A. (2002). A Geoenvironmental model for polymetallic vein deposits: a case study in the Coeur D'Alene Mining District and comparisons with drainage from mineralized deposits in the Colorado Mineral Belt and Humboldt Basin, New Nevada. In: R. R., Seal II, Foley, N. K. (Eds), *Progress on Geoenvironmental Models for Selected Mineral Deposit Types*. USGS Open-File Report 02-195, Washington DC, USA, 176-195.

Barcelona, M., Gibbs, J. P., Helfrich, J. A. & Garske, E. E. (1985). Practical Guide for Ground Water Sampling. *Illinois State Water Survey IDWS contract report*, 374, 85.

D. Chapman, (Ed) (1996). Water Quality Assessment: A guide to the use of biota, sediments and water in environmental monitoring, 2nd edit, EandFN SPON, 621.

Henriksen, A. (1979). A simple approach for identifying and measuring acidification of freshwaters. *Nature*, 278, 542-545.

Henriksen, A. (1980). Acidification of freshwater- a large scale titration. In: Proceedings of International Conference on Ecological Impact of Acid Precipitaiton. *Norway*, 1980 SNSF Project, PP68-74.

Jacks, G., Lnutsson, g., Maxe, L. & Fylker, A. (1984). Effect of acid rain on soil and groundwater in Swede. In: B. Yaron, et al. (Eds), *Ecological studies Volume*, 47, spromger, 94-144.

Claridge, P. G. & Dowing, B. W. (1993). Environmental geology and geochemistry at the Windy Craggy massive sulphide deposit, *northwestern British Columbia CIM bulletin*, 86 (966), 50-57.

COX, D. P. & BAGBY, W. C. (1986). Descriptive models of Au-Ag-Te veins, in D. P. Cox, & D. A. Singer, (Eds.), Mineral deposit models: *US Geological Survey Bulletin*, 1693. 124.

Carlos, D., Da Rosa ACD & Lyon, J. S. (1997). *Golden Dreams, Poisoned Stream*s (How Reckless Mining Pollutes America's waters and How we can stop it). Mineral Policy Centre, Washington D.C

Caritat, D P. (1995). Intensifying groundwater acidification at Birkenes, South Norway, *Journal of Hydrology*, 170, 47-62

Edmunds, W M. & Kinniburgh, D. G. (1986). The susceptibility of UK groundwaters to acid deposition. *J.Geol. Soc.*, 143, 667-672.

Grimvall, A., Cole, C A., Allard, B. & Sanden, P. (1986). Quality trends of public water supplies in Sweden. Acid precipitation and human health- part 1. *Water quality bulletin, Volume 11*, No. 1. 6-11.

Downing, B W., Grvel, J. & Mills, C. (2000). *Trace element Geochemistry in acid rock drainage*. Electronic copy.

http://www.infomine/tecnology/environmine/ard/introduction/Trace.htm

Eisenlohr, B. N. & Hirdes, W. (1992). The structural development of the early Proterozoic Birimian and Tarkwaian rocks of southwest Ghana, West Africa. *J Afr Earth Sci.*, *14(3)*, 331-325.

Eppinger, R. G., Briggs, P. H., Rosenkrans, D. & Ballestrazze, V. (1999). Environmental Geochemical Studies of Selected Mineral Deposits in Wrangell, St. *Elias National Park and Preserve*, Alaska. USGS Professional Paper 1619, Washington, DC, USa. 47.

Evangelou, V. P. (1995). Pyrite Oxidation and its Control: Solution Chemistry, Surface Chemistry, Acid Mine Drainage (AMD), Molecular Oxidation Mechanisms, Microbial Role, Kinetics, Control, *Amelioration and Limitations*, Microencapsulation. CRS Press Inc, 293.

Ferguson, W. A. & Errington J. C. (1998). Guidelines for Metal Leaching and Acid Rock Drainage at Minesites in British Columbia. B. C. Ministry of Energy, *Mines and Petroleum Resources*, 79.

Ficklin, W. H. & Mosier, E. L. (1999). Field methods for sampling and analysis of environmental samples for unstable and selected stable constituents. In: G. S. Plumlee, & M. J. Logsdon, (Eds), *The Environmental Geochemistry of Ore Deposits*, Part A, Processes, Techniques, and Health Issues, Soc Econ Geol, Littleton CO, USA, Vol 6A, 249-260.

Foose, M. P., Zientek, M. L. & Klein, D. P. (1986) Magmatic Sulphide deposits.

Freeze, R. A. & Cherry, J. A. (1979). Ground Water. *Prentice Hall Inc*, New Jersey, 238-302.

Hem, J. D. (1985). Study and Interpretation of the Chemical characteristics of Natural Water. *USGS Water Supply paper*, 2254, 263.

Henderson, P. (1982). *Inorganic Geochemistry*. Pergamon Press, Headington Hill Hall. England, 255-271.

Jamieson, H. E. & Pryzbylowicz, W. J. (1997). Incorporation of toxic elements in iron sulfates precipitated from acid mine waters. *Geological Association of Canada/Mineralogical Association of Canada Abstracts*, 22, 1997, A73.

Junner, N. R. (1935). Gold in the Gold Coast. *Gold Coast Geol. Surv*. Memo. 4. 5, 20.

Jonah, S. E. (1987). The Impact of the ERP on the Mining Industry in Ghana. In: *Proceedings of a Seminar on the Mineral Sector in Ghana*, Accra. Minerals Commission.

Jones, D. R. (1998). Treatment: Dealing with an Acid Problem. *Groundwork*, September 1998 contents page. Web posted and Copyright © 1998, Australian Minerals and Energy Environment Foundation.

Kuma, J. S. & Younger, P. L. (2000). *Conceptual ground water model and related environmental concerns in the Tarkwa area*, Ghana. Ghana Mining J 6, 42-52

environmental concerns in the Tarkwa area, Ghana. *Ghana Mining Journal*, 6, 42-52.

Kuma, J. S., Younger, P. L. & Bowell, R. J. (2002). Hydrogeological framework for assessing the possible environmental impacts of are scale mines. In P. L., Younger, & N. S. Robins, (Eds). *Mine Water Hydrogeology and Geochemistry*. Geological Society, London, Special Publication, 198, 121-136.

Kuyucak, N. (2001). Acid Mine Drainage (AMD) prevention and control. *J of Environmental Management*. January, 12-16.

Canadian Mine Environment Neutral Drainage (MEND), 2001. MEND Manual *Volume 2, Sampling and Analysis*. MEND 5.4.2b. 111.

Lawrence, R. W. & Scheske, M. (1997). A method to calculate the neutralization potential of mine waste. *Environ. Geol.*, 32, 100-106.

Lee, G., Bigham, J. M. & Faure, G. (2002). Removal of trace metals by co-precipitation with Fe, Al, and Mn from natural waters contaminated with acid mine drainage in the Ducktown Mining District, Tennessee. *Applied Geochemistry, 17(2002)*, 569-581.

Marston, R J., Woolrich, P. & Kwesi, J. (1992). Closely associated stock work and paleoplacer gold mineralisaiton in the early Proterozoic Tarkwaian Sytem of Ghana. Regional Trends in Africa Geology. Proceedings of the 9th International Geological

Coference, Accra. 2-7th November, 1992. *Geological Society of Africa/Geological Survey of Ghana.*

Miliesi, J. P., Ledru, P., Feybesse, J. L., Dommanget, A. & Marcoux, E. (1992). Early Proterozoic ore deposits and tectonics of Birimian orogenic belt, West Africa. *Precambrian research, 58*, 305-344.

Miller, S. (1998). Prediction: Predicting Acid Drainage. *Groundwork*, Australian Minerals and Energy Environment Foundation.

Miller, W. R. (2000). Influence of rock composition on the geochemistry of streams and spring waters from mountainous watersheds in the Gunnison, Uncompahgre, and Grand Mesa National Forest, Colorado. *US Geological survey Professional Paper*, 1667. 30.

Miller, J. R., Lechler, P. J., Rowland, J., Desilets, M. & Hsu, L. C. (1995). An integrated approach to the determination of the quantity, distribution, and dispersal of mercury in Lahontan Reservoir, Nevada, USA. In: R. J. Allan, & W. Salomons, (Eds.) *JGE 52*, 45-55.

Minerals Commission, (1991). Draft. *Study on the effect of mining on the environment.* CR574/1.

Mogollon, J. L., Bifano, C. & Davies, B. E. (1995). Distribution of metals in mechanical fractions of soils from a lake catchment in Venezuela. In: *Environmental Geochemistry and Health, 17*, 103-111.

J. O. Nriagu, (Ed.) 1978. *Sulfur in the Environment.* John Wiley and Sons. 314-325.

Plumlee, S G., Smith, K. S., Montour, W. H., Ficklin, W. H. & Mosier, E. L. (1999). Geologic controls on the composition of natural waters and mine waters draining diverse mineral deposits types. In: Filipek HL, Plumlee SG (eds), The Environmental Geochemistry of Mineral deposits. Part B: Processes, Techniques and Health Issues, *Reviews in Economic Geology*, Soc Econ Geol, Littleton CO, USA, vol 6B, 373-409.

Rollinson, H. R. (1993). *Using Geochemical Data: Evaluation, Presentation,* Interpretation. Longman Singapore Publishers. 1-172.

Salami, M. B. (1999[1]). Environmental Impact Assessment policies, their effectiveness or otherwise for mining sector environmental management in Africa. Paper presented at the Conference on Mining, *Development and Social conflicts in Africa.* 15- 18 November, Accra, Ghana.

Salami, M. B. (1999b). *Environmental Monitoring and Training*, Law and Practice. Paper presented at the Seminar on Social and Environmental Management in the Mines and its effects on the community. 26- 29 October, Agona- Ahanta.

R. R. Seal II, & K. N. Foley, (2002). (Eds). *Progress on Geoenvironmental Models for Selected Mineral Deposits.* U.S. Geological Survey Open File 02-195.213.

Salomons, W. (1995). Environmental impacts of metals derived from mining activities: Processes, Predictions and Prevention. In: R. J. Allan, & W. Salomons, (Eds.) *JGE, Vol. 52*, Nos 1, 2 , 5-21.

Schreck, P. (1998). Environmental Impacts of uncontrolled waste disposal in the mining and industrial areas in central Germany. *Environmental Geology, 35(1)*, 66-72.

Seyire, J. S. (2001). The Environmental Impact of Acid Mine Drainage (AMD) from the abandoned Konongo Mine. *Unpublished Masters Thesis*, University of Ghana, Legon p 25-50.

Smedley, P. L. & Kinniburgh, D. G. (2002). A review of the source, behaviour and distribution of arsenic in natural waters. *Applied Geochemistry*, 17 (2002) 517-568.

Smedley, P. L., Edmunds, W. M. & Pelig- Ba, K. B. (1993). *Mobility of trace elements* (As, I and F) *in groundwaters from Ghana and their impact on human health.*

Smith, R. M., Grube, W. E. Arkle, T. & Sobek, A. (1974). Mine Spoil Potentials for Soil and Water Quality, Prepared for U.S. *Environmental Protection Agency*, Cincinnati, Ohio, EPA-670/2-74-070.

Smith, M. W. & Skema, V. W. (2001). Evaluating the potential for acid mine drainage remediation through remining in the Tangascootack Creek watershed, Clinton County, Pennsylvania. *Mining Engineering.* February, 41-48.

Smith, L. (2001). The geology of the Tarkwa mine resources, Gold Fields Ghana. *Goldfields (GH) Ltd Report.* February.

Sobek, A. A., Schuller, W. A., Freeman, J. R. & Smith, R. M. (1978). Filed and laboratory methods applicable to overburdens and mine soils. *Report EPA.*

Songsore, J., Yankson, P. W. K. & Tsikata, G. K. (1994). Mining and the Environment: Towards a Win-Win strategy (*A study of the Tarkwa – Aboso – Nsuta Mining complex in Ghana*). 185.

Walker, S. R., Jamieson, H. E. & Andrade, C. F. (2004). Determining arsenic speciation in roaster-derived iron oxides: Application of synchrotron micro-XRD and micro-XANES at the grain scale. *Submitted to Canadian Mineralogist*, March 2004 44.

Nyame, F. K., Beukes, N. J., Kase, E. K. & Yamamoto, M. (2002). Compositional variations in manganese carbonate micronodules from the Lower Proterozoic Nsuta deposit, Ghana: product of authigenic precipitation or post-formational diagenesis? *Sediemntary geology, 154,* 159-175.

Kaiser, H. P. (1960). The application of electronic computers to factor analysis, *Educational and psychological measurements, 20,* 141-151.

Yager, D. B., Mast, M. A., Wrigh, W. G. & Hageman, P. L. (2000). Natural versus mining – related water quality degradation to tributaries draining Mount Molly, Silverton, Colorado. Proc, 5th International Conf on.

Acid Rock Drainage. (2000). *ICARD, Vol 1,* 535-547.

USDA FOREST SERVICE, (1993). Acid Mine Drainage From Mines on the National Forests, A management Challenge. *Program Aid,* 1505, 10-13.

USEPA, (1994). Technical Document on Acid Mine Drainge Prediciton. *EPA530-r-94-036.* 48.

In: Mine Drainage and Related Problems
Editor: Brock C. Robinson, pp. 189-210

ISBN: 978-1-60741-285-4
© 2010 Nova Science Publishers, Inc.

Chapter 5

CHARACTERIZATION, DEWATERING, DISPOSAL AND BENEFICIAL USE OF ACID MINE DRAINAGE SLUDGE

Xinchao Wei[*]
Department of Civil and Environmental Engineering, West Virginia University, Morgantown, WV, USA

ABSTRACT

The formation of acid mine drainage (AMD) has been an environmental problem for decades in coal mining and hard rock mining regions throughout the world. Typical AMD treatment processes are designed to raise the pH of the effluent to within the range of 6.0-9.0 and to precipitate various metals. A major problem in the AMD treatment is the generation of large volumes of sludge through pH neutralization and subsequent metal precipitation. The sludge typically contains a low concentration of solids, resulting in associated handling and disposal difficulties. Good sludge properties can increase the effective throughput rate of AMD treatment facilities, reduce the capital costs of settling ponds, alleviate the task of sludge dewatering and lower the transportation cost of sludge to ultimate disposal sites. In this chapter, detailed physical and chemical characterization is provided for AMD sludge generated from different neutralization processes. Factors affecting sludge properties include chemical composition of AMD, chemical agents applied for active AMD treatment, passive treatment systems, and AMD treatment operations. AMD sludge treatment can include conditioning, thickening, and dewatering. Sludge disposal can be achieved by deep mine disposal, retained-in-pond disposal, or disposal at coal refuse areas. Finally, the potential beneficial uses of AMD sludge are discussed.

Key Words: acid mine drainage, sludge characterization, sludge dewatering, sludge disposal

---

[*] Corresponding author: Email: xinchao.wei@mail.wvu.edu, Phone: 304-293-9954, Fax: 304-293-7109.

1. INTRODUCTION

The term "acid mine drainage" (AMD) refers to the water flowing from or caused by surface mining, deep mining or coal refuse piles that is typically high in acidity and dissolved metals. AMD forms when sulfide minerals in rocks are exposed to oxidizing conditions. AMD has been an environmental problem for decades in coal mining areas such as the *Mid Appalachian* region and the western United States. In West Virginia alone, AMD has impacted about 3,200 km (~2,000 mi) of streams (Demchik and Garbutt, 1999). Problems associated with AMD include acidic water with elevated concentrations of dissolved metal ions, which can be toxic to aquatic organisms, corrode infrastructure, damage benthic habitats, and affect the overall aesthetics of receiving streams (Tabak *et al.*, 2003; DeNicola and Stapleton, 2002; Soucek *et al.*, 2000). In order to minimize the negative impacts of AMD, various active and passive treatment processes have been implemented to increase alkalinity of AMD waters and subsequently precipitate metal ions (Hammarstrom *et al.*, 2003; Dempsey and Jeon, 2001; Zinck and Aube, 2000; Gazea *et al.*, 1996). Active treatment involves the continuous addition of neutralizing agents to AMD waters. Hydrated lime, sodium hydroxide, limestone, ammonia and soda ash are commonly used in active AMD treatment (Johnson and Hallberg, 2005). Typical AMD treatment processes are designed to raise the pH of the effluent to within the range of 6.0-9.0 and reduce the total iron concentration to less than 3 ppm (Hesbach and Lamey, 2001). Constructed wetlands, anoxic limestone drains (ALD), vertical flow systems such as successive alkalinity producing systems (SAPS), limestone ponds, and open limestone channels (OLC) are passive treatment systems to increase alkalinity. No matter which method is used for AMD treatment, dissolved metal ions are precipitated from AMD as a loose, open-structured mass of tiny grains called floc or sludge. The sludge typically contains a low concentration of solids, resulting in associated handling and disposal difficulties.

2. AMD SLUDGE PROPERTIES

2.1. Physical Properties of Sludge

The physical properties may include dewaterability, percent solids, sludge density, particle size, particle surface properties (e.g. net surface charge), and sludge flow properties such as viscosity. Sludge dewaterability refers to the ability of sludge to dewater, or its resistance to dewater. It plays an important part in selecting dewatering process and equipment and affects the dewatering efficiency. Percent solids content is another physical property of interest. It is usually expressed by percent solids by weight in sludge. The percent solids content of AMD sludge reported in the literature varies from 1 to 20%, with a typical value of 1 to 5% (Johnson and Hallberg, 2005; Kuyucak, 1998; Ackman, 1982; Keefer, 1979).

The physical properties of representative AMD sludge samples are listed in Table 1 (Dempsey and Jeon, 2001). Sample 1 is sludge formed from AMD active treatment system with addition of NaOH, while sample 2 is AMD sludge produced in passive treatment system. It should be noted that in Table 1 the dewaterability is expressed in term of specific resistance to filtration. Sludge from passive treatment is much denser than sludge created through active treatment. For AMD sludge produced from ammonia neutralization, the physical characteristics are presented in Table 2 (Viadero, et al., 2006).

2.2. Chemical Properties of Sludge

The chemical composition of sludge generally consists of hydrated ferrous or ferric oxides, gypsum, hydrated aluminum oxide, calcium carbonate and bicarbonate, with trace amounts of silica, phosphate, manganese, copper, and zinc. In general, chemical composition of sludge AMD is highly variable and non-uniform in nature. A typical chemical analysis of AMD sludge from hydrated lime treatment system is shown in Table 3 (Coal Research Bureau, 1971). For AMD sludge produced from ammonia neutralization, the chemical characteristics are presented in Table 4 (Viadero, et al., 2006).

Table 1. Physical properties of some sludge samples
(Dempsey and Jeon, 2001)

Characteristics	Sample 1 (active treatment)	Sample 2 (passive treatment)
pH	7.79	6.44
Weight Density (kg/l)	0.03	0.20
Settling Rate (cm/min)	0.01	1.84
Reduced Viscocity (cm^3/g)	43.24	7.96
Zeta-potential (mV)	-13.5	-10.1
Specific Resistance to Filtration (10^9 m/kg)	26-125	2.16-4.16
Coefficient of Compressibility	0.86	0.38

Table 2. Physical characteristics of AMD sludge samples from
ammonia neutralization (average ± standard deviation)

Parameters	Values
Specific Conductance	3572 ± 450 µS/cm
Temperature	18.1 ± 6.3 °C
Density	1.008 ± 0.011 g/mL
% Solids	0.72 ± 0.24 % (g/g)
Total Solids	7.31 ± 2.53 g/L
Dissolved Solids	2.1 ± 0.49 g/L
Specific Conductance	3572 ± 450 µS/cm
Viscosity	11.8 ± 1.1 cP
Specific Resistance to Filtration	1.31 ± 0.32 × 10^{11} m/kg
Coefficient of Compressibility	1.07

Table 3. Typical compositions of AMD sludge from hydrated lime treatment.

Species	% by weight
Calcium sulfate ($CaSO_4$)	~40%
Magnesium sulfate ($MgSO_4$)	~5%
Free lime (CaO)	~3%
Magnesia (MgO)	~1%
Ferric oxide (Fe_2O_3)	~15%
Manganese oxide (Mn_2O_3)	~4%
Silica (SiO_2)	~20%
Aluminum oxide (Al_2O_3)	~12%

Table 4. Chemical characteristics of AMD sludge samples from ammonia neutralization (average ± standard deviation; unit in mg/L except pH)

Parameters	Values
pH	8.54 ± 0.67
Alkalinity (as $CaCO_3$)	1983 ± 995
Total Fe	1401 ± 692
Total Al	380 ± 172
Total Mn	13.3 ± 4.5
Total Ca	68.5 ± 33.0
Total Mg	60.3 ± 10.3
Sulfate, SO_4^{2-}	1294 ± 145
TAN**	429.7 ± 95.9
Nitrate, NO_3^-	0.78 ± 0.14
Nitrite, NO_2^-	BDL***
Total Phosphorus	0.23 ± 0.14

3. FACTORS AFFECTING AMD SLUDGE PROPERTIES

AMD sludge properties are influenced by two main factors: 1) influent characteristics; 2) AMD treatment systems. It is to note, for any AMD treatment system, AMD nature and treatment methods commonly determine the sludge precipitation and its properties.

3.1. Effects of AMD Composition on Sludge Properties

Since AMD treatment method is determined by raw acid mine water, the characteristics of AMD itself greatly affect the sludge formation and its properties. Ziemkiewicz and Skousen (1996) summarized acid mine drainage into 4 types (Table 5). However, in terms of the consideration of the effects of raw AMD characteristics on sludge properties, the classification by Coal Research Bureau (1971) is more relevant to sludge disposal. AMD was categorized into three types depending on chemical characteristics (Table 6).

Coal Research Bureau (1971) summarized the effects of AMD type on sludge properties. Sludge formed from Type I AMD water is usually light and fluffy in character and settles to 0.5 percent of its original treated volume. This type of water does not necessarily need neutralization due to the lack of present acidity. However, when there is a high concentration of ferrous (Fe^{3+}) iron present, lime may be needed to raise the pH to handle potential acidity resulting from the oxidation of ferrous iron. If lime addition is not needed, the volume of the sludge is totally dependent upon the concentration of iron and other precipitable metals present in the water.

Sludge from Type II AMD water can vary depending on the raw water constituents. Iron content will be a major contributor to sludge volume. To treat this type of water, neutralization processes have to be introduced, which dramatically changes sludge properties. Sludge from this type of AMD normally has a relatively tight floc formation. When this type of water contains a large amount of ferrous and varying concentrations of bicarbonate and is treated with lime and aerated, extended retention is required for complete sedimentation. The problem can be solved by eliminating the aeration step and allowing the iron to remain in the unoxidized ferrous hydroxide form. In this case, the bicarbonate content of the raw water played a deciding role in affecting sludge settling characteristics.

Table 5. Classification of acid mine drainage (Ziemkiewicz and Skousen, 1996)

AMD Type	Characteristics
Type 1	AMD having no alkalinity, high metals and acidity (oxidized).
Type 2	AMD having high total dissolved solids containing metals, no or low oxygen content, but high pH.
Type 3	Neutralized AMD with high total suspended particulates where settling of metal hydroxides has not occurred.
Type 4	Neutralized AMD with high total dissolved solids containing hardness, but no metals.

Table 6. Classification of acid mine drainage (Coal Research Bureau, 1971)

AMD Type	Characteristics
Type I	AMD usually has a pH of 6.5 to 7.5 or greater, very little or no acidity, and contains iron, usually in ferrous state (Fe^{2+}), that varies from less than 60 mg/l up to 1000 mg/l.
Type II	AMD is partially oxidized water that can have a pH in the range of 3.5 to 6.5, acidity values that range from 0 to 1000 mg/l, iron in both ferric and ferrous state and small amount of aluminum. This type of water can contain both acidity and alkalinity and the relative amounts of ferric and ferrous iron dependent upon the pH.
Type III	This type of AMD water is highly acidic. This water frequently contains high concentration of acidity (1,000-15,000 mg/l), large concentration of iron (500-10,000 mg/l), mostly in the ferrous state, and aluminum (0-200 mg/l).

The volume of the sludge formed from Type III is generally high. The high iron content will naturally contribute to the sludge volume, but the presence of aluminum in the form of hydrous oxide adds to the sludge volume. The total effect of aluminum on the sludge volume will be dependent upon the treatment pH. Silica present with alumina also aids in the sludge sedimentation due to the formation of large flocs which in turn aids in the coagulation of iron hydroxide.

In general, iron, aluminum, sulfates, bicarbonate, silicon, calcium and acidity are the main species in AMD that influence the sludge properties. From the AMD formation mechanism, it can be seen that iron and its various derivatives are very important to sludge properties. Ferrous iron will not precipitate from solution unless the pH is raised above 8.5, forming a blue-green iron hydroxide. Ferric iron, the oxidized form, reacts with hydroxides and precipitates as "yellow boy" at a pH above 3.5 (Ziemkiewicz and Skousen, 1996).

3.2. Effects of AMD Treatment Systems on Sludge Properties

Since there are various AMD treatment methods and a number of neutralizing agents, it is difficult to consider every aspect of reagent effect on sludge properties. However, an overview of representative reagents is made for the active and passive AMD treatment systems.

3.2.1. Effects of active treatment with chemicals on sludge properties

Active treatment systems involve treating mine drainage with alkaline chemicals to raise water pH, neutralize acidity, and precipitate metals. Active chemical treatment of AMD to remove metals and acidity is often effective, but expensive due to the cost of equipment, chemicals, and manpower (Skousen *et al.* 1998). Addition of chemicals also presents a future liability.

There are six primary chemicals used to treat AMD: (1) limestone, (2) hydrated lime, (3) pebble quicklime, (4) soda ash, (5) caustic soda and (6) ammonia. Other industrious wastes, such as fly ash, flue gas desulfurization sludge and fluidized bed combustion waste, may also be used in AMD treatment to raise alkalinity. Also, in order to improve the settling of sludge floc, coagulants and flocculants are also often added to the treatment systems. Among these chemicals, each has its own characteristics that make it more or less appropriate for a particular circumstance, and thus, the choice of chemicals depends on both technical and economic factors.

Generally, a low-density sludge is produced in active treatment of AMD with neutralization reagents. Ackman (1982) studied the chemical and physical properties of floc formed from active chemical treatment of AMD, and found that the sludge had relatively high volume with low total solids content and was difficult to dewater. Brown *et al.* (1993) studied floc generation in AMD treatment with calcium hydroxide, ammonium hydroxide, sodium carbonate and sodium hydroxide. All the sludge samples contained 0.5-1.5% of solids. However, different neutralizers and other chemicals have various effects on sludge properties and their effects are reviewed respectively as follows.

Limestone

The feasibility of AMD treatment limestone by limestone was established by Tracy (1922) as early as 1913. Limestone $(CaCO_3)_{(s)}$ has been used for decades to raise pH and precipitate metals of AMD. It has the lowest material cost, is the safest chemical for AMD treatment, and is easy to handle. From the perspective of sludge disposal, active AMD treatment with limestone produces the most compact and easy to handle sludge material (Skousen et al. 1998). In another study on the sludge settling rates and volumes of settled solids of AMD treatment with limestone, lime and a combination of lime and limestone, it was found that limestone and the combination process produced a sludge of more compact solids that was easier to settle (Coal Research Bureau, 1971).

Hydrated lime

Hydrated lime, $Ca(OH)_2$, is the most commonly-used chemical for treating AMD. It is generally easy and safe to use, effective, and is relatively inexpensive. Payette et al. (1991) used scanning electron microscope analysis to study the AMD floc and found that AMD floc neutralized by hydrated lime exists in the formation of crystalline gypsum as well as various amorphic metal hydroxides. Ackman (1982) stated that calcium hydroxide generates a granular, dense floc by comparison with a more gelatinous, loose floc generated by addition of sodium hydroxide or ammonia. Similarly, Sanmarful (1969) investigated sludge properties formed from a synthetic mine water using various alkaline agents and found that sludge from hydroxide lime treatment was light, gelatinous, and voluminous in nature as compared to sludge generated by treatment with $CaCO_3$, Na_2CO_3 and $NaOH$. Compared to limestone, AMD treatment with hydrated lime produces the voluminous sludge. Since hydroxide lime does not contain unburned limestone or "core" in the quicklime, considerably less sludge is produced from hydroxide lime treatment as compared to quicklime neutralization.

Pebble quicklime

Sometimes pebble quicklime, CaO, is used in conjunction with limestone due to inefficiency of $CaCO_3$ in the treatment of AMD containing large amounts of ferrous iron. Carbonate compounds do not raise pH of AMD above 8.5 while hydroxide compounds can raise pH to above 10.0 (Ziemkiewicz and Skousen, 1996). Limestone is added to raise the pH of AMD to about 5.0 and then lime is added to raise pH to the required level. Ferric iron converts to a yellowish-orange precipitate at a pH of 5.5 or greater and ferrous converts to the solid bluish-green ferrous hydrous hydroxides at a pH of 9.0 or greater (Ziemkiewicz and Skousen, 1996). Thus, the chemical properties of sludge are changed through adjustment of pH. The optimum combination of limestone and lime is expected to produce superior AMD treatment results, good sludge properties and reduced costs.

Soda ash

Soda ash, Na_2CO_3, is especially effective for treating AMD in remote areas with low flow and low acidity and metals concentration, but its use is declining (Ziemkiewicz and Skousen, 1996). The major disadvantage of Na_2CO_3 is relatively high reagent cost compared to limestone. Generally, AMD sludge treated with soda ash has poor settling properties and is difficult to dewater (Skousen et al., 1998).

Caustic soda

Caustic soda (NaOH) is often used for treating low flow mine water (<100 gpm) with high acidity situations in remote locations (Skousen et al. 1998). It is especially effective in treating AMD with a high Mn concentration in the AMD because caustic can raise the pH of AMD water to 13.0. The AMD sludge from caustic soda treatment usually has poor sludge properties, such as low settling rate and high sludge volume, which make the disposal of sludge difficult (Skousen et al., 1998).

Ammonia

Ammonia, the common term for anhydrous ammonia ($NH_{3(s)}$), behaves as a strong base and can easily raise the pH of receiving water to 9.2. At pH 9.2, it buffers the solution to further pH increase (Skousen et al., 1998). It is effective in AMD treatment and is one of the quickest ways to raise pH of AMD. However, it is a material that is difficult and dangerous to use and can affect biological conditions after addition.

Fossil fuel combustion products

Fossil fuel combustion products are solid wastes generated from coal-fired power plants, such as fly ash, flue gas desulfurization sludge and fluidized bed combustion (FBC) waste. The compositions of fly ash are highly variable, and commonly consist of oxides of Si, Al, Fe, and Ca, and of these elements Na, P, K and S (Lau and Wong 2001). Hesbach and Lamey (2001) studied the AMD treatment with FBC ashes, and recommended that the method as a way to combine two waste products (AMD sludge and coal combustion ash) to economically address a problem of the mining industry. The addition of either fly ash or FBC ash reduced the volume of the resulting sludge to much less than that produced by adding lime to acid mine drainage. The solids percent was increased by a factor of 10 to over 40%. With addition of either FBC ash or fly ash, the sludge settled much more quickly than the settling rate produced by lime addition. Thus, sludge properties were greatly improved.

Coagulants and flocculants

Once metal hydroxide flocs are formed by the addition of alkaline reagents, adequate retention time is needed for the sludge to settle. In order to improve the settling efficiency of sludge, a variety of chemicals, mainly coagulants and flocculants, are added. Coagulants reduce the net repulsive forces at particle surfaces, thereby promoting consolidation of small particles into large particles. By comparison, flocculants aggregate or combine small particles by bridging the space between particles to form (Svarovsky, 2000). From these floc formation mechanisms, it is found that normally the sludge formed by coagulation has a relatively high solids content than by flocculation. According to Ziemkiewicz and Skousen (1996), ferric sulfate coagulates particles over a large pH range than ferrous sulfate and the precipitate is a heavy, quick settling floc. Activated silica can also be used to strengthen flocs and to reduce the potential deterioration. The resulting sludge is larger, denser, faster-settling, and more stable.

3.2.2. Effects of passive treatment on AMD sludge properties

Active chemical treatment of AMD to remove metals and acidity is often expensive and results in a long term liability. Therefore, a variety of passive treatment systems have been developed. These systems take advantage of naturally occurring chemical and biological processes to cleanse contaminated mine waters, so that continuous chemical inputs can be avoided. The primary passive technologies include constructed wetlands, anoxic limestone drains (ALD), vertical flow systems such as successive alkalinity producing systems (SAPS), limestone ponds, and open limestone channels (OLC) (Skousen et al., 1998). Compared to active chemical treatment, passive systems generally require longer retention times and greater space.

Physical properties of some field AMD sludge samples at C&K coal site are listed in Table 7 (Dempsey and Jeon, 2001). In Table 7, all samples are passive treatment sludge except CK Pond #1 sludge, which is AMD sludge from active treatment with NaOH addition. The wet density of the passive treatment sludge (0.12-0.28 kg/l) is considerably higher than AMD active treatment sludge (0.03kg/l). Data of specific resistance to filtration (SRF) indicate that passive sludge is relatively easy to dewater and low specific viscosity also shows that AMD passive sludge has good dewaterability.

AMD Sludge from passive treatment systems contains iron oxy-hydroxides, such as goethite, lepidocrocite, and phases that convert to hematite upon heating (Kirby, 1999). Hsieh (1993) studied the dewaterability of Fe_2O_3 sludge and found that the sludge from passive treatment of mine drainage was easily dewatered compared with ferric oxide sludge that was prepared by addition of chemical reagents.

3.2.3. Effects of some AMD treatment processes on sludge properties

Aeration

Aeration is the process of introducing oxygen into water as air. Oxidation occurs when oxygen in air reacts with metals in the water. Without the introduction of oxygen, ferrous iron will keep its state and reacts with neutralization reagents to form sludge of ferrous hydroxide. In aeration process, ferrous iron is oxidized into ferric iron and precipitated as ferric hydroxides. Pudlo (1970) studied the effects of aeration on sludge volume and settling characteristics. Results showed that non-aerated sludge settled faster than aerated sludge, but produced a larger settled volume, which meant flocs of the nonaerated sludge contain more water content.

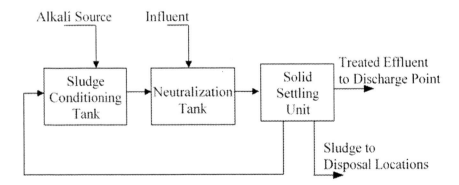

Figure 1. Schematic of dense sludge process (Zick et al., 1999)

Table 7. Characteristics of the field sludge samples (Dempsey and Jeon, 2001)

Sample	pH	Wet density (kg/L)	Settling rate (cm/min)	Reduced viscosity (cm^3/g)	Zeta-potential (mV)	SRF ($\times 10^{10}$) (m/kg)	S_o
C&K, inlet	5.96	0.25	1.51	6.58	-11.13	5.63-10.3	0.28
Bend #1	5.86	0.28	1.93	6.64	-8.5	0.89-1.46	0.29
Bend #1-2	6.14	0.23	1.55	5.23	-10.9	2.85-5.65	0.38
Bend #2	6.43	0.14	0.85	15.71	-13.2	3.44-10.1	0.62
Bend #3	6.18	0.12	1.23	13.96	-13.3	3.83-8.43	0.44
Pond #1	7.79	0.03	0.01	43.24	-13.5	26.0-125	0.86
H.B. Pond #1	6.44	0.20	1.84	7.96	-10.1	2.16-4.16	0.38
Pond #2, Infl.	3.71	0.17	1.16	15.99	1.0	6.73-15.2	0.46
Pond #3, infl.	3.40	0.20	1.68	10.38	-5.5	4.73-5.0	0.65
Pond #3	2.51	0.21	1.91	11.67	1.13	3.11-6.41	0.39

High-density sludge process (HDSP)

Developed by Bethlehem Steel Corp. in the early 1970s, the high-density sludge process was originally intended for use on AMD and diluted waste pickle liquor discharges requiring continuous chemical-treatment. High density sludge process combines conventional AMD treatment with sludge recirculation. According to Zick et al. (1999), there are approximately 15 operating systems that have been installed to treat mine drainage in the United States. The simplified schematic of high-dense sludge process is shown in Figure 1.

In the high-density sludge process, recycled sludge is mixed with an alkali source before the neutralization process. The alkali source can be lime, caustic, ammonia or any other neutralization agent that reacts readily and can be continuously metered. The sludge particles react with the alkali to provide attraction sites for the removal of metals and cause gypsum crystals to grow (Zick et al. 1999). The continued recirculation of sludge ultimately converts the metal hydroxides to oxides through a series of steps. The properties of AMD sludge generated through HDSP include: 1) the dense sludge particles settle faster, dewater more readily, are more easily pumped; 2) the sludge density of the clarifier underflow increased from 1% to 2% solids to approximately 15% solids; 3) actual sludge volume for disposal was reduced by approximately 90% (Zick, et al., 1999). In another study, Herman et al. (1984) stated that high-density sludge process produced 20-30% solids after sedimentation and 40-50% solids after storage.

HTP process

Developed by HPT Research, Inc., HPT process provides a viable economic and environmental solution to the myriad of problems associated with AMD treatment. The HPT process used for AMD treatment relies upon newly developed technology that provides the means of altering the chemical and physical properties of metal contaminants within aqueous solutions. According to Mining Engineering magazine (2000), this process produces a minimum amount of metal hydroxide sludge. This process produces a pure anhydrous calcium sulfate void of any metal contamination with potential value as an agricultural soil amendment or for safe on-site disposal.

4. SLUDGE DEWATERING

After AMD is treated with active or passive systems, a large volume of sludge is produced. Due to the loose structure of flocs, the sludge is primarily composed of water. Typical sludge solids concentrations produced at site range from 1.5-2.0 %wt. for active treatment systems. The high water content creates various problems for the final sludge disposal or application. In general, sludge dewatering is needed prior to any final disposal schemes. In order to facilitate or improve the full-scale dewatering performances, sludge conditioning and sludge thickening are required before dewatering processes.

4.1. Water States within Sludge Flocs

Characterization of water states within flocs is important to decide its dewaterability and to select the appropriate dewatering technology. AMD sludge contains various metallic hydroxide flocs, which have different floc structures. These structures determine the water distribution within flocs. According to Waite (1997) and Vesilind (1974), the different physical states of water in sludge can be described as: bulk water, interstitial water, vicinal water and water of hydration (Table 8).

Bulk water is the easiest to remove from the sludge. Dewatering of this portion of water can be achieved by drainage, settling, thickening or mechanical dewatering. Since interstitial water is trapped within flocs, it can be eliminated from the floc only by either the destruction of floc structures or by the expenditure of sufficient mechanical energy to squeeze the water out. This can be accomplished by mechanical dewatering such as pressure filtration. According to Tsang and Vesilind (1990), it was found that the dewatering procedures could only remove part of the free and interstitial moisture; surface and bound moisture were not affected. Part of the free and interstitial moisture also remained with the sludge cake, which could be considered as the inherent inefficiency of the dewatering process. The complete removal of all the free moisture can be considered as the theoretical best performance any mechanical dewatering device can achieve.

Table 8. Water states within sludge flocs (Waite, 1997 and Vesilind, 1974)

Water State	Description
Bulk water	Water is not associated with and not influenced by suspended particles
Interstitial water	Water is trapped in the crevices and interstitial spaces of flocs and organisms.
Vicinal water	Multiple layers of water molecules are held tightly to the particle surface by hydrogen bonding.
Water of hydration	The water is chemically bound to particles and removable only upon the addition of thermal energy.

4.2. Sludge Conditioning

Sludge dewatering is usually improved if the sludge is previously treated or conditioned. Sludge conditioning is a process whereby sludge solids are treated with chemicals or various other means to prepare the sludge for dewatering processes. Chemical conditioning usually reduces the specific resistance of the sludge cake to filtration or drainage, and hence renders the dewatering process more economical.

Mechanisms of sludge conditioning include: reduction of charge on sludge particulate, binding of dispersed particles into aggregates, and auxiliary factors (Stanley, 1969). Reduction of charge is achieved mainly by addition of a conditioning agent that supplies opposite charge radicals to the charged dispersed particles. A reduction in charge on dispersed particles alters the equilibrium between electrostatic forces of repulsion and forces of mass attraction so as to promote aggregation. Since small particles normally possess the highest charge per unit mass, they are aggregated preferentially by the action of conditioning agent. The aggregation of the fines yields sludge with improved dewatering properties. Binding of sludge particles into aggregates may be effected by the forces of mass attraction or by chemical binding mechanisms of conditioning agents. The binding is particularly effective if the conditioning agent forms a polymer possessing long length, good fiber strength, and an appropriate chemical structure to bond to the surface of sludge particles. The stem and branches of the polymer may contain functional groups that are capable of forming electrostatic or chemical bonds. Auxiliary factors in sludge conditioning consist of the side effects of addition of conditioning agents, such as pH. Some conditioning agents can alter the pH of sludge system, while pH affects the absorption and ionization equilibrium of both the dispersed sludge particles and the condition agents. The pH also influences solubility of substances and the degree of curl of polymer. Control of pH is often necessary in the sludge and conditioning agent system. Other auxiliary factors include substances such as phosphates and carbonates that may precipitate or inactivate sludge conditioning agents and aid in the stabilization of sludge dispersion. In practice, the combined effects of these mechanisms are needed to be maximized in order to achieve the best conditioning results. In the operations of sludge dewatering, the performance is affected as a consequence of variations in the three basic parameters: 1) dosage of chemical; 2) mixing time; and 3) mixing intensity (Christensen, 1995).

Ferric and aluminum compounds, especially ferric chloride and aluminum sulfate, are the common inorganic chemicals used in sludge conditioning. The double layer theory assumes that the ionized Fe^{3+} and Al^{3+} will enter the outer layer of the particle, neutralize it, and make it possible for the particles to flocculate (Vesiliand, 1974). Organic polymer coagulants or flocculants have been developed and are now widely used in sludge conditioning due to their ease in handling, small storage space requirements, and their effectiveness. The important characteristics of these polymers for sludge conditioning include molecular weight and the nature of functional group (Kim, 1990). These compounds can be characterized according to their ionized form as cationic, anionic or nonionic. Lotito and Spinosa (1990) studied the influence of polyelectrolyte dosage and characteristics on dewaterability parameters. Results showed the negligible influence of polyelectrolyte molecular weight, as compared to other characteristics considered, and that medium and high charge density polymers are preferable in dewatering by filtration, while medium ones are best in centrifugation. Chang et al. (1998) studied the conditioning behaviors of kaolin sludge with different cationic polyelectrolytes.

The rate of dewatering did not significantly differ from polyelectrolyte molecular weight above 6.6×10^6, and as well as a charge density exceeding 15%. The moisture content of dewatered sludge cake depended on polyelectrolyte dosage, molecular weight and charge density. When the dosage approached a value such that the streaming current became zero, the moisture content of dewatered kaolin sludge cake reaches the maximum of comparable values for different polyelectrolytes.

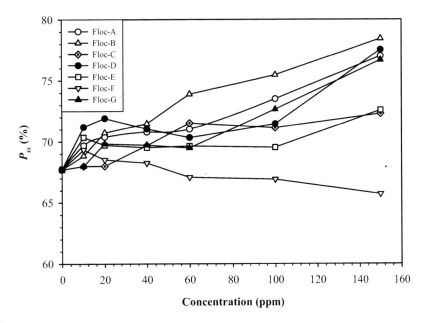

Figure 2. Percent settled sludge at different concentrations of anionic polymers (Viadero et al., 2006).

Table 9. Polymeric flocculants and coagulants (Viadero et al., 2006)

	Code	Chemical Type
Flocculants	Floc-A	Anionic proprietary polymer, high charge
	Floc-B	Anionic, modified polyacrylamide, high charge
	Floc-C	Anionic, ammonium polyacrylate, high charge
	Floc-D	Anionic, high molecular weight polyacrylamide, high charge
	Floc-E	Anionic, very high molecular weight polyacylamide, medium charge
	Floc-F	Anionic acrylamide copolymer
	Floc-G	Anionic acrylic emulsion copolymer
	Floc-H	Cationic polyacrylamide copolymer, high charge
	Floc-I	Cationic polyacrylamide copolymer, low charge
	Floc-J	Cationic acrylic copolymer, synthetic polymeric demulsifer
	Floc-K	Cationic acrylic copolymer, synthetic polymeric flocculant
Coagulants	Coag-A	Cationic, medium molecular weight polyDADMAC, high charge
	Coag-B	Cationic, high molecular weight polyDADMAC, high charge

Coagulation and flocculation experiments were conducted using ammonia-treated AMD sludge (Viadero, et al., 2006). The polymeric flocculants and coagulants used are presented in Table 9. Settling results for anionic polymers applied at different concentrations are shown in Figure 2. With the exception of Floc-F, none of the anionic polymers resulted in a reduced sludge volume. In fact, an increase in P_{ss} (Percent settled sludge) was recorded for the other anionic polymers at high polymer concentrations. Floc-F reduced sludge volume at high applied concentrations ($\geq$ 60ppm). However, the volume reduction was less than 3% when compared to controls – even at the highest applied concentration of 150 ppm. Therefore, it was concluded that none of anionic polymers effectively reduced sludge volume. However, based on visual observations made during testing, the addition of most anionic polymers resulted in large flocs during slow mixing and the sludge settled much more rapidly than sludge in which floc formation was not visible. Consequently, the anionic polymers did not reduce the sludge volume, though the sludge settling rate had been enhanced. Results of settling tests performed using four cationic polymers are presented in Figure 3. Floc-I, a low charge cationic polyacrylamide copolymer, resulted in sludge volume reduction of 9% at concentrations $\geq$ 60 ppm, when compared with controls. None of the other cationic polymers produced meaningful sludge volume reduction. Consequently, coagulation and flocculation were not effective in reducing AMD sludge volume.

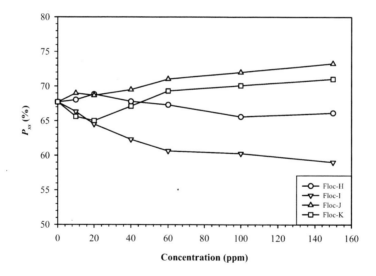

Figure 3. Percent settled sludge at different concentrations of cationic polymers (Viadero et al., 2006)

Table 10. Vacuum Filter Variables (Vesilind, 1974)

Machine Variable	Operational Variable
Vacuum pressure Drum submergence Drum speed Agitation Filter medium	Sludge type Condition of sludge Chemical conditioners Sludge characteristics

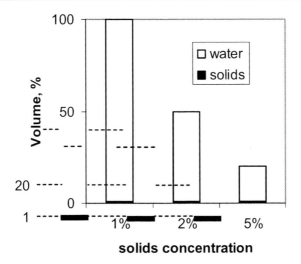

Figure 4. Volume Reduction by Solids Concentration (Vesiliand, 1974)

4.3. Sludge Thickening

Sludge thickening refers to the processes of removing water from sludge after initial separation from the treated water. The objective of thickening or concentrating sludge is to reduce the volume of liquid sludge to be handled, so that the efficiency of further treatment such as dewatering process can be increased greatly. Since typical AMD sludge is low in solids content, the volume reduction of liquid sludge is considerable even with modest increase of its solids concentration. Figure 4 illustrates the volume reduction of AMD sludge with solids concentration from 1% to 2% and 5%. Volume reduction of liquid sludge makes sludge thickening economically attractive.

The simplest and most economical method of thickening is gravity thickening, which can be accomplished by natural or manmade lagoons, settling tanks and mechanical thickeners. Sludge thickening in lagoons is used in the treatment of coal mine drainage sludge (Coal Research Bureau, 1971). In most cases, sludge is perpetually stored in large lagoons where compaction occurs. Sludge also can be removed by batches at set time intervals, so that lagoons can be reused.

Sedimentation of coal mine drainage sludge was described as three phases: free settling and floc growth, a transition zone, and compression. After sludge is prepared, free settling starts along with floc formation and growth. Then, clear liquid layer in the upper part appears and a layer of sludge settles down on the bottom. Between them is the transition layer. When all the solids are settled, transition layer disappear and then compression begins to occur. In compression, part of the liquid that was entrapped by the solids is forced out the weight of the deposit breaks down the structure of the flocs. Equilibrium finally occurs when the weight of solids equals to the strength of the sludge flocs, and settling stabilizes with a fixed sludge height.

Gravity thickening in conventional mechanical thickeners is accomplished in a tank equipped with a slowly rotating rake mechanism. The sludge slurry is usually pumped from a centrally located discharge pipe into the center of the thickener where slurry moves radially allowing the solids to settle to the bottom of the tank and clear water to overflow around the perimeter. The rake arms rotate slowly and move the settled sludge to the center of the tank, where the sludge flows through an opening to the inlet of a sludge pump. Movement of rakes aid the compaction of the settled particles by breaking up the bridge between sludge flocs, thereby producing a more concentrated sludge underflow.

Farrow *et al.* (2000) studied consolidation and densification of kaolin slurry during gravity thickening. They found that rake action, not compression, was the dominant dewatering mechanism. Microscopic analysis showed that the dewatering induced by the rake action occurred not only by removal of inter-aggregate liquor but also by densification of the aggregates (i.e. removal of intra-aggregate liquor).

4.4. Sludge Dewatering

The objective of AMD Sludge dewatering is to increase the solids content so that it can facilitate the sludge handling, reduce transportation costs, and produce concentrated sludge for final disposal, such as land application. Sludge dewatering can be accomplished by many processes, which can be classified into two general processes: mechanical dewatering processes using pressure and gravity dewatering using sludge drying beds (Cheremisinoff, 1998).

4.4.1. Mechanical dewatering

Mechanical dewatering involves processes in which water is forced out of the sludge through mechanically induced pressure. These processes mainly include vacuum filtration, pressure filtration.

Vacuum filtration

Vacuum filtration uses negative pressure to pull the water to the interior of the filter media while the sludge adheres to the exterior. The vacuum filter for dewatering sludge is a drum over which is laid the filtering medium consisting of a cloth of cotton, wool, nylon, dynel, fiber glass or plastic, or a stainless steel mesh, or a double layer of stainless steel coil springs. The drum with horizontal axis is set in a tank with about one quarter of the drum submerged in conditioned sludge. Valves and piping are so arranged that, as a portion of the drum rotates slowly in the sludge, a vacuum is applied on the inner side of the filter medium, drawing out water from the sludge and holding the sludge against it. The application of the vacuum is continued as the drum rotates out of the sludge and into the atmosphere. This pulls water away from the sludge, leaving a moist mat or cake on the outer surface. This mat is scraped, blown or lifted away from the drum just before it enters the sludge tank again. The chemicals used to condition the sludge and, thus, make it easier to filter are combined with the feed sludge. The important variables that affect the filtration performances of vacuum filters are listed in Table 10 (Vesilind, 1974). Vacuum filters have been used in wastewater treatment for decades to dewater the sludge. In digested sludge dewatering, vacuum filters are typically able to form cakes of 20-40% solids. The filtrate quality can vary anywhere from 10-20,000 mg/l of solids, corresponding to solids recoveries of 99-50%. One of the first attempts to dewater lime treated coal mine drainage sludge by vacuum filtration was conducted using a unit that had a 3.2 square meter filter surface area. Optimum conditions were obtained with a special nylon filter cloth and a drum speed that corresponded to a 20 second drying time with a vacuum of 0.6 atm at immersion. The vacuum filter produced a filter cake that averaged approximately 23 percent solids as compared to the original 0.6 percent solids.

"Operation Yellowboy" studies included filtration tests on thickener underflow sludge of coal mine drainage treated with hydrated lime (Coal Research Bureau, 1971). Filter cakes were formed with solids contents as high as 29.8 percent. Glover (1967) studied the vacuum filtration of AMD sludge from biochemical oxidation and limestone neutralization treatment and found this AMD sludge has very good vacuum filtration possibility. The sludge was dewatered on a one square foot model rotary vacuum drum filter. Filter cake had a thickness of 3/32 inch with a 45% solids content and exhibited good discharge and handling characteristics.

Pressure filtration

Pressure filtration differs from vacuum filtration in that the liquid is forced out through filter media by a positive pressure instead of a vacuum. Currently, there are various commercial pressure filters available. Among them, filter press is the most widely used in the chemical process industry. It is also widely used in wastewater treatment in Europe (Cheremisinoff, 1998). Filter press consists of a series of horizontally arranged vertical filter elements, each dressed with a synthetic woven material, paper, or felt medium. The filter press also consists of a skeleton to support the filter plates and a closure mechanism to provide the necessary force on the sealing faces of the plates to counteract the applied squeezing force. Filter press operates by pumping the sludge between the plates that are covered with filtration medium. The filter medium retains suspended solid particulates while the liquid seeps through. As a filter cake builds in the chamber, liquid flows through the particulate bed, through the filter medium, across the drainage surface of the filter plate, and is collected from the filter through drainage outlets. When the filtration cycle is completed, the closing force is released and the filter plates are separated, permitting the filter cake to be discharged from the chambers. Filter pressing is a cyclic operation and different designs enable automatic cake removal. Other filter presses include chamber filter press and belt filter press.

The dewatering of sewage sludges using filter presses indicated that the solids content of the cake can be in excess of 30%. Day (2002) conducted sludge dewatering with belt filter press and found that the solids contents of sludge cakes were above 20% for various sludge types. According to Coal Research Bureau (1971), AMD sludge dewatering by pressure filtration showed that the solids content of the filter cake ranged from 20-30% with a feed sludge of 1.2 % solids.

Bench-scale belt filter press was used to dewater AMD sludge generated from ammonia neutralization. The initial total solids content of the sludge was ~0.4% solids. Ultimately, a sludge cake of ~10% solids was achieved, which was a conveyable, stackable material. Since the total solids content was very low (~0.4%) in the feed sludge, the performance of the belt filter press was expected to improve if the total solids content of the feed could be raised to ~1%.

4.4.2. Gravity dewatering

Gravity dewatering refers to the dewatering processes that utilize gravity and/or natural evaporation or induced drainage. Typical gravity dewatering processes are drying beds and lagoons.

Drying beds

Drying beds are generally used for dewatering of well digested sludge. They consist simply of shallow ponds with sand bottoms and tile drains. The granular filter materials that can be used as beds include silica sand, crushed anthracite, crushed magnetite and garnet sands (Coal Research Bureau, 1971). Sludge is pumped to these beds at a depth of 6-12 inches. The time required for the sludge to dewater to a liftable consistency ranges from several weeks to several months (Vesilind, 1974). Partitions around and between the drying beds are generally open to the weather, but may be covered with ventilated green-house type enclosures where it is necessary to dewater sludge in wet climates. The removal of water from sludge in drainage beds is in two steps (Vesilind, 1974). Initially, the water is drained through the sludge into the sand and out tile drains. This process might last a few days, until the sand is clogged with fine particles and/or all the free water has drained away. Further dewatering occurs by evaporation. Nishimura et al. (1994) found that the cake layer formed above media has poor permeability, and dewatering rate was limited. The destruction of this cake layer by aeration and/or mechanical stirring to maintain high dewatering rate was the principle of high speed sludge drying bed.

According to Cheremisinoff (1998), dry beds can produce a dryer sludge than mechanical dewatering and up to 40% solids under normal operation and over 60% solids with additional drying time or with the use of underdrainage systems. However, this method requires a larger land area and is more labor intensive than mechanical dewatering.

Sludge lagoons

Sludge lagoons are nothing but excavated areas in which sludge is allowed to drain and dry over a period of months or even a year or more. They are usually dug out by bulldozers, or other dirt-moving equipment. The lagoons can serve three purposes in the treatment of AMD sludge: as a clarification basin, as a sludge dewatering area, and finally as a sludge storage area. However, when a single lagoon performs all three functions the sludge dewatering is served the least efficiently. A series system of lagoons can achieve better sludge treatment result (Coal Research Bureau, 1971). The first lagoon catches most of the sludge as it precipitates and settles. The second lagoon "polishes" the treated water allowing for complete sedimentation. The solids content of the coal mine sludge from lagoons ranges from 12-20% (Coal Research Bureau, 1971).

5. AMD Sludge Disposal

The precipitated colloidal sludge created through active or passive treatment of AMD must be collected and removed regularly from settling ponds, and must be disposed of properly to avoid any further negative environmental consequences, or after dewatering it may be exploited as useful resources for other applications.

AMD sludges are typically disposed of by the following methods: (1) deep mine disposal, (2) permanent retention in pond, (3) haulage to and disposal at coal refuse area, and (4) onsite burial (Ackman, 1982). The most commonly used method is deep mine disposal with sludge either pumped or trucked to boreholes drilled into underground abandoned deep mines or inactive portions of producing mines. For the retention-in-pond disposal, the ponds serve as permanent sludge impoundment sites while serving as dewatering and storage areas. The pond construction varies from simple excavation to earthen dams placed across valleys or hollows. Disposal of AMD sludge in the coal refuse area includes the transport of sludge to the sites where sludge was disposed of with coal refuse. Onsite burial is the least used method in sludge disposal.

6. BENEFICIAL USE OF AMD SLUDGE

AMD sludge from treatment facilities are currently considered a waste material which must be disposed of. However, AMD sludges are rich in Fe, which gives rise to the ochreous colors, and these sediments may represent a vast untapped resource if they can be recovered economically in a usable form. Hedin (2003) and Kirby *et al.* (1999) studied on the use of AMD sludge as a raw material in the manufacture of pigments.

AMD sludge generally has a high specific surface area and numerous functional groups which are chemically active in an aqueous environment, which indicates that the AMD sludge containing a mixture of iron and aluminum hydroxide precipitates would be a suitable medium for the adsorption of dissolved contaminants in the water or wastewater.

AMD sludge was tested to remove synthetic dye from wastewater by adsorption and precoat filtration (Wei and Viadero, 2007). Adsorption of Congo Red onto AMD sludge followed the *Langmuir* isotherm model with a maximum adsorption capacity of 389.1 mg/g. Based on thermodynamic studies, adsorption was found to be exothermic, and an increase in temperature led to a decline in dye removal. Dye removal decreased with an increasing pH. Through metal leaching tests, it was observed that most metals associated with AMD sludge remained insoluble when adsorption occurred at pH 6 to 10. A rather rapid process was observed for the adsorption of Congo Red onto AMD sludge with more than 80% adsorption taking place within 5 minutes. Adsorption kinetics followed pseudo-second-order model. Precoat filtration continuously removed Congo Red from aqueous solution with success, and dye removal mechanism via precoat filtration was adsorption. The adsorption and precoat filtration studies showed dye removal could be a beneficial use of AMD sludge prior to its final disposal.

AMD sludge was also used as an adsorbent to develop a cost-effective treatment approach to phosphorus removal from municipal secondary effluents (Wei et al., 2008). Batch tests were carried out to study the effects of pH, temperature, concentration, and contact time for phosphorus removal from wastewater. Batch tests were followed by continuous flow tests using a continuous stirred tank reactor (CSTR). Adsorption of orthophosphate onto AMD sludge particles followed the *Freundlich* isotherm model with an adsorption capacity ranging from 9.89 mg/g to 31.97 mg/g when the final effluent concentration increased from 0.21 mg-P/L to 13.61 mg-P/L. P adsorption was found to be a rather rapid process and neutral or acidic pH enhanced phosphorus removal. Based on a thermodynamic assessment, P adsorption by

AMD sludge was found to be endothermic; consequently, an increase in temperature could also favor phosphorus adsorption. Results from batch tests showed that leaching of metals common to AMD sludges was not likely to be a major issue of concern over the typical pH range (6-8) of secondary wastewater effluents. CSTR tests with three types of water (synthetic wastewater, river water, and municipal secondary effluent) illustrated that P adsorption by AMD sludge was relatively independent of the presence of other ionic species. In treating municipal secondary effluent, a phosphorus removal efficiency in excess of 98% was obtained. Results of this chapter indicated that it was very promising to utilize AMD sludge for phosphorus removal from secondary effluents and may be relevant to future efforts focused on the control of eutrophication in surface waters.

7. Conclusion

AMD sludge generated from treatment facilities typically contains a low concentration of solids, resulting in associated handling and disposal difficulties. The physical properties of AMD sludge may include dewaterability, percent solids, sludge density, particle size, particle surface properties (e.g. net surface charge), and sludge flow properties such as viscosity. The chemical composition of sludge generally consists of hydrated ferrous or ferric oxides, gypsum, hydrated aluminum oxide, calcium carbonate and bicarbonate, with trace amounts of silica, phosphate, manganese, copper, and zinc. AMD sludge properties are mainly influenced by influent characteristics, the neutralization reagents, and AMD treatment systems. In general, sludge dewatering is needed to reduce the sludge volume prior to any final disposal. In order to facilitate or improve the dewatering performances, sludge conditioning and sludge thickening are required before dewatering processes. Dewatering can be achieved either by gravity (drying beds and sludge lagoons) or mechanical dewatering (vacuum filtration or pressure filtration). AMD sludges are typically disposed of by deep mine disposal, permanent retention in pond, haulage to and disposal at coal refuse area, and onsite burial. AMD sludge might be potentially used to make pigments, or used as adsorbents to remove contaminants from wastewater.

References

Ackman, T. *Sludge Disposal from Acid Mine Drainage Treatment* (Report of Investigation 8672); US Bureau of Mines: Pittsburgh, PA, 1982.

Brown, H; Skousen, J; Renton, J. *Green Lands*, 1993, 23, 44-51.

Chang, MC; Lin, HL; Huang, CL; Wang, YY; Wan, CC. *Colloid. Surface. A*, 1998, 139, 75-80.

Cheremisinoff, NP. *Liquid Filtration*, Butterworth-Heinemann: Boston, MA, 1998, 211-228.

Christensen, JR; Christensen, GL; Hansen, JA. *J. Environ. Eng-ASCE*, 1995, 121, 236-244.

Coal Research Bureau. *Dewatering of Mine Drainage Sludge*, Project 14010 FJX; U.S. Environmental Protection Agency: Washington, DC, 1971.

Day, P. *Filtr. Separat*, 2002, 39, 18-20.

Demchik, M; Garbutt, K. *J. Environ. Qual*, 1999, 28(1), 243-248.

Dempsey, B; Jeon, BH. *Geochemistry: Exploration, Environment, Analysis*, 2001, 1, 89-94.
DeNicola, DM; Stapleton, MG. *Environ. Pollut*, 2002, 119(3), 303-315.
Farrow, JB; Johnston, RRM; Simic, K; Swift, JD. *Chem. Eng. J*, 2000, 80, 141-148.
Gazea, B; Adam, K; Kontopoulos, A. *Minerals Eng.*, 1996, 9(1), 23-42.
Glover, HG. *Proceedings of the Twenty-Second Industrial Waste Conference*; Purdue University: West Lafayette, IN, 1967, 2, 823-847.
Hammarstrom, JM; Sibrell, PL; Belkin, HE. *Appl. Geochem*, 2003, 18(11), 1705-1721.
Hedin, RS. *Land Contam. Reclam*, 2003, 11(2), 93-97.
Herman, ST; Pfeiffer, JB; Sewald, RT; Sterner, CJ. United States Patent, patent number 4, 465, 597, *Bethlehem Steel Corp*, Bethlehem, PA, 1984.
Hesbach, PA; Lamey, SC. *SME Trans*, 2001, 310, 111-117.
Hsieh, Y. *Water Sci. Technol*, 1993, 28, 23-30.
Johnson, DB; Hallberg, KB. *Sci. Total Environ.*, 2005, 338(1-2), 3-14.
Keefer, GB. *Utilization and Recycle of Acid Mine Drainage Neutralization Sludge*, Ph.D. Dissertation, West Virginia University, Morgantown, WV, 1979, 4-23.
Kim, YH. in Solid/Liquid Separation: *Waste Management and Productivity Enhancement*, Importance of Flocculants Preparation for Use in Solid/Liquid Separation; Battelle Memorial Institute: Columbus, OH, 1990, 289-304.
Kirby, CS; Decker, SM; Macander, NK. *Environ. Geol*, 1999. 37(3), 243-254.
Kuyucak, N. *Int. J. Environ. Pollut*, 1998, 10(2), 315-325.
Lau. SSS; Wong, JWC. *Water Air Soil Poll*, 2001, 128, 243-254.
Lotito, V; Spinosa, L. *Filtr. Separat*, 1990, 27, 122-125.
Mining Engineering, *Min. Eng.*, 2000, 52(9).
Nishimura, O; Gotoh, K; Sato, A. *Proceedings of the Japan Society of Civil Engineers*; 1994, 497(2-2), 119-126.
Payette, C; Lam, W; Angle, C; Mikula, R. in *Proceedings of the Second International Conference on the Abatement of Acidic Drainage*; Evaluation of Improved Lime Neutralization Processes; Montreal, Canada, 1991.
Pudlo, GH. *Sludge Volum from Treatment of Acid Mine Drainage*; M.S. Thesis in Mining Engineering, West Virginia Univeristy, 1970.
Sanmarful, IC. *Evaluation of Common Alkalies in Neutralizing Acid Mine Drainage*; M.S. Thesis in Civil Engineering, The Pennsylvania State University, 1969.
Skousen, J; Rose, A; Geidel, G; Foreman, J; Evans, R; Hellier, W. *Handbook of Technologies for Avoidance and Remediation of Acid Mine Drainage*; The National Mine Land Reclamation Center: Morgantown, WV, 1998.
Soucek, DJ; Cherry, DS; Currie, RJ; Latimer, HA; Trent, GC. *Environ. Toxicol. Chem.*, 2000, 19(4), 1036-1043.
Stanley, WE. *Sludge Dewatering*; WPCF Manual of Practice NO. 20, Water Pollution Control Federation: Alexandria, VA, 1969.
Svarovsky, L. *Solid-Liquid Separation*, CRC Press: Baco Raton, FL, 2000, 104-128.
Tabak, HH; Scharp, R; Burckle, J; Kawahara1, FK; Govind, R. *Biodegradation*, 2003, 14, 423-436.
Tracy, LD. *Trans. AIME*, 1922, 66, 609-623.
Tsang, KR; Vesilind, PA. *Water Sci. Technol.*, 1990, 22, 135-142.
Vesilind, PA. *Treatment and Disposal of Wastewater Sludges*; Ann Arbor Science Publishers: Ann Arbor, MI, 1974.

Viadero, Jr., RC; Wei, X; Buzby KM. *Environ. Eng. Sci.*, 2006, 23(4), 734-743.
Waite, TD; Wang, T; Kurucz, CN; Cooper, WJ. *J. Environ. Eng-ASCE*, 1997, 123, 335-344.
Wei, X; Viadero, Jr., RC. *J. Environ. Eng-ASCE.* 2007, 133(6), 633-640.
Wei, X; Viadero, Jr., RC; Bhojappa, S. *Water Res.*, 2008, 42(13), 3275-3284.
Zick, RL; Leon, MH; Finn, DC. *Min. Eng.*, 1999, 51(11), 46-50.
Ziemkiewicz, PF; Skousen, JG. *Acid Mine Drainage Control and Treatment* (2nd Edition); National Mine Land Reclamation Center: Morgantown, WV, 1996.
Zinck, JM; Aube, BC. *CIM Bull*, 2000, 93(1043), 98-105.

NEUTRALISATION OF ACID MINE DRAINAGE WITH FLY ASH IN SOUTH AFRICA

Vernon Somerset[a,], Michael Klink[d],*
Leslie Petrik[c] and Emmanuel Iwuoha[b]

[a]NRE, Council for Scientific and Industrial Research (CSIR),
Stellenbosch, 7599, South Africa
[b]SensorLab, Department of Chemistry, University of the Western Cape,
Bellville 7535, South Africa
[c]Environmental and Nano Sciences Group, Department of Chemistry,
University of the Western Cape, Bellville 7535, South Africa
[d]School of Mathematical and Physical Sciences, Faculty of Agriculture, Science and Technology, North-West University, Mafikeng Campus, Mmabatho, South Africa

ABSTRACT

This paper reports the results obtained when FA is neutralised with AMD, using different FA:AMD ratios. XRF analysis of South African FA has shown that it consists of three main phases of SiO_2, Al_2O_3 and Fe_2O_3. These three phases form more than 70% of the composition of the FA source, while the other major components are CaO and MgO representing the relative soluble bases present in the FA and related materials. The neutralisation process was investigated by adding different volumes of AMD to a constant mass of FA, using ratios of 1:1; 1:2; 1:3; 1:4; 1:5; and 1:10. The results obtained have shown that the FA in FA:AMD ratios of 1:10; 1:5; 1:4 and 1:3 were able to raise the pH of the AMD from pH 2.4 to pH 6.5. The use of FA:AMD ratios of 1:1 and 1:2 have shown that the FA was able to raise the pH to 11.7 and 11.2 respectively. For ratios 1:10, 1:5 and 1:1 a good plateau was observed between 60 and 210 minutes, indicating that a limited alkalinity was available after this period to influence the pH. Analysis and determination of the sulphate, iron and aluminium concentrations for the co-disposal reactions have shown clear increasing and decreasing trends for the different FA:AMD ratios investigated. Investigation of the B and As concentrations have shown no clear

---
[*] Corresponding author: Email: vsomerset@csir.co.za.

increasing or decreasing trends, while for the Co and Cu concentrations evaluated, it was observed that a clear increasing trend was observed as the FA:AMD ratio increased and the pH of the co-disposal reaction dropped from pH 12 to 5.

Keywords: Fly Ash; AMD; Co-disposal reaction; Sulphate; Iron; Aluminium;

1. INTRODUCTION

Acid mine drainage (AMD) is one of the most serious environmental problems facing the coal- and metal-mining industry. Weathering causes sulphide minerals, pyritic material and coal seams to be exposed to groundwater and the atmosphere, resulting in mine water discharges from the mine workings. Low pH-values and high levels of sulphates and metals characterize this discharge (Sheetz and Earle, 1998; Burke and Banwart, 2002; Christensen *et al.*, 1996; Somerset, 2003; Cheng *et al.*, 2007; Neculita and Zagury, 2008).

Several studies have indicated that AMD is a concentrated wastewater, which contains considerable amounts of heavy metals (Burgess and Stuetz, 2002; Suteerapataranon *et al.*, 2006; Gitari *et al.*, 2008). These metals can be released as a result of direct solubilisation of metal sulphides and also by the acidic extraction of metals adsorbed on mineral surfaces. The environmental impact of AMD produced by a single source can be felt for up to a hundred years and even a considerably longer time, if no corrective measures are taken (Somerset, 2003).

AMD forms where pyritic (FeS_2) material is in contact with both water and atmospheric influences. When iron pyrite (FeS_2) reacts with oxygen, water and in the presence of the oxidising bacteria *Thiobacillus ferrooxidans*, AMD is formed according to the following set of reactions (Sheetz and Earle, 1998; Burke and Banwart, 2002; Christensen *et al.*, 1996; Gray, 1997; Somerset, 2003):

$$FeS_2 \text{ (s)} + {}^7/_2 O_2 + H_2O \rightarrow Fe^{2+} + 2\ SO_4^{2-} + 2\ H^+ \quad (1)$$

$$Fe^{2+} + \tfrac{1}{4} O_2 + H^+ \rightarrow Fe^{3+} + \tfrac{1}{2} H_2O \quad (2)$$

$$Fe^{3+} + 3\ H_2O \rightarrow Fe(OH)_3 \text{ (s)} + 3\ H^- \quad (3)$$

$$FeS_2 \text{ (s)} + 14\ Fe^{3+} + 8\ H_2O \rightarrow 15\ Fe^{2+} + 2\ SO_4^{2-} + 16\ H^+ \quad (4)$$

In equation 1, iron pyrite combines with oxygen and water, to produce sulphuric acid, hydrogen and ferrous ions. These ferrous ions then react with oxygen, hydrogen and the *Thiobacillus ferrooxidans* bacteria. The ferrous ions are then oxidized to produce ferric ions and water, as shown in equation 2. In equations 3 and 4 the ferric ions are reduced by the iron pyrite and more ferrous ions are released, along with additional acid, and AMD forms (Sheetz and Earle, 1998; Rios *et al.*, 2008).

According to Gray (1997) the above reactions indicate that pyritic material can remain in a stable state, in undisturbed coal strata, as long as conditions are anaerobic. During AMD formation, the rate-limiting step has been shown by Singer and Strumm (1970) to be the oxidation of the ferrous ion in equation 2. There is a propagation cycle between reactions 2 and 4. The ferric (Fe^{3+}) ion of reaction 2 acts as an oxidant of the pyrite in reaction 4, and the ferrous ion (Fe^{2+}) produced by this reaction, can be used as reductant in reaction 2. Since this process is limited by the oxidation of pyrite, the surface area available for oxidation determines the rate of the reaction (Singer and Strumm, 1970; Gray, 1997).

The major minerals associated with AMD are listed below with its chemical formula indicated in brackets. These minerals are arsenopyrite ($FeS_2 \cdot FeAs$), bornite ($CuFeS_4$), chalcocite (Cu_2S), chalcopyrite ($CuFeS_2$), covelite (CuS), galena (PbS), millerite (NiS), mobybdenite (MoS_2), pyrite (FeS_2), pyrhhdite ($Fe_{11}S_{12}$), sphalerite (ZnS) (Gray, 1997). This list gives a clear breakdown of all the minerals associated with AMD and the possible metal ions that can be present in AMD. It is therefore necessary to look at the effect and influence of what AMD discharges can have on the surrounding environment (Burke and Banwart, 2002; Somerset, 2003).

Since the 1980s there has been increasing public awareness of the potential environmental hazards arising from mining activities, in particular the pollution associated with AMD. Acid mine drainage is a multi-factor pollutant. Major impact areas are rivers, lakes, estuaries, and coastal waters. However, AMD affects different aquatic ecosystems in different ways Acidic mine drainage can affect a river system in numerous and interactive ways. This causes multiple pressures, both direct and indirect, on all the organisms comprising the community structure of the ecosystem. The four main categories, in which an ecosystem is affected, are chemical, physical, biological and ecological (Gray, 1997).

The chemical effects of AMD on a river system include: increased acidity of the river system, reduction in pH, destruction of the river's bicarbonate buffering system, an increase in the soluble metal concentrations, and an increase in particulate metals in the river system. Some of the physical effects associated with AMD include: substrate modification, an increase in river stream velocity, changes in the turbidity, sedimentation of the river system, adsorption of metals onto river sediment, reduction in turbulence due to sedimentation increasing laminar flow, and a decrease in light penetration into the river water (Gray, 1997).

Some of the biological effects associated with AMD include: the behavioural patterns of species, the respiratory mechanism of the river system, reproduction of species, osmoregulation of the system, acute and chronic toxicity of all species, ultimate death of sensitive species in the ecosystem, an acid-base balance failure in organisms, and migration or avoidance of species (Gray, 1997).

The total ecological effects of AMD on the environment include: a total habitat modification for some species, niche loss and total degradation of surroundings, bioaccumulation of toxins within the food chain, loss of food source or prey in the food chain, elimination of sensitive species in the flora and fauna, a reduction in primary productivity for certain species, and a total food chain modification (Gray, 1997). From these effects it is clear that AMD needs to be treated with effective methods in order to reduce the severe impact it can have on the environment (Somerset, 2003).

Since it is known that AMD is characterised by a low pH, high sulphate content and various dissolved metal ions such as Fe, Al, Mn, Zn, Cu, and Ni, this study was designed to investigate the potential use of fly ash (FA) as a liming agent. Several studies have shown that South African FA contains relatively high concentrations of SiO_2, Al_2O_3, and CaO, with CaO considered as a liming agent to neutralize AMD (Foner *et al.*, 1999; Somerset *et al.*, 2004; Somerset *et al.*, 2005a; Somerset *et al.*, 2008 ;Gitari *et al.*, 2008). Conventional techniques for AMD treatment involves the addition of neutralising agents such as hydrated lime, anhydrous ammonia, limestone, sodium hydroxide and sodium carbonate to raise the pH and then precipitate the metal ions from the AMD solution. In this study a co-disposal reaction was designed that involved the addition of AMD to FA in order to neutralise the acidic drainage and to raise the pH of the resulting solution to a circumneutral value. Different AMD:FA ratios were investigated to determine a pH and EC profile for the co-disposal reaction and to investigate what happens to the metal ion concentrations during the co-disposal process.

2. MATERIALS AND METHODS

2.1. Sample Collection

All FA samples used in this study were obtained directly from a coal-fired power station in the Mpumalanga province, South Africa. The FA samples were kept in tightly locked poly(vinyl chloride) (PVC) buckets to prevent any ingress of CO_2 into the buckets, since it leads to the loss of alkalinity. All AMD samples were collected from holding dams and 5 litre high density polyethylene (HDPE) containers were used during sampling. The HDPE sample containers were rinsed thoroughly with the AMD before sampling took place, while the containers were kept under water to fill during sampling. AMD samples were kept on ice at 4 °C in cooler boxes while transported to the laboratory, followed by refrigeration in the laboratory (Klink, 2003; Somerset, 2003; Gitari et al., 2008).

2.2. Batch Experiment Description

Co-disposal reactions were performed wherein FA was reacted with AMD in a specific FA:AMD ratio (e.g. 1:2; 1:4; 1:5). The AMD was first stirred with an overhead stirrer before the FA was added to it. The readings for pH and electrical conductivity (EC) of the co-disposal reaction mixture were taken at regular time intervals (e.g. 10, 30, 60, 90, 150, 180, 210 min.) using a Hanna HI 991301 portable pH/EC/TDS/Temperature probe. The overhead stirrer was stopped briefly at the designated time intervals to record the pH and EC readings (Somerset, 2003; Klink, 2003; Somerset et al., 2005b; Gitari et al., 2008).

2.3. Chemical Characterisation of AMD

Prior to analysis the AMD samples were filtered through a 0.45 μm nucleopore membrane and the filtrate was prepared for major and trace element analysis, and sulphate ion concentration. Cation analysis of the AMD samples was determined by inductively coupled mass spectrometry (ICP-MS) using a Perkin-Elmer Elan ICP-MS unit, equipped with ChemStation Software and an autosampler unit. Analysis of the Fe^{2+}/Fe^{3+} ions was done with a colorimetric method using 2,2-bypyridal as the complexing agent, while the SO_4^{2-} content of the AMD samples was done by ion chromatography, using a Dionex DX-120 system (Petrik et al., 2003; Somerset et al., 2004; Yang and Swami, 2007; Gitari et al., 2008).

2.4. Elemental Characterisation of FA Using X-Ray Fluorescence (XRF) Spectrometry

The chemical composition of the raw FA and co-disposal solids were analysed using XRF spectrometry. A Phillips 1404 XRF Wavelength Dispersive Spectrometer equipped with an array of six analysing crystals and fitted with a rhodium X-ray tube target was used. A vacuum was used as the medium of analyses to avoid interaction of X-rays with air particles (Somerset et al., 2004; Somerset et al., 2008).

2.5. X-Ray Diffraction (XRD) Analysis of FA Material

The mineralogy of the FA materials was evaluated with a Phillips Analytical XRD spectrometer. This instrument was equipped with a graphite monochromator and Cu-Kα radiation samples were scanned for 2θ ranging from 7 to 70. The data files presented by X'Pert Graphics & Identify data collection software were used to identify the minerals present in the samples (Somerset et al., 2004; Somerset et al., 2008; Gitari et al., 2008).

3. RESULTS AND DISCUSSION

3.1. Characterisation of AMD and FA Samples

AMD is generally characterised by high acidity (pH 2–4), also containing high sulphate concentrations and high concentrations of heavy metals such as Fe, Mn, Al, Cu, Ca, Pb, Mg, Na and Ni. In order to establish these characteristics for the Navigation AMD used in this study, the AMD was analysed and the results are shown in Table 1.

From the results in Table 1, it can be seen that laboratory results obtained for the AMD sample have shown that the sample is strongly acidic with an electrical conductivity of 9.45 mS/cm. The high sulphate content of the AMD can be attributed to the sulphide rich coal mine tailings of the mining area from which the AMD originates. Several trace metal ion concentrations in the AMD was analysed, including the calcium and iron concentrations. The results have shown that the boron (B) concentration is higher than the total arsenic (As), cadmium (Cd), chromium (Cr), lead (Pb) and mercury (Hg) concentrations with the concentration ranging from 0.001 to 0.125 mg/L for this group of metal ions. Relatively higher concentrations of copper (Cu), cobalt (Co), nickel (Ni) and zinc (Zn) were obtained with Cu having the lowest (0.320 mg/L) and Zn the highest concentration (4.911 mg/L) for these four metal ions. It was further observed that high concentrations of aluminium (Al), calcium (Ca) and total iron (Fe_{Tot}) were present in the AMD sample.

The elemental composition of the FA and co-disposal solid samples was evaluated using XRF spectrometry and the results are shown in Table 2.

Table 1. Chemical characterisation of the Navigation AMD source, using ICP-MS, Hg-vapour and IC analysis (Somerset, 2003; Somerset et al., 2005b; Somerset et al., 2008)

Constituent	Conc. (mg/L)
pH	2.64
EC (mS/cm)	9.45
B	0.125
As_{Tot}	0.009
Cd	0.002
Cr	0.018
Co	0.640
Cu	0.320
Pb	0.003
Hg	0.001
Ni	0.922
Zn	4.911
Al	13.84
Ca	497.22
Fe_{Tot}	3,522.92
SO_4	18,888.62

Table 2. XRF results for the chemical composition of the fresh FA and co-disposal solid samples

Oxide (wt%)	Co-disposal solid samples						FA
	1:1	1:2	1:3	1:4	1:5	1:10	
SiO_2	53.95	54.63	52.85	53.92	53.93	52.54	55.08
TiO_2	1.29	1.19	1.23	1.22	1.24	1.26	1.30
Al_2O_3	22.39	18.91	20.79	19.34	21.37	22.76	22.63
Fe_2O_3	5.71	8.13	6.94	8.02	6.62	6.62	5.40
MnO	0.06	0.08	0.08	0.07	0.06	0.05	0.06
MgO	2.69	2.87	2.69	2.70	2.51	2.26	2.67
CaO	7.77	8.48	7.89	8.31	7.58	7.14	8.03
Na_2O	0.20	0.14	0.16	0.19	0.16	0.17	0.17
K_2O	0.52	0.46	0.49	0.48	0.50	0.51	0.52
P_2O_5	0.19	0.18	0.18	0.18	0.19	0.19	0.20
Cr_2O_3	0.03	0.03	0.03	0.03	0.03	0.03	0.03
NiO	0.01	0.01	0.01	0.01	0.01	0.01	0.01
V_2O_5	0.02	0.02	0.02	0.02	0.02	0.02	0.02
ZrO_2	0.05	0.05	0.05	0.05	0.05	0.05	0.05
LOI	4.19	3.45	4.74	4.19	4.64	4.52	2.23
Total	99.07	98.62	98.13	98.71	98.89	98.13	98.39

From the results in Table 2, it can be seen that three major phases are present in the FA and the co-disposal solids, i.e. SiO_2, Al_2O_3 and Fe_2O_3. This three phases form more than 70% of the composition of the FA source, while the other major components are CaO and MgO representing the relative soluble bases present in the FA and related materials. No clear increasing or decreasing trends were observed in the SiO_2, Al_2O_3 and Fe_2O_3 concentrations, as the ration of FA:AMD for the co-disposal reactions were increased (Somerset, 2003; Klink, 2003; Somerset et al., 2004).

XRD spectroscopy results of the FA and a representative co-disposal solid sample (FA:AMD = 1:4) were collected to report on the mineralogical characterisation of the samples, as shown in Figure 1.

The XRD spectra in Figure 1 show that two predominant mineralogical phases are present in the FA and co-disposal solid sample. These two phases are (SiO_2) and mullite ($Al_6Si_2O_{13}$). Previous research results reported by the authors of this study has shown that the mineralogy of the raw FA and co-disposal solid samples are relatively similar, also providing a rich source of SiO_2 and Al_2O_3 for hydrothermal zeolite synthesis (Somerset et al., 2004; Somerset et al., 2005b; Somerset et al., 2008).

3.2. Neutralisation of AMD with FA

One of the main investigations of this study was the neutralisation of AMD with the addition of FA as an alternative neutralisation agent. AMD is commonly neutralised with the addition of chemicals such as CaO, $Ca(OH)_2$, $CaCO_3$, NaOH and Na_2CO_3. The neutralisation process was investigated by adding different volumes of AMD to a constant mass of FA (e.g. 1:1; 1:2; 1:3; 1:4; 1:5; 1:10) (Somerset et al., 2005a; Somerset et al., 2008; Gitari et al., 2008).

In Figure 2 the results obtained for the pH readings of the co-disposal reaction mixture taken at regular time intervals (e.g. 10, 30, 60, 90, 150, 180, 210 min.) are shown.

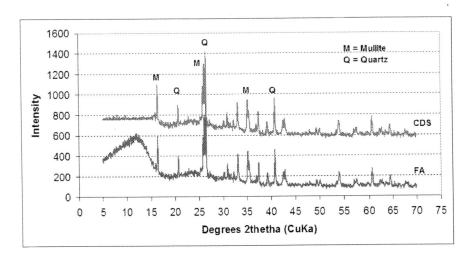

Figure 1. XRD spectroscopy results collected for raw fly ash (FA) and a co-disposal solid (CDS) sample, showing that mullite and quartz are the predominant mineralogical phases present (Somerset et al., 2005b; Somerset et al., 2008)

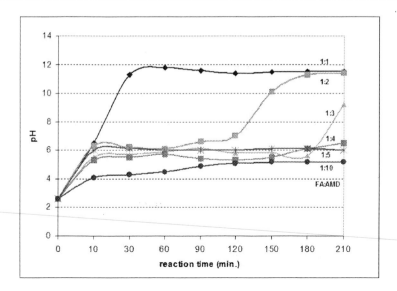

Figure 2. Results obtained for the pH measurements of the co-disposal reaction for different FA:AMD ratios investigated (Petrik *et al.*, 2003; Somerset *et al.*, 2004)

From the results in Figure 2 it can be seen that the FA in FA:AMD ratios of 1:10; 1:5; 1:4 and 1:3 were able to raise to pH of the AMD from pH 2.4 to pH 6.5. Only when the FA:AMD ratios of 1:1 and 1:2 were used, the FA was able to raise to pH to 11.7 and 11.2 respectively. For ratios 1:10, 1:5 and 1:1 a good plateau was observed between 60 and 210 minutes, indicating that pH was only altered in the first hour of the co-disposal reaction and thereafter the FA was unable to induce any changes. Only for ratio 1:2 it was observed that a significant change in the pH from 6.5 to 10.2 was possible after 120 minutes of the reaction, while for ratio 1:3 a considerable change in the pH occurred after 180 minutes when the value changed from 5.5 to 9.2 (Petrik *et al.*, 2003; Somerset *et al.*, 2004).

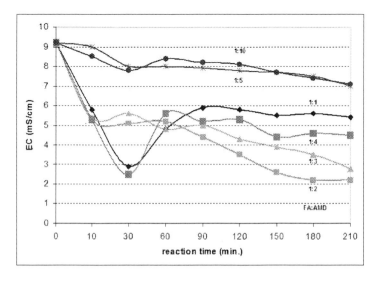

Figure 3. Results obtained for the EC measurements of the co-disposal reaction for different FA:AMD ratios investigated (Petrik *et al.*, 2003; Somerset *et al.*, 2004)

In Figure 3 the results obtained for the EC readings of the co-disposal reaction mixture taken at regular time intervals (e.g. 10, 30, 60, 90, 150, 180, 210 min.) and collected simultaneously with the pH readings, are shown.

The results in Figure 3 shows that for FA:AMD ratios of 1:10 and 1:5, it was observed that relatively small changes in the EC readings were observed as the value decreased from 9.3 to 7.2 mS/cm. For ratios 1:4 and 1:1 a considerable decrease in the EC readings from 9.3 to 2.6 and 2.9 mS/cm respectively, were observed in the first 30 minutes of the co-disposal reaction, followed by an increase in the readings. Only the ratios 1:3 and 1:2 showed a continuous decrease in EC over the entire duration of the co-disposal reaction with the EC readings decreasing from 9.3 to 2.7 and 2.3 mS/cm respectively.

The results for the co-disposal reactions in which different FA:AMD ratios were investigated, have shown that neutralisation is dictated by the FA:AMD ratio and contact time of the reaction. The final pH of the solution is reached after 120 minutes and the solutions with ratios 1:5 and 1:4 attained neutral pH values of between 6 and 7. For these ratios, ratio 1:4 also attained a relatively low EC reading, suggesting that ratio 1:4 can be employed for further experimentation. From the results in Table 2 it is seen that FA ash contains significant quantities of total alkalinity in the form of CaO, MgO, K_2O and Na_2O, thereby increasing the neutralisation potential of FA. The results in Figures 2 and 3 have thus confirmed that FA may therefore be used as a substitute for limestone or lime treatment in the neutralisation of AMD.

3.3. AMD Neutralisation and Trends in Sulphate and Iron Concentrations

During the co-disposal reaction the final sulphate ion (SO_4^{2-}) concentrations for the different ratios investigated were determined and the results are shown in Figure 4.

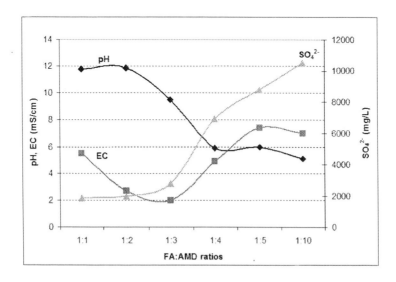

Figure 4. Results obtained for the final SO_4^{2-} concentrations of the co-disposal reactions for different FA:AMD ratios investigated, shown in comparison with the pH and EC readings for each ratio (Petrik et al., 2003; Klink, 2003)

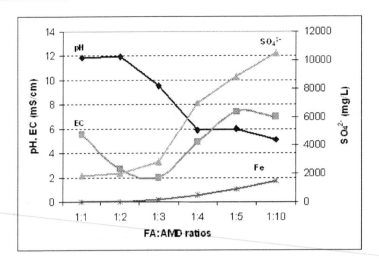

Figure 5. Results obtained for the SO_4^{2-} and Fe concentrations of the co-disposal reactions for different FA:AMD ratios investigated, shown in comparison with the pH and EC readings for each ratio (Petrik et al., 2003; Klink, 2003).

The results in Figure 4 are that obtained for the final sulphate concentrations for the different FA:AMD ratios investigated. The samples for sulphate ion concentration determination were collected after the co-disposal reaction was completed and the final pH and EC results for the different reactions are also displayed in Figure 4. The results in Figure 4 shows that FA:AMD ratio 1:10 had the lowest final pH, while ratio 1:3 had the lowest final EC reading. A clear trend was observed for the sulphate concentration with the results showing that the concentration increased as the FA:AMD ratio was increased. Ratio 1:1 had a SO_4^{2-} concentration of 1850 mg/L with the concentration for ratio 1:10 at 10500 mg/L. It was further observed that for FA:AMD ratios of 1:1, 1:2 and 1:3, characterised by pH values between 9 and 12, the corresponding sulphate concentrations were relatively low as compared to the other ratios investigated. For ratios 1:4, 1:5 and 1:10, the pH values was between 5 and 6, with the corresponding sulphate concentrations at high values for this pH range (Petrik et al., 2003; Klink, 2003).

In the treatment of mine effluent and waters influenced by mining activities, it is crucial that the sulphate concentration is measured. This is also a requirement and an influential step in environmental monitoring, regulatory inspection and compliance determination. The determination of the sulphate concentration also allows the assessment of the redox state of the water system involved, since sulphur is a crucial element in controlling the fate and solubility of polluting elements in the aquatic environment (Reisman et al., 2007).

In Figure 5 the results for the sulphate and iron concentration determined for the different FA:AMD ratios are shown.

The results in Figure 5 shows that the FA:AMD ratios of 1:1, 1:2 and 1:3 have relatively low Fe concentrations, ranging from 10 to 150 mg/L and the corresponding sulphate are also at low levels. On the other hand, the sulphate and Fe concentrations are higher for ratios 1:4, 1:5 and 1:10. During the co-disposal reaction, it was observed that an orange precipitate was formed in the solution in the form of flocculants. This precipitation formation can be explained as follows. When AMD emerges as surface waters, it usually has a high chemical oxygen demand. Under these conditions it is often found that Fe^{2+} precipitate as iron oxide and hydroxysulphate species, with the pH of the solution greater than 5. It is also observed that the composition of the surface sediments plays an important role and it is observed that intermediate iron species precipitate as oxyhydroxy, oxyhydroxysulphate or amorphous to crystalline oxyhydroxides with Schwertmannite ($Fe_8O_8(OH)_6SO_4$) and Goethite ($Fe(OH)_3$) being prominent phases (Kalin et al., 2006).

3.5. Trends in Aluminium and Sulphate Concentrations

In Figure 6 the Fe and Al concentrations for the different FA:AMD ratios investigated are shown. These individual concentrations are shown in comparison with the pH and EC readings for each ration investigated during the co-disposal reactions.

The results in Figure 6 indicate that an increasing trend in the Al concentrations from 0.2 to 12.4 mg/L were observed as the FA:AMD ratio was increased. The Al concentrations were also low for ratios 1:1, 1:2 and 1:3 that was characterised by high pH values of between 9.5 and 12. On the contrary, for ratios 1:4, 1:5 and 1:10 the Al concentration was higher with the pH values ranging between 5 and 6. It was further observed that the Al concentrations increased as the sulphate concentrations also increased. These observations and results obtained for the Al and sulphate concentrations of the co-disposal reaction can be explained as follows. In AMD the primary sources of aluminum include potassium feldspar ($KAlSi_3O_8$), muscovite ($KAl_2(AlSi_3O_{10})(F,(OH))_2$) and kaolinite ($Al_2Si_2O_5(OH)_4$). When potassium feldspar is exposed to acidity, kaolinite and silicic acid are produced as shown in equation 5. The kaolinite then undergoes further degradation in an acidic environment, resulting in mobile aluminum ions being formed as shown in equation 6.

$$KAlSi_3O8 + H^+ + {}^9\!/_2 H_2O \rightarrow 2\ H_4SiO_4 + \tfrac{1}{2}\ Al_2Si_2O_5(OH_4) \tag{5}$$

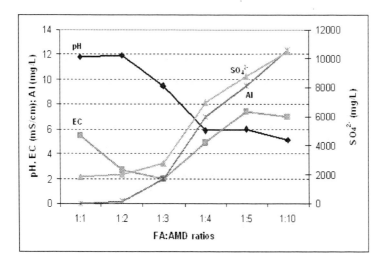

Figure 6. Results obtained for the Fe and Al concentrations of the co-disposal reactions for different FA:AMD ratios investigated, shown in comparison with the pH and EC readings for each ratio (Petrik et al., 2003; Klink, 2003).

$$Al_2Si_2O_5(OH)_4 + 6\ H^+ \rightarrow 2\ Al^{3+} + 2\ H_4SiO_4 + H_2O \tag{6}$$

This will result in the aluminum precipitating in the presence of water at a pH of approximately 5 and produces acidity that can best be represented by equation 7 (Younger et al., 2002; Watzlaf et al., 2003).

$$Al^{3+} + 3\ H_2O \rightarrow Al(OH)_3 + 3\ H^+ \tag{7}$$

The above equations are this key to explain the behaviour of the sulfate and Al concentrations as the co-disposal reaction is performed.

3.4. Trends in Some Trace Metal Ion Concentrations Determined

Figure 7 shows the B and As concentrations obtained for the different FA:AMD ratios investigated during the co-disposal reaction. All readings are shown in comparison with the pH and EC readings for the different ratios investigated.

The results in Figure 7 show that the no clear increasing or decreasing trends in the B and As concentrations were observed. In the case of the B concentrations, it was found that the concentration dropped from 7500 µg/L for ratio 1:1, to 7400µg/L for ratio 1:2 and was then back at µg/L for ratio 1:3. The B concentration then remain relatively constant for ratios 1:3, 1:4 and 1:5, while the lowest concentration was obtained for the highest FA:AMD ratio of 1:10.

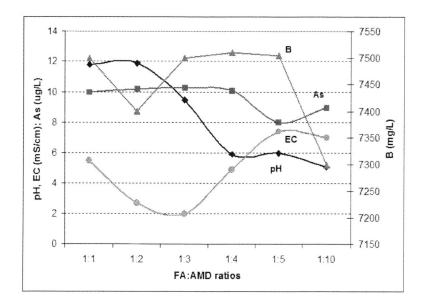

Figure 7. Results obtained for the B and As⁻ concentrations of the co-disposal reactions for different FA:AMD ratios investigated, shown in comparison with the pH and EC readings for each ratio (Petrik et al., 2003; Klink, 2003).

Metal ions have different degrees of mobility in various environmental conditions, but factors such as pH, solubility, sorption and redox conditions have the biggest influences (Smith, 2007). The degree of mobility of metal ions under oxidising and reducing conditions are as follows. Under oxidising conditions accompanied by a pH < 3, it is commonly observed that Cd, Co, Cu, Ni and Zn are very mobile species. Under the same conditions, metal ions such as Al, As, Ca, Fe, Hg, K, Mg, Mn, Na and Se are less mobile. The degree of mobility decreases even further for metal ions such as Ag, Ba, Bi, Cr, Cs, Li, Mo, Ti and W (Smith, 2007).

Under oxidising conditions with pH between 5 and 7 and no iron substrates available, it is found that metal ions of Cd and Zn are very mobile. The mobility of metal ions of Ca, Mg, Mo, Na, Se and Sr are decreased, while metal ions such as As, Ba, Bi, Co, Cr, Cs, Cu, Hg, K, Li, Mn, Ni and Sb are only slightly mobile. On the contrary, under oxidising conditions with pH between 5 and 7 and abundant iron substrates available, it is found that metal ions of Ca, Cd, Mg, Na, Sr and Zn are moderately mobile. Under the same conditions, the metal ions of Ba, Bi, Co, Cs, Hg, K, Li, Mn, Ni and Se have decreased mobility (Smith, 2007).

Under reducing conditions with the pH between 5 and 7 in the absence of hydrogen sulphide, it is found that the metal ions of Ca, Cd, Cu, Fe, Mg, Mn, Na, Ni, Pb and Zn have moderate mobility. Under the same conditions, the metal ions of As, Ba, Co, Cr, Cs, Hg, K and Li have decerased mobility. Contrary to the above, under reducing conditions with the pH between 5 and 7 in the presence of hydrogen sulphide, it is found that the metal ions of Ca, Mg, Mn, Na and Sr are moderately mobile. This mobility is even more decreased for the metal ions of Ba, Cs, K and Li (Smith, 2007).

Relating the mobility of metal ions discussed in the above paragraphs with the results shown in Figure 7, it can be concluded that B and As confirmed to these conditions and are moderately mobile, as the B concentration dropped sharply to 7300 µg/L at a pH value of 5 for FA:AMD ratio of 1:10. The As concentration was lowest for ratio 1:5 at a pH of 6 and was followed by a moderate increase in concentration.

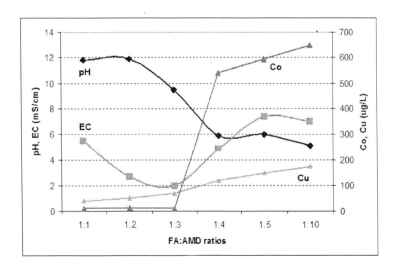

Figure 8. Results obtained for the Cu and Co concentrations of the co-disposal reactions for different FA:AMD ratios investigated, shown in comparison with the pH and EC readings for each ratio (Petrik et al., 2003; Klink, 2003).

In Figure 8 the results obtained for the Cu and Co concentrations obtained for the different FA:AMD ratios investigated during the co-disposal reaction, are shown in comparison with the pH and EC readings for each ratio investigated.

Figure 8 shows that clear trends in the Co and Cu concentrations were observed for the different FA:AMD ratios investigated in the co-disposal reactions performed. For ratios 1:1 to 1:3 the Co concentration was between 10 and 12 µg/L, when the pH was between 9 and 12. A very big increase in the Co concentration was observed when the ratio of FA:AMD was changed from 1:3 to 1:4, and the corresponding concentrations changed from 10 to 540 µg/L. This change in concentration was experienced when the pH changed from 10 to 6, indicating that the Bo concentration was affected by the pH of the co-disposal reaction solution. For ratios 1:4 to 1:10, a further increase in the Co concentration was observed to the highest value of 650 µg/L for ratio 1:10, with associated pH of 5.

In the case of the Cu concentrations, a more gradual increase in the concentration was observed as the FA:AMD ratio increased. For ratio 1:1 the Cu concentration was 40 µg/L, while for ratio 1:10 it was 175 µg/L. This result indicates that the Cu concentration increased as the pH decreased from 12 to 5, also indicating that the Cu was released into solution during the duration of the co-disposal reaction. Relating the mobility of the metal ions of Co and Cu with the trends discussed in the above paragraphs, the following conclusions can be made for the experimental results obtained. Both metal ions of Co and Cu have shown increased mobility under the experimental conditions for the co-disposal reaction, since both experienced an increase in concentration at pH values between 4 and 6.

4. CONCLUSIONS

The results obtained in this study have shown that FA can be neutralised with AMD, using different FA:AMD ratios. XRF analysis of South African FA has shown that it consists of three main phases of SiO_2, Al_2O_3 and Fe_2O_3. These three phases form more than 70% of the composition of the FA source, while the other major components are CaO and MgO representing the relative soluble bases present in the FA and related materials. The neutralisation process was investigated by adding different volumes of AMD to a constant mass of FA, using ratios of 1:1; 1:2; 1:3; 1:4; 1:5; and 1:10. FA:AMD ratios of 1:10; 1:5; 1:4 and 1:3 were able to raise to pH of the AMD from pH 2.4 to 6.5, while for ratios of 1:10, 1:5 and 1:1 a good plateau was observed between 60 and 210 minutes. Analysis of the Fe and sulphate concentrations for FA:AMD ratios of 1:1, 1:2 and 1:3 have shown relatively low Fe concentrations, ranging from 10 to 150 mg/L and the corresponding sulphate concentrations are also at low levels. For ratios 1:4, 1:5 and 1:10, the sulphate and Fe concentrations were higher. Further analysis have shown that the Al concentrations were also low for ratios 1:1, 1:2 and 1:3 that was characterised by high pH values of between 9.5 and 12. On the contrary, for ratios 1:4, 1:5 and 1:10 the Al concentration was higher with the pH values ranging between 5 and 6. No clear increasing or decreasing trends in the B and As concentrations were observed. On the contrary, for the Co and Cu concentrations evaluated, it was observed that a clear increasing trend was observed as the FA:AMD ratio increased and the pH of the co-disposal reaction dropped from pH 12 to 5.

ACKNOWLEDGMENT

The authors wish to express their gratitude to the Water Research Commission (WRC), Coaltech 2020 Consortium, and the National Research Foundation (NRF) for funding and financial support to perform this study. The assistance provided by the CSIR, Pretoria and Eskom, Witbank in the collection of FA samples, iThemba labs for XRD analysis, and the University of Cape Town (UCT) for ICP-MS analysis. A special thank you goes to the Chemistry Department of the University of the Western Cape for their assistance and guidance.

REFERENCES

Burgess, J. E. & Stuetz, R. M. (2002). Activated sludge for the treatment of sulphur-rich wastewaters. *Minerals Engineering, 15*, 839-846.

Burke, S. P. & Banwart, S. A. (2002). A geochemical model for the removal of iron(II)(aq) from mine water discharges. *Applied Geochemistry, 17*, 431-443.

Cheng, S., Dempsey, B. A. & Logan, B. E. (2007). Electricity Generation from Synthetic Acid-Mine Drainage (AMD) Water using Fuel Cell Technologies. *Environmental Science & Technology, 41*, 8149-8153.

Christensen, B., Laake, M. & Lien, T. (1996). Treatment of Acid Mine Water by Sulfate-Reducing Bacteria; Results form a Bench Scale Experiment. *Water Resources, 30(7)*, 1617-1624.

Foner, H. A., Robl, T. L., Hower, J. C. & Graham, U. M. (1999). Characterization of fly ash from Israel with reference to its possible utilization. *Fuel, 78*, 215-223.

Gray, N. F. (1997). Environmental Impact and remediation of Acid Mine drainage: A Management Problem. *Environmental Geology, 30(1/2)*, 62-71.

Gitari, W. M., Petrik, L. F., Etchebers, O., Key, D. L. & Okujeni, C. (2008). Utilization of fly ash for treatment of coal mines wastewater: Solubility controls on major inorganic contaminants. *Fuel, 87*, 2450-2462.

Klink, M. (2003). The potential use of South African coal fly ash as a neutralization treatment option for acid mine drainage. *Unpublished MSc Thesis.* University of the Western Cape, Bellville, RSA, 205.

Neculita, C. M. & Zagury, G. J. (2008). Biological treatment of highly contaminated acid mine drainage in batch reactors: Long-term treatment and reactive mixture characterization. *Journal of Hazardous Materials, 157*, 358-366.

Petrik, L. F., White, R. A., Klink, M. J., Somerset, V. S., Burgers, C. L. & Fey, M. V. (2003). Utilization of South African fly ash to treat acid coal mine drainage, and production of high quality zeolites from the residual solids, in: *Proceedings of the International Ash Utilization Symposium*, Centre for Applied Energy Research, University of Kentucky, 1-26.

Reisman, D. J., Sundaram, V., Al-Abed, S. R. & Allen, D. (2007). Statistical validation of sulfate quantification methods used for analysis of acid mine drainage. *Talanta, 71*, 303-311.

Rios, C. A., Williams, C. D. & Roberts, C. L. (2008). Removal of heavy metals from acid mine drainage (AMD) using coal fly ash, natural clinker and synthetic zeolites. *Journal of Hazardous Materials, 156*, 23-35.

Scheetz, B. E. & Earle, R. (1998). Utilization of fly ash. *Solid States & Materials Science, 3*, 510-520.

Singer, P. C. & Strumm, W. (1970). Acidic mine drainage: the rate-determining step. *Science, 167*, 1121-1123.

Smith, K. S. (2007). *Geochemical and Geological Factors that Influence Metal Speciation, Transport, and Aquatic Toxicity.* Notes from ERAC Workshop, Kuopio, Finland, February 7th. U.S. Geological Survey, Denver.

Somerset, V. S. (2003). The Preparation and Characterisation of High Capacity Ion Exchange Adsorbents, Made by the Co-disposal of Fly Ash and Acid Mine Drainage for Their Possible Use in Electrochemical Systems for Water Purification. Unpublished M.Sc. Thesis. University of the Western Cape, Bellville, RSA, 305.

Somerset, V. S., Petrik, L. F., White, R. A., Klink, M. J., Key, D. & Iwuoha, E. (2004). The use of X-ray fluorescence (XRF) analysis in predicting the alkaline hydrothermal conversion of fly ash precipitates into zeolites. *Talanta, 64*, 109-114.

Somerset, V. S., Petrik, L. F., White, R. A., Klink, M. J., Key, D. & Iwuoha, E. I. (2005a). Alkaline hydrothermal zeolites synthesized from high SiO_2 and Al_2O_3 co-disposal fly ash filtrates. *Fuel, 84*, 2324-2329.

Somerset, V., Petrik, L. & Iwuoha, E. (2005b). Alkaline hydrothermal conversion of fly ash filtrates into zeolites 2: utilization in wastewater treatment. *Journal of Environmental Science and Health, A40(8)*, 1627-1636.

Somerset, V., Petrik, L. & Iwuoha, E. (2008). Alkaline hydrothermal conversion of fly ash precipitates into zeolites 3, The removal of mercury and lead ions from wastewater. *Journal of Environmental Management, 87*, 125-131.

Suteerapataranon, S., Bouby, M., Geckeis, H., Fanghanel, T. & Grudpan, K. (2006). Interaction of trace elements in acid mine drainage solution with humic acid. *Water Research, 40*, 2044-2054.

Watzlaf, G., Schroeder, K., Kleinmann, R., Kairies, C. & Nairn, R. (2003). *The passive treatment of coal mine drainage.* 72 pp. Information Circular, National Energy Technology Laboratory, US Department of Energy.

Younger, P., Banwart, S. & Hedin, R. (2002). *Mine water: hydrology, pollution, remediation.* 442. Kluwer Academic, London.

In: Mine Drainage and Related Problems
Editor: Brock C. Robinson, pp. 227-238
ISBN: 978-1-60741-285-4
© 2010 Nova Science Publishers, Inc.

Chapter 7

HOW MANY DATA ARE ENOUGH? ESTIMATION OF AN OPTIMUM SAMPLING DENSITY IN MINING AND ENVIRONMENTAL APPLICATIONS

Konstantinos Modis[*] *and Konstantinos Komnitsas*

[1] School of Mining and Metallurgical Engineering, National Technical University of Athens, Greece

[2] Dept of Mineral Resources Eng., Technical Univ. of Crete, Greece

ABSTRACT

In mining, environmental and geochemical applications sampling is considered as an extremely important factor to accurately evaluate reserves, spatial heterogeneity of trace elements and contaminants and reliably assess risk as well as environmental impacts. Various sampling campaigns are designed and huge numbers of samples are often analyzed resulting thus in high costs and delays. It would have been therefore useful if a critical sampling density was established for the analysis of earth related space-distributed natural variables.

In the present chapter a case study that predicts the net neutralization potential (NNP) and therefore the capacity of acid generation of the rock formations in a mixed sulphide mine in northern Greece is discussed. A critical sampling grid that extracts maximum information is proposed, so that sampling above this threshold provides limited improvement in the mapping results. The study also revealed that in some parts of the irregular ore body, the initial sampling grid used was denser than required.

Keywords: Optimum sampling density, geostatistics, acid mine drainage

---

[*] Corresponding author: Email: kmodis@mail.ntua.gr.

1. Introduction

Most mining, geochemical, and environmental studies are quite complex since they involve analysis of multivariable data. The number of samples required in these studies for the prediction among others of mining reserves and the assessment of environmental impacts is an extremely important parameter (François-Bongarçon, 2004). Excessive sampling causes considerable delays and results in high costs.

Mine exploration and feasibility studies consider the formation process of ores, the relationships between geological environments and structures, as well as the geological factors affecting ore geometry and quality (Heriawan and Koike, 2008). The uncertainty involved in mining applications represents often a high risk during assessment of a deposit. Therefore, in every phase of a mining project a quantitative value should be assigned to each parameter that is used for the evaluation of a deposit; if the uncertainty associated with each value is taken into account then decision making becomes faster and more accurate (Bastante et. al., 2008).

In geochemical studies, geochemical maps are of great interest and constitute an effective tool for environmental planning, for establishing healthcare policies and for promoting sustainable development. Reliable geochemical maps are based on systematic sampling and should take into account background concentrations of chemical elements in all affected media to distinguish between "natural" and "man-made" levels of element concentration (Martinez et al., 2007).

In environmental applications, e.g in waste disposal sites, a number of parameters should be taken into account in order to accurately assess environmental impacts; some of these parameters are the type of wastes, the number, concentration, bioavailability and fate of contaminants of interest, the vicinity of the site with aquatic sources, the presence of hot spots and the climatic conditions (Komnitsas and Modis 2006).

The prediction of acid generation potential of geological formations and waste disposal sites enables the selection and implementation of an environmentally sound exploitation strategy and closure plan. Such predictions are normally based on static and/or kinetic evaluation of samples collected from the site; in most cases, several experimental studies are required due to the mineralogical nature of the materials involved. Acid–base accounting (ABA) is a widely applied static test that determines the balance between acid generating and acid consuming phases. The reliability of this test is sometimes criticized, since sulphur may be partially present in forms that are not amenable to oxidation. If the mineralogy of the formation varies widely, numerous samples must be examined and assumptions must be made to predict the actual overall acid generation potential (Adam et al., 1997; Komnitsas et al., 1998).

Modeling and geostatistical techniques are often applied to improve the quality of mining and environmental applications, in issues related with ore grade, presence of trace elements, simulation of geological domains, spatial heterogeneity between wells, training images for vain-type formations, degree of soil contamination, health and risk assessment as well as fate of contaminants in affected media. (Modis et al., 1998; Watson et al., 2001; Diko et al., 2001; Taboada et al., 2002; Marinoni, 2003; Emery, 2007; Zhang, 2008; Boisvert et al., 2008).

The accuracy of geostatistical estimators is principally defined by the quality of the sampling campaign. An important question that should be always asked and answered, regardless of the estimation procedure used, is how representative the samples are. The usual practice in order to determine the optimum grid size up to date, was to employ an estimation variance as a criterion of efficiency (David 1976; Dowd and Milton 1987); various grid setups may be considered while the average estimation variance may be plotted as a function of sampling density. The optimal grid size is the threshold beyond which no further improvement of the estimation variance is practically seen. This process is awkward and may lead to subjective results. (Modis and Papaodysseus, 2006; Modis and Komnitsas, 2007).

In the present chapter a case study that predicts the Net Neutralization Potential (NNP) and therefore the capacity of acid generation of the rock formations in a mixed sulphide mine in Chalkidiki peninsula, northern Greece, is presented and discussed. Past mining and waste disposal activities as well as the continuous generation of Acid Mine Drainage (AMD) have affected the ecological balance of the area (Gaidajis, 2003; Lazaridou et al., 2004; Kelepertsis et al., 2006).

Therefore, the prediction of AMD potential of all materials involved is extremely important to enable future sustainable mining and mitigate environmental impacts. Based on the information theory, a mathematical approach is proposed in this chapter in order to estimate an optimum sampling density for mining and environmental applications. This methodology is then applied to estimate the appropriate sampling grid and predict AMD generation at the Madem Lakkos and Mavres Petres mines. The proposed approach enables the establishment of a theoretical optimum sampling grid size with a minimum mean square error that overcomes the limitations of the subjective and awkward process of iteratively calculating kriging estimation error for various sampling grids by progressively increasing the sampling density until some improvement is seen.

2. SAMPLING THEORY

The information theory that was originally developed for the deterministic electrical signals, deals with an estimation problem of a signal that is reconstructed from its samples. Reconstruction involves filtering of the sampled signal by an interpolation filter; if the conditions of the sampling theorem are satisfied, then the original signal is exactly reproduced (Papoulis and Pillai, 2002).

It is known that in a bandlimited function, the spectrum vanishes above a certain frequency limit. The spectrum of an arbitrary sampled bandlimited function is a scaled, periodic replication of the spectrum of the original function (Pratt 1991). In case of no spectrum overlapping, in other words if the sampling frequency is greater than a critical value defined as the Nyquist rate, the original function can be reconstructed from its samples by linear filtering. The transfer function of the filter must be such that when applied to the samples sequence spectrum, the spectrum of the resulting signal is identical to the original. Thus, the filter aims to clear the spectrum from all replicas and retain only one. Based on this criterion, the square pulse is the only appropriate transfer function of the interpolation filter that is used in this application.

A product in the frequency domain is a convolution in time or space domain. The sinc function, seen below

$$\operatorname{sinc}(x) = \begin{cases} 1, & x = 0 \\ \dfrac{\sin(\pi x)}{\pi x}, & x \neq 0 \end{cases} \qquad (1)$$

is the continuous inverse Fourier transform of the rectangular pulse Π(x) of width 2π and height 1:

$$F[\Pi(x)] = \frac{1}{2\pi} \int_{-\pi}^{\pi} e^{i\omega x} dx = \operatorname{sinc}(x) \qquad (2)$$

By taking into account that the Fourier transform of the sinc function is the square pulse function, it is deduced that convolution of the sinc function with the samples produces the original function (Figure 1a). The sinc function is therefore the ideal interpolation filter. The critical Nyquist frequency for band limited signals is two times higher than the signal bandwidth.

In practice, apart from the square pulse, other easier to realize functions may be used as approximate transfer functions to interpolate the sampled function in one or two dimensions. Appropriate transfer functions may be selected to define all deterministic interpolation methods (Figure 1b and 1c). When the sampling rate is lower than the Nyquist limit, then overlapping of the infinite copies of the initial spectrum is seen and therefore all aforementioned approximations are intrinsically problematic. In such cases, instead of causal approaches stochastic approximations must be considered.

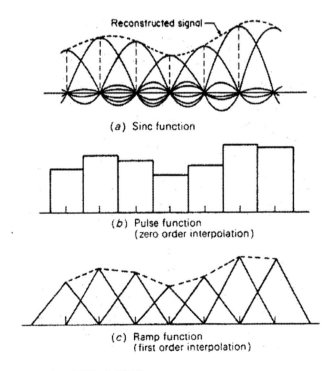

Figure 1. Signal reconstruction (Niblack 1986)

When the information theory is applied to earth related sciences, the natural variables under study are usually modelled as random fields. In mining applications, the most commonly used covariance (variogram) model is the spherical scheme; its formula is based on the geometric covariogram (Matheron, 1970) of a sphere of radius a. The corresponding correlation function is expressed by equation (3), after being normed to 1, i.e. corresponding to random function with unit a priori variance.

$$R(h) = \begin{cases} 1 - \dfrac{3|h|}{2a} + \dfrac{|h|^3}{2a^3} & |h| \leq a \\ 0 & |h| > a \end{cases} \quad (3)$$

By using the Fourier transform of (3), the power spectrum of the underlying random function model is expressed by equation (4):

$$S(\omega) = \frac{3}{a\omega^2} + \frac{12}{a^3\omega^4}\sin^2\left(\frac{\omega a}{2}\right) - \frac{6}{a^2\omega^3}\sin\omega a \quad (4)$$

The spherical correlation function is not strictly bandlimited but again it vanishes asymptotically. In Figure 2, presenting the power spectrum of the underlying RF as a function of 'a', the value $\omega = 2\pi/a \approx 6$ can be approximately considered as the vanishing point for the spectrum. This is seen by applying the 'rule of thumb' (Modis and Papaodysseus, 2006) which is true for every value of the range of influence 'a'. Based on the sampling theorem it is again seen that the sampling lag size must be less or equal to half the size of the range of influence:

$$\Delta s = a/2 \quad (5)$$

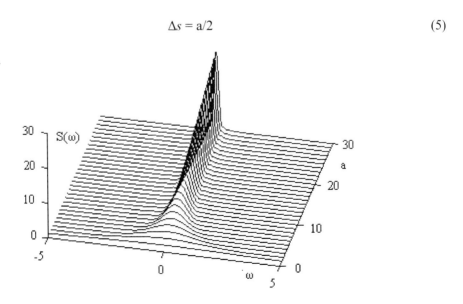

Figure 2. Fourier transform of the spherical correlation model as a function of range of influence 'a' (Modis and Papaodysseus 2006)

It is therefore deduced that if the structural analysis of an ore body reveals an underlying structure with a certain range of influence, then the optimum exploratory grid size is defined as half the value of this range.

3. ESTABLISHMENT OF AN CRITICAL SAMPLING GRID FOR NNP MODEL ESTIMATION

Kassandra is located 100 km east from Thessaloniki; it belongs to the Chalkidiki prefecture, and is the largest mixed sulphide mine site in the country. The polymetallic deposits of Olympias and Stratonion, which contain gold, lead, silver, and zinc, and the copper-gold porphyry deposit referred to as Skouries are spread over an area of 317 km^2.

The Olympias mine is located 2.3 km W of the coastal town of Olympias and 22 km N of Stratonion. The Stratonion mines of Mavres Petres and Madem Lakkos are located 6 km to the south and 3 to 3.5 km W-NW from the port and loading facilities of the village of Stratonion (latitude: 40° 31' 0 N, longitude: 23° 49' 60 E). The Skouries site is located about 11 km SW of the Stratonion mines, 11 km south of the town of Paleohori, and 3 km NE of the village of Megali Panagia.

The orebodies are mainly developed in the marbles or in the contacts between marbles and other formations. The form of the deposits is often irregular; however, it is controlled by the nature of the contact between marbles and other rocks.

The origin of the Madem Lakkos and Olympias ores is subject to controversy. Nikolaou and Kokonis (1980) based on geological evidence and ore textures support a metasomatic replacement origin for the ores at the Olympias deposit. Nebel et al. (1992) suggest a pre-metamorphic syngenetic origin for the Olympias and Madem Lakkos ores, which were subsequently regionally metamorphosed and later reworked by Tertiary hydrothermal activity related to igneous rocks in the area. Gilg and Frei (1994) favor an epigenetic origin of the ores and a genetic relationship to the emplacement of Tertiary porphyritic stocks in the area of eastern Chalkidiki. Finally, Kilias et al. (1996) have studied the microthermometry of synore gangue material from undeformed and deformed Olympias ores.

The Mavres Petres orebody is a hydrothermal to mesothermal complex sulphide deposit located in marble with a relatively small thickness, or at the contact between marble and surrounding schists or gneisses. The Madem Lakkos orebody is also a hydrothermal complex sulphide deposit, mainly located at the upper contact of a lenticular body of marble of considerable thickness with the overlying gneiss. A typical cross section of the Madem Lakkos sulphide deposit is shown in Figure 3.

The lead-rich ores from the Madem Lakkos mine at Stratonion were smelted for silver while the Olympias arsenopyrite ores were processed for their high gold content. Ancient mining reached a peak in the area between 350 and 300 BC. It has been estimated, from the volume of ancient slags, that about one million tonnes of ore was extracted from each site during that period.

The ore in the Stratonion mines was mined underground by the cut-and-fill method. Earlier, sub-level caving was employed and this resulted in the generation of extensive cracks in the overlying strata, facilitating water infiltration and subsequently acid generation.

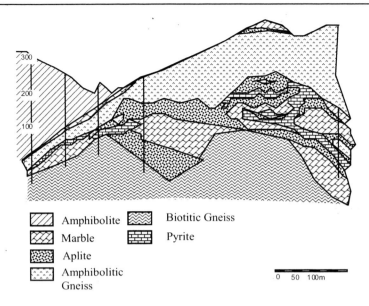

Figure 3. A typical cross section of the Madem Lakkos sulphide deposit (Modis et al., 1998)

Estimated proven reserves at Olympias are 10 Mt of ore grading 8.1 g/t gold, 108 g/t silver, 3.5% lead, and 4.6% zinc, while probable reserves are 4 Mt grading 9.7 g/t gold, 148 g/t silver, 4.9% lead, and 6.6% zinc. Estimated proven reserves at Stratonion are 1 Mt grading 191 g/t silver, 8% lead, and 10% zinc, while probable reserves are 862,000 t grading 189 g/t silver, 8% lead, and 11.7% zinc. Estimated probable reserves at Skouries gold-copper porphyry deposit are 129 Mt grading 0.89 g/t gold and 0.56% copper (Newman 2004).

The NNP model estimation was based on the analysis of multiple samples collected from 171 exploratory drill-holes, with a depth up to 900 m. In total 1820 samples from the Madem Lakkos and Mavres Petres mining areas were collected while due to time and budget limitations only 89 mineralized samples were chemically analyzed. Mean values for the different formations were used to correlate the drill-hole logs (Modis and Komnitsas, 2007; Modis et al., 2008;).

The geostatistical structural analysis of the NNP spatial distribution involves estimation of the NNP variogram function as a measure of the variability in different directions in space. The mean experimental variograms were calculated for five spatial directions, using sample pairs to calculate particular points (Figure 4). No significant difference was observed in each specific direction, so isotropy can be reasonably assumed. Therefore, the fitted function, $\gamma(r)$, is an isotropic spherical model with sill $C_0 = 40,000$ (kg $CaCO_3$ equiv/t sample)^2, and range of influence, $a = 90$ m. The model was validated using the back estimation technique: a variogram model was used to estimate each data value, which was in turn removed. The best model was the one exhibiting the minimum average estimation error.

The location of each drill hole is presented in Figure 5. It can be easily seen from this figure that drilling did not follow any grid configuration and that the distance between drill-holes varies widely. The shortest distance between two successive drill holes is 20 m while the longest is 250 m; the average distance is about 150 m. A higher density of drill holes is seen in the south-central (Madem Lakkos mine) and northwest (Mavres Petres mine) areas, where the distance between drill holes is between 20 and 120 m. Thus, a separate distance distribution with different averages can be identified for each site.

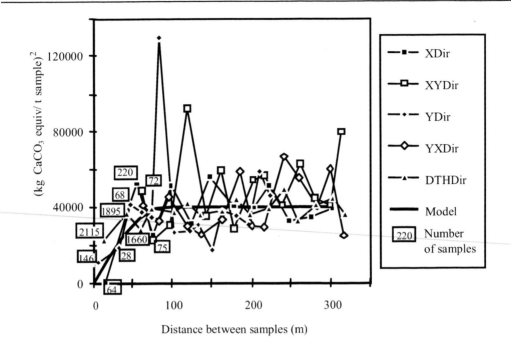

Figure 4. Experimental mean variograms and fitted isotropic model for sample pairs; DTH refers to "down the hole"

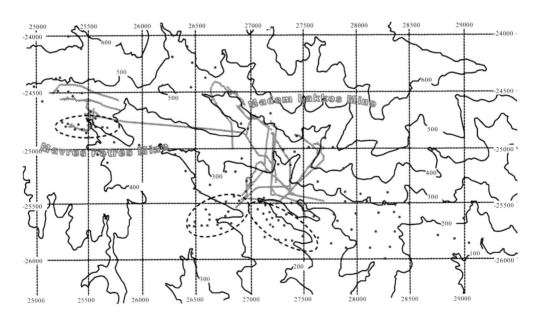

Figure 5. Drill-hole location and over sampled areas (seen in ellipses) in the Stratonion mining area

If the range of influence as defined by the structural analysis of the NNP values (Figure 4) is inserted into formula (3), the resulting size of the ideal square sampling grid becomes 45 m. However, the drill-holes are distributed unevenly, indicating over-sampling in some parts and under-sampling in others (Figure 5). The southwestern part of the Madem Lakkos mine and the southern part of the Mavres Petres mine were over-sampled, since the drill-hole distance was as short as 20 m, which is shorter than the recommended distance of 45 m. The rest of the mining area was under-sampled, since no drill-holes are close to the optimum distance. As previously explained this may cause problems during the NNP modeling process and subsequently affect the accuracy of AMD prediction for the formations under study.

4. CONCLUSIONS

A number of geostatistical estimation techniques may be used in mining, geochemical and environmental applications to analyze information extracted from drill cores or other means of sampling in order to

- establish the natural variability in mineralization and thus evaluate the financial viability of mining deposits
- eliminate uncertainty, generate accurate geological models in terms of shape, extension and distribution of geological units and design optimum mining technologies
- define geometrical parameters of geological layers and improve safety calculations
- significantly reduce smoothing effect around zero values
- generate reliable geochemical maps that may be used as effective tools for environmental planning, for establishing health care policies and for promoting sustainable development
- establish background concentrations that may be used as soil clean-up criteria; also establish whether anomalies related with trace element concentrations in soil are geogenic or anthropogenic
- take into account hot spots and assess risk in waste disposal and other affected sites and implement appropriate rehabilitation options

It is well known that dense sampling is often not realistic and that geological units are not always logged in blast holes. In environmental applications on the other hand, the design of an appropriate sampling campaign becomes complex due to the presence of hot spots and the heterogeneity of wastes. Thus, the design of a drilling campaign should be better considered as an optimization problem. The selection of an appropriate drill hole network relative to the grid size can assist in the extraction of maximum information related with the distribution of several parameters.

In order to increase the probability of accurately estimating NNP potential using values deriving from the implementation of standard static or kinetic tests, the optimum sampling density can be better established if issues such as topography and geology of the site (e.g. faults, lenses of different materials, permeability of the overlying formations), as well as mineralogy are taken into account. For example, the reliability of the model can be definitely improved if only the percentage of sulphur that is amenable to oxidation (sulphidic instead of total sulphur) is considered in calculations.

In this chapter, a variogram model generated by structural analysis of NNP spatial distribution was used to establish a critical sampling grid. If the density of the sampling grid is close to or greater than the critical value, the NNP numerical model can be obtained using simpler interpolation algorithms, such as the inverse distance square, and the accuracy would be equal to that derived by geostatistics. However, if the available data density is less than the critical sampling density, geostatistical analysis is recommended. In this specific case, the variogram model generated by structural analysis of NNP spatial distribution indicates that in some parts of the ore body, the sampling grid is denser than required.

Finally, it is mentioned that a similar approach could be followed to identify a critical sampling grid for a) prediction of AMD potential in waste disposal sites, especially if different layers of wastes have been disposed of over the years, b) assessment of soil contamination in the vicinity of mining and waste disposal sites, and c) other environmental applications involving earth-related space-distributed data.

REFERENCES

Adam, K., Kourtis, A., Gazea, B. & Kontopoulos, A. (1997). *Evaluation of static tests used to predict the potential for acid drainage generation at sulphide mines*, T I Min Metall A *106*, A1-A8.

Bastante, F. G., Ordóñez, C., Taboada, J. & Matías, J. M. (2008). Comparison of indicator kriging, conditional indicator simulation and multiple-point statistics used to model slate deposits, *Eng Geol*, *98*, 50-59.

Boisvert, J. B., Leuangthong, O., Ortiz, J. M. & Deutsch C. V. (2008). A methodology to construct training images for vein-type deposits, *Comput Geosci*, *34*, 491-502.

David, M. (1976). What Happens If? - Some remarks on useful geostatistical concepts in the design of sampling patterns. Proc, Symp on Sampling Practices in the Minerals Industry. *Australasian Institute Mining and Metallurgy*, 1-15.

Diko, L., Vervoort, A. & Vergauwen, I. (2001). Geostatistical modelling of lateritic orebodies in Surinam: effect of the vertical dimension, *J Geochem Explor*, *73*, 131-153.

Dowd, P. A. & Milton, D. W. (1987). Geostatistical estimation of a section of the perseverance nickel deposit, in: G. Matheron, M. Armstrong, (Eds), *Geostatistical Case Studies*, D Reidel Publ Co, Norwell, MA, USA, 39- 67.

Emery, X. (2007). Simulation of geological domains using the plurigaussian model: New developments and computer programs, *Comput Geosci*, *33*, 1189-1201.

François-Bongarçon, D. (2004). Theory of sampling and geostatistics: an intimate link, *Chemometr Intell Lab*, *74*, 143- 148.

Gaidajis, G. (2003). Ambient concentrations of total suspended particulate matter and its elemental constituents at the wider area of the mining facilities of TVX Hellas in Chalkidiki, Greece, *J Environ Sci Heal, A*, *38(11)*, 2509-2520.

Gilg, H. A. & Frei, R. (1994). Chronology of magmatism and mineralization in the Kassandra mining area, Greece. The potentials and limitations of dating hydrothermal illites, *Geochim Cosmochim Acta*, *58*, 2107-2122.

Heriawan, M. N. & Koike, K. (2008). Identifying spatial heterogeneity of coal resource quality in a multilayer coal deposit by multivariate geostatistics, *Int J Coal Geol*, *73*, 307-330.

Kelepertsis, A., Argyraki, A. & Alexakis, D. (2006). Multivariate statistics and spatial interpretation of geochemical data for assessing soil contamination by potentially toxic elements in the mining area of Stratonion, north Greece, *Geochem-Explor Env., A, 6(4)*, 349-355.

Kilias, S. P., Kalogeropoulos, S. I. & Konnerup-Madsen, J. (1996). Fluid inclusion evidence for the physicochemical conditions of sulfide deposition in the Olympias carbonate-hosted Pb-Zn (Au,Ag) sulfide ore deposit, E. Chalkidiki peninsula, N. Greece, *Miner Deposita, 31*, 394 406.

Komnitsas, K., Kontopoulos, A., Lazar, I. & Cambridge, M. (1998). Risk assessment and proposed remedial actions in coastal tailings disposal sites in Romania, *Miner Eng., 11*, 1179-1190.

Komnitsas, K. & Modis, K. (2006). Soil risk assessment of As and Zn contamination in a coal mining region using geostatistics, *Sci Total Environ, 371*, 190-196.

Lazaridou-Dimitriadou, M., Koukoumides, C., Lekka, E. & Gaidagis, G. (2004). Integrative evaluation of the ecological quality of metalliferous streams (Chalkidiki, Macedonia, Hellas), *Environ Monit Assess, 91*, 59-86.

Martínez, J., Llamas, J., de Miguel, E., Rey, J. & Hidalgo, M. C. (2007). Determination of the geochemical background in a metal mining site: example of the mining district of Linares (South Spain), *J Geochem Explor, 94*, 19-29.

Matheron, G. & (1970). La Theorie des Variables Regionalisees et ses Application: Les Cahiers du Centre de Morphologie Mathematique, Fasc. 5, *CGMM Fontainebleau, 212*.

Marinoni, O. (2003). Improving geological models using a combined ordinary–indicator kriging approach, *Eng Geol, 69*, 37-45.

Modis, K., Adam, K., Panagopoulos, K. & Kontopoulos, A. (1998). Development and validation of a geostatistical model for prediction of acid mine drainage in underground sulphide mines, *T I Min Metall A, 107*, 102-107.

Modis, K. & Papaodysseus, K. (2006). Theoretical estimation of the critical sampling size for homogeneous orebodies with small nugget effect, *Math Geol, 38(4)*, 489-501.

Modis, K., Papantonopoulos, G., Komnitsas, K. & Papaodysseus, K. (2008). Mapping optimization based on sampling size in earth related and environmental phenomena, Stoch Env Res Risk A, *Springer, Vol 22*, 83-93.

Modis, K. & Komnitsas, K. (2007). "Optimum sampling density for the prediction of acid mine drainage in an underground sulphide mine", *Mine Water Environ, Springer, 26*, 237- 242.

Nebel, M. L., Hutchinson, R. W. & Zartman, R. E. (1992). Metamorphism and polygenesis of the Madem Lakos polymetallic sulfide deposit, Chalkidiki, Greece - a reply, *Econ Geol, 87*, 1187-1190.

Newman, H. R. (2004). The Mineral Industry of Greece. *US Geological Survey Minerals Yearbook, 13*, 1-13.8.

Niblack, W. (1986). An Introduction to Digital Image Processing. Prentice Hall, *Upper Saddle River*, NJ, USA, *215*.

Nikolaou, M. & Kokonis, I. (1980). Geology and development of the Olympias mine, eastern Chalkidiki, Macedonia, Greece. In: M. J. Jones, (Ed), *Complex Sulphide Ores*, Institute Mining Metallurgy, London, UK, 260-270.

Papoulis, A. & Pillai, U. (2002). *Probability, Random Variables and Stochastic Processes.* McGraw Hill, NY, *852*.

Pratt, W. K. (1991). Digital image processing. *Wiley-Interscience*, New York, *698*.

Watson, W. D., Ruppert, L. F., Bragg, L. J. & Tewalt, S. J. (2001). A geostatistical approach to predicting sulfur content in the Pittsburgh coal bed, *Int J Coal Geol, 48*, 1-22.

Zhang, T. (2008). Incorporating Geological Conceptual Models and Interpretations into Reservoir Modeling Using Multiple-Point Geostatistics, *Earth Sci Frontiers, 15(1)*, 26-35.

Taboada, J., Vaamonde, A., Saavedra, A. & Ordóñez, C. (2002). Geostatistical study of the feldspar content and quality of a granite deposit, *Eng Geol, 65*, 285-292.

In: Mine Drainage and Related Problems
Editor: Brock C. Robinson, pp. 239-263

ISBN: 978-1-60741-285-4
© 2010 Nova Science Publishers, Inc.

Chapter 8

TREATMENT OF ACID MINE DRAINAGE BY A COMBINED CHEMICAL/BIOLOGICAL COLUMN APPARATUS: MECHANISMS OF HEAVY METAL REMOVAL[*]

Francesca Pagnanelli [a,†], Ida De Michelis [b], Michele Di Tommaso [b], Francesco Ferella [b], Luigi Toro [a], and Francesco Vegliò [b]

[a] Department of Chemistry, Sapienza University of Rome,
P. le Aldo Moro 5, 00185 Rome, Italy
[b] Department of Chemistry, Chemical Engineering and Materials, University of L'Aquila,
Monteluco di Roio 67040, L'Aquila, Italy

ABSTRACT

Natural oxidation of sulphide minerals, exposed to the combined action of oxygen and water, results in the worst environmental problem associated with mining activities, i.e. acid mine drainage (AMD). Waters polluted by AMD are often characterised by low pH, elevated concentrations of iron, sulphates and toxic metals.

Biological remediation options in passive systems (permeable reactive barriers, PRB) usually exploit sulphur production by sulphate reducing bacteria, SRB.

In this report a combined chemical-biological treatment was tested for decontamination of synthetic AMD containing iron, arsenic, copper, manganese and zinc.

Particular attention was paid to the investigation of the mechanisms involved in pollutant removal (chemical precipitation, sorption, bioprecipitation and biosorption) as a fundamental preliminary step for permeable reactive barrier design and long term performance estimation.

---

[*] A version of this chapter was also published in Causes and Effects of Heavy Metal Pollution, edited by Mikel L. Sànchez published by Nova Science Publishers, Inc. It was submitted for appropriate modifications in an effort to encourage wider dissemination of research.
[†] Corresponding author: Email: francesca.pagnanelli@uniroma1.it; fax: +39 06 490631.

Experimental tests were performed both in batch reactors and in a two-column apparatus for sequential treatment by chemical precipitation (first column filled with natural limestone) followed by bioprecipitation/biosorption (second column filled with a natural organic mixture inoculated by sulphate reducing bacteria).

Distinct mechanisms of removal for each metal were identified by combining theoretical data of metal solution chemistry, and results obtained from independent experimental tests: batch and column tests, blank tests using natural organic mixture as biosorbing materials, acid digestions, and selective extractions of metals using solid samples of filling material after column dismantlement.

This analysis allowed isolating metal-specific mechanism of abatement and denoted the relevant contribution of biosorption phenomena in metal removal in biological column. This contribution, generally neglected in biological PRB design with respect to bioprecipitation, should be taken into account in order to avoid misleading estimation of SRB performance and also to better estimate PRB duration.

Keywords: water treatment; heavy metals; sulphate reducing bacteria; biosorption, fixed bed column reactor.

1. INTRODUCTION

1.1. Acid Mine Drainage

Natural oxidation of sulphide minerals, exposed to the combined action of oxygen and water, results in the worst environmental problem associated with mining activities, i.e. acid mine drainage (AMD) (Johnson and Hallberg, 2005). Waters polluted by AMD are often characterised by low pH, elevated concentrations of iron, sulphates and toxic metals.

AMD can be generated in underground and open pit operating mines, but also abandoned sites can remain active generation points for decades or even centuries after mine closure. The other main source of AMD is the huge amount of mill tailings (often about the 90% of treated ore): about 18 billion m^3 are produced every year, stored in impoundments or left exposed to natural weathering (Sheoran and Sheoran, 2006).

AMD is generated by biologically-accelerated oxidation of iron pyrite (FeS_2), the most abundant sulphide mineral on the planet, generally present in metal ores and coals. AMD originates from a complex series of reactions, which result in pyrite dissolution and release of sulphates and protons:

$$FeS_2 + 3.5O_2 + H_2O \rightarrow Fe^{2+} + 2SO_4^{2-} + 2H^+ \qquad (1)$$

Nevertheless, this global reaction can be misleading not showing that, in most situations, the main sulphide oxidant is Fe(III) rather than oxygen (Johnson and Hallberg, 2005; Baker and Banfield, 2003; Peppas et al., 2000):

$$FeS_2 + 14Fe^{3+} + 8H_2O \rightarrow 15Fe^{2+} + +2SO_4^{2-} + 16H^+ \qquad (2)$$

Ferrous iron regeneration by O_2 at low pH is the rate-limiting step of the whole process:

$$14Fe^{2+} + 3.5O_2 + 14H^+ \rightarrow 14Fe^{3+} + 7H_2O \qquad (3)$$

The rate of this reaction can be augmented by several orders of magnitude by the action of iron oxidising bacteria (such as *Thiobacillus ferooxidans*) resulting in the global acceleration of pyrite dissolution (reaction 1, which is the sum of reactions 2 and 3).

The same mechanisms of oxidation by O_2 and Fe(III) are responsible for dissolution of other metal sulphides releasing toxic metals in soil solution system (such as As, Zn, Cd, Pb, Cu, Ni and Mn).

The true scale of AMD pollution is of difficult assessment: an estimate of 1989 spoke about 19˙300 Km of rivers and 72˙000 ha of lake damaged by AMD (Johnson and Hallberg, 2005).

The long-term activity of pollution sources along with the predicted doubling of mine tailing production in the next 20-30 years (Sheoran and Sheoran, 2006) indicate a need for the application of an integral approach based on both prevention and treatment.

1.2. Prevention and Treatment of AMD Pollution

Operative strategies for AMD pollution involve both prevention and treatment options.

The preclusion or attenuation of AMD generation can be obtained by different ways (Peppas et al., 2000; Johnson and Hallberg, 2005; Sheoran and Sheoran, 2006; Pagnanelli et al., 2007):

- minimization of oxygen diffusion (flooding/sealing of underwater site, underwater storage of mine tailings, storage in sealed waste heap with covers of organic materials and clays);
- control of pH of mineral wastes by blending with solid material (lime, limestone, phosphates, fly ash, paper mill waste) in order to precipitate Fe(III) and reduce its oxidant action;
- total solidification of wastes;
- inhibition of iron and sulphur oxidising bacteria by such biocides as surfactants.

These prevention technologies can be integrated with the treatment of AMD polluted waters performed according to both abiotic and biological strategies.

In abiotic approaches various chemicals can be used to promote heavy metal removal by decreasing their solubility. Generally AMD polluted waters are treated by limestone in order to neutralise AMD pH and to precipitate iron:

$$2Fe(HCO_3)_2 + 1/2O_2 + H_2O \rightarrow 2Fe(OH)_3 \downarrow + 4CO_2 \qquad (4)$$

$$2Fe_2(SO_4)_3 + 6CaCO_3 + 3H_2SO_4 \rightarrow 2Fe(OH)_3 \downarrow + 6CaSO_4 \downarrow + 6CO_2 \qquad (5)$$

Iron removal is a fundamental step to preclude AMD mechanism going on (equation 2), but also toxic metals with low solubility in basic conditions can be removed.

Biological strategies exploit H₂S production by sulphate reducing bacteria, SRB, promoting metal precipitation as sulphides (Johnson, 2006; Neculita et al., 2007):

$$Me^{2+} + H_2S \rightarrow MeS \downarrow + 2H^+ \qquad (6)$$

Precipitation of heavy metals as sulphides seems to be a better alternative with respect to metal precipitation for pH rise. In fact sulphide precipitation occurs almost regardless of wastewater characteristics, is less pH-dependent and leave lower residual metal concentrations than hydroxide precipitation (Tunay and Kabdasli, 1994). However, chemically produced H₂S is relatively expensive to handle and to use safely. The precipitation of metals with biologically produced H₂S by SRB has been then proposed as an alternative process (Foucher et al., 2001).

As for the application of abiotic and biological strategies, both active and passive systems can be adopted. Active systems require ex-situ treatment of polluted streams (by filtration, precipitation and adsorption onto active carbons), while passive systems are based on in situ technologies using permeable reactive barriers (PRB) which intercept polluted streams and determine pollutants abatement within the barrier. PRB are especially advantageous for widespread pollution sources as in the case of AMD in mine districts (Gibert et al., 2002).

Treatment in abiotic PRBs can exploit neutralizing agents, adsorbents and zero-valent iron as reactive filling materials (Table 1).

Table 1. Some examples of abiotic reactive barriers for heavy metal pollution (http://www.rtdf.org/public/permbarr/PRBSUMMS/)

Location	Installation	Scale	Reactive medium	Pollutants
Durango, CO (USA)	1995	Pilot	Fe(0)	As, Mo, Se, U, V, Zn
Elizabeth City, NC (USA)	1996	Pilot	Fe(0)	TCE, Cr(VI)
Hanford, WA (USA)	1997	Full	Na₂S₂O₄	Cr(VI)
Ontario, Canada	1998	Full	Zeolites	Sr-90
Ponticello, UT (USA)	1999	Full	Fe(0)	U, As, Mn, Se, V
Kolding, Denmark	1999	Full	Fe(0)	TCE, Cr(VI)
Nesquehoning, PA (USA)	1998	Full	Limestone	Pb, Cd, As, Zn, Cu

Biological PRB exploiting SRB activity are generally made up of organic mixtures as electron donor in the dissimilatory reduction of sulphate to sulphide, which generates alkalinity and promotes metal precipitation:

$$SO_4^{2-} + 2CH_2O + 2H^+ \rightarrow H_2S + 2H_2CO_3 \qquad (7)$$

Organic components used in PBR are a mix of biological materials chosen on the base of the local availability: biodegradable materials (mushroom compost, manure of cow, horse and sheep, municipal compost) are generally mixed with more recalcitrant ones (sawdust, peat, straw, leaf compost) to ensure long term growth of SRB (Alvarez et al., 2007; Chang et al., 2000; Christiensen et al., 1996; Cocos et al., 2002; Gibert et al. 2003; Gibert et., 2004; Hammack et al., 1992; Peppas et al., 2000; Waybrant et al., 1998; Waybrant et al., 2002). Full scale applications of organic-carbon based sulphate reducing PRB are also characterised by the addition of gravel to improve barrier permeability and limestone to increase pH and favour SRB growth (Benner et al., 1999; Jarvis et al., 2006; Ludvig et al., 2002).

Table 2. Reactive mixtures used in lab- and in full-scale biological permeable reactive barriers for the treatment of AMD and heavy metal's contaminated wastewaters

Composition	Type	Abatement	Reference
Municipal compost Sawdust Manure Cellulose Sediments with SRB Silica sand Limestone	Batch	SO_4^{2-} ~ 100% Fe 99% Ni ~ 100% Cd 99%	Waybrant et al., (1998)
Wood chips (3%) Composted leaves (30%) Chicken manure (20%) Silica sand (5%) Sediments with SRB (37%) Limestone (2%) Urea (3%)	Batch	SO_4^{2-} 97% Ni 72 % Zn 88%	Cocos et al., (2002)
Pirite Silica sand Leaves Chips Sawdust Biological sludge Sediments with SRB	Column	SO_4^{2-} 20-60%	Waybrant et al., (2002)
Limestone (50%) Compost (45%) Sediments with SRB (5%)	Column	Fe 99% Zn 55% Cd 80% Cu 97%	Gibert et al. (2003)
Compost limestone Sheep manure	Column	SO_4^{2-} 18-27%	Gilbert et., (2004)
Gravel (50%) Municipal compost (20%) Soil (20%) Wood chips (9%) Limestone (1%)	Full-Scale PRB Ontario (Canada) 1995	SO_4^{2-} 60% Fe 85%	Benner et al. (1999)
Module 1 Limestone (50%) Municipal compost (30%) Sludge (20%) *Module 2* Limestone (50%) Municipal compost (50%) *Module 3* Zero-valent iron (0) (1%) Limestone (66%) Municipal compost (33%)	Full-Scale PRB Aznalcòllar (Spagna) 1998	Metals 90%	Carrera et al. (2001)
Composted leaves (15%) Gravel (84%) Limestone (1%)	Full-Scale PRB Vancouver (Canada) 2000	Metals 80%	Ludwig et al. (2002)
Manure and straw (25%) Municipal compost (25%) Limestone (50%)	Full-Scale PRB Northumberland (UK) 2003	SO_4^{2-} 67% Fe 95% Al 87%	Jarvis et al. (2006)

Table 2 shows some reactive mixtures reported in the literature, which have been used in lab-scale (batch and column) studies and in full-scale permeable reactive barriers for treatment of AMD and heavy metal's contaminated wastewaters.

These reducing and alkalinity-producing systems (RAPS) (Johnson and Hallberg, 2005) can be inadequate for highly iron-concentrated AMD due to limestone armouring (reduction of reactivity by deposition of iron oxides precipitates) (Simon et al., 2005) and barrier plugging for iron precipitates. In such cases, an alternative engineering configuration can be adopted by using a two-step procedure: chemical precipitation of iron by limestone (step I, pre-treatment of chemical precipitation) and then a refinement of heavy metal removal as sulphide precipitates mediated by sulphate reducing bacteria (step II, bioprecipitaton).

Chemical pretreatments of AMD aim firstly to obtain an increase of pH, which allows the precipitation of Fe(III) (involved in sulphide oxidation and then in AMD generation) and the partial removal of other toxic components. The adjustment of pH and the reduction of iron concentration are also necessary for the successive biological treatment in permeable reactive barriers to avoid the rapid plugging of the barrier (for the large amounts of iron precipitates) and to ensure neutral pH conditions for bacterial growth.

1.3. Aim of the Work

Experimental results reported in this work denote some crucial points that should be specifically addressed in order to develop new biotechnological applications for the treatment of heavy metal pollution. According to this, the research specifically focused on the investigation of the chemical and biological mechanisms responsible of heavy metal removal in biological permeable reactive barriers.

Biological permeable reactive barriers are complex continuous flow multiphase reactors in which different mechanisms can act simultaneously in heavy metal removal (Sheoran and Sheoran, 2006; Johnson and Hallberg, 2005; Johnson, 2006; Whitehead et al., 2005).

Generally bioprecipitation and chemical precipitation were addressed as the main mechanism operating. In fact different metals simply precipitate in the conditions of pH typical of SRB growth as oxyhydroxides (ferric iron, copper and aluminium) and as carbonates (copper, manganese and zinc) (Sheoran and Sheoran, 2006; Peppas et al., 2000). Mechanism investigations were generally developed according to geo-chemical studies considering both chemical equilibria in solution and metal speciation in solid phase (Herbert et al., 2000; Morrison et al., 2002; Wilkin and McNeil 2003; Johnson and Hallberg, 2005b; Swash and Monhemius 2005).

Experimental data using zero-valent iron as reactive medium denoted the importance of adsorption as initial and rapid metal uptake mechanism (Wilkin and McNeil 2003). In the same way adsorption onto organic matter (biosorption) used as reactive media in biological PRB should be accounted for in lab-studies for further development. Bioprecipitation by SRB is the long term active mechanism that should operate in metal removal. Nevertheless abatement estimates for biological permeable reactive barriers can not neglect the relevant contribution of biosorption onto organic reactive materials generally used as long term carbon source for SRB. In fact, bioprecipitation remains active in removing metals and sulphates until SRB are alive, while biosorbents tend to be saturated and undergo biodegradation processes causing the release of initially sorbed pollutants.

According to these observations the development and design of new biological PRB can not neglect the identification of the specific mechanisms involved in metal removal as key-step of lab-study research.

In this view knowledge, isolation and quantification of the different mechanisms operating in biological PRB are the primary goal of this work. To this aim experimental results of batch and column tests of SRB growth on solid media were combined with theoretical metal speciation, blank tests of biosorption and analysis of pollutants speciation in solid phase after column dismantlement.

2. MATERIALS AND METHODS

2.1. Synthetic Acid Mine Drainage (AMD)

A synthetic solution was used to have a standard average composition of the influent stream to be treated. Synthetic AMD solution was obtained by dissolving weighted amounts of reagent grade chemicals ($FeSO_4$, $Fe_2(SO_4)_3$, $MnSO_4$, $CuSO_4$, $ZnSO_4$, As_2O_5) in distilled water.

Average values ± standard deviations of measured chemical composition in synthetic AMD were reported in Table 3.

2.2. Sulphate Reducing Bacteria (SRB)

Bacterial biomass was kindly furnished by the research group of Professor Groudev (Department of Engineering Geoecology, University of Mining and Geology, Sofia, Bulgaria), who collected it in the Curilo mine district located near Sophia (Groudev et al., 2001).

Table 3. Average values of chemical composition of synthetic AMD in the feed, after chemical pre-treatment (10th output of pre-treatment for PV=70) and after biological treatment (10th output of biological treatment column for PV=60)

Species	Feed	After Chemical Pre-Treatment	After Biological Treatment
Ph	2.9 ± 0.2	6.1 ± 0.2	8.1 ± 0.3
So_4^{2-}	800 ± 50 Mg/L	700 ± 50 Mg/L	600 ± 50 Mg/L *
Fe	400 ± 50 Mg/L	3 ± 1 Mg/L	N.D.
Cu	50 ± 10 Mg/L	0.2 ± 0.1 Mg/L	N.D.
Zn	50 ± 10 Mg/L	46 ± 5 Mg/L	N.D.
Mn	50 ± 10 Mg/L	48 ± 5 Mg/L	N.D.
As	2.0 ± 0.5 Mg/L	N.D.	

*Pre-treated samples of AMD after chemical treatment were upgraded to 2000±100 mg/L of sulphates; n.d.: not detectable, i.e. <0.01 ppm

Bacteria used in batch and column experiments were cultivated in closed shaken flasks using standard procedures for SRB reported in the literature (Postgate, 1979). In particular medium cultivation (defined as C Medium) was used for bacterial growth and acclimatising in column tests: KH_2PO_4 0.5 g/L; NH_4Cl 1 g/L; Na_2SO_4 4.5 g/L; $CaCl_2*6H_2O$ 0.06 g/L; $MgSO_4*7H_2O$ 0.06 g/L; sodium lactate 6 g/L; yeast extract 1 g/L; $FeSO_4*7H_2O$ 0.004 g/L; sodium citrate*$2H_2O$ 0.3 g/L.

2.3. Solid Mixture

Solid mixture used both in batch and column tests was made up of 80% v/v of compost, 15% v/v of cow manure, 5% v/v of straw and traces of limestone.

2.4. Batch Tests

Preliminary batch tests were performed for both chemical precipitation and sulphate removal by SRB.

Batch tests of chemical precipitation were performed by limestone addition to synthetic AMD (250 g in 500 mL). Quarry quality limestone used in this work was mainly made up of calcium carbonates and silica (mean particle size 6 mm). Limestone-bearing suspensions for batch tests of chemical precipitation were kept under stirring and monitored during time for iron concentration and pH.

Batch tests of SRB growth in liquid medium were performed according to the cultivation procedure reported above (paragraph 2.2) and monitored during time for cell, lactate and sulphate concentration.

Batch tests of SRB growth on solid media were performed according to the following procedure: a sample (20 g) of solid mixture (see 2.3) was added in each flask, filled with 80 mL of C Medium prepared without sodium lactate and yeast extract. Therefore the flasks were sealed and 20 mL inoculum of bacteria cultivated in C Medium (in exponential growth phase) were added by a syringe through the sampling port. All experiments were conducted at 37 °C under shacking conditions. Medium characteristics (pH, E_h, SO_4^{2-} concentration and H_2S production) were monitored during the experiments, which lasted 22 days.

Measurements of pH (by CRISON GLP22), E_h (by CRISON GLP22) and H_2S (by lead acetate paper) were determined immediately after sample collection. Samples were then filtered through 0.45 µm cellulose acetate filters and used for sulphate analysis (see 2.6.1). Each test was performed twice both in presence and in absence of bacteria. Average values ± standard deviations were considered.

Sulphate removal onto single organic components (compost, olive pomace and leaves) without inoculum were carried out as those previously described, using the same amount of component as in the mixture.

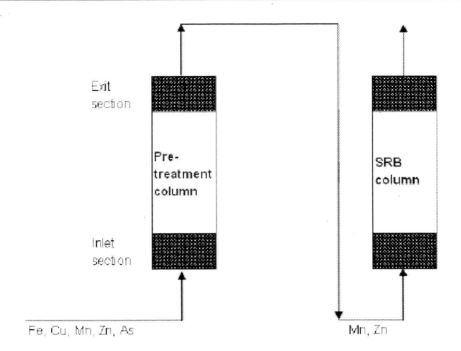

Figure 1. Schematic representation of the two-column systems for the combined chemical-biological treatment of AMD

2.5. Column Tests

Column tests of chemical precipitation and biological precipitation were performed in two distinct fixed bed columns. Both columns were made of Plexiglas (height 1 m; diameter 0.2 m; column volume, $V_b=6.65*10^{-3}$ m^3) with 10 equally distant outputs (0.1 m) along the axial length, numbered from the bottom to the top of the column.

Figure 1 reported a schematic representation of the 2-column system used in this work.

For chemical precipitation tests, the first column was filled with limestone and fed with synthetic AMD from the bottom (30 mL/h). The total pore volume was evaluated by pumping a known volume of water through the dry column and collecting the effluent. The difference is the total column pore water (2 L). Column effluents from the different outputs were monitored during time for pH, iron and metal's concentration. For biological precipitation tests, the second column was filled with a solid medium containing 80% v/v of compost, 15% v/v of cow manure, 5% v/v of straw and traces of limestone (total column pore water 0.9 L). SRB were inoculated along the whole length of the column through the ten outputs. After inoculation, the column was preliminary fed (2 months) with liquid C Medium with and then without lactate for bacteria acclimatisation. After acclimatisation, pre-treated sample of AMD coming from the 10th output of the first column were upgraded to 2000 mg/L of sulphates, and fed to the second column (10 mL/h) (Figure 1).

2.6. Analytical Methods

2.6.1. Sulphate determination

Sulphates were determined by a turbidimetric method: 30 mL of sample to be analysed were placed in a 100 mL Erlenmeyer Maier flask and mixed under magnetic stirring with 10 mL of a solution of glycerine and sodium chloride. This solution was prepared mixing two volumes of a glycerine solution (pure glycerine diluted at 50% in distilled water) with one volume of NaCl solution (200 g of NaCl + 40 mL of concentrated HCl diluted to 1 L by distilled water). After the addition of glycerine/NaCl solution, each sample was added for 5 mL of a solution of $BaCl_2$ (prepared with 90 g of $BaCl_2*2H_2O$ in 1 litre of distilled water). After two minute stirring, samples were analysed by spectrophotometer at 420 nm. Instrument calibration was performed by standard solutions of anhydrous Na_2SO_4 (dried for 1 hour in an oven at 110 °C).

2.6.2. Metal and metalloid determination

Concentrations of Fe, As, Cu, Mn and Zn in feed and treated solutions were determined by an Inductively Coupled Plasma Optical Emission Spectrometer (Varian Vista-MPX CCD Simultaneous ICP-OES). Radiation wavelengths used to identify the elements were 238.204 nm for Fe, 188.89 nm for As, 327.395 nm for Cu, 257.61 nm for Mn, 213.86 nm for Zn. Solutions for calibration were prepared by diluting 1 g/L standard solutions (Fluka) with 1% nitric acid solution. Concentrations were determined by radiation intensity of three replicates at the characteristic wavelengths.

2.6.3. Cell growth monitoring: Turbidimetry and lactate determination

Qualitative information about cell growth in liquid medium during batch tests were obtained by turbidimetry measuring the absorbance change of the suspension by a spectrophotometer (440 nm).

A lactate testing kit (K-D/LATE Astori) was used for the colorimetric measurements of lactate.

2.6.4. Chemical composition by fluorescence analysis

Chemical compositions of the native limestone and the precipitated solid obtained by chemical precipitation in pre-treatment column were determined by X-Ray fluorescence. After column emptying and drying of the filling material, the finest size fractionated sample (<355 micron) obtained by sieving of the filling material of the chemical treatment was taken as representative of precipitated solid. 3 g samples of grounded material (<66 micron) were pelletised using a press with an additive as a binder. The pellets were analysed in a SPECTRO XEPOS bench-top XRF spectrometer.

2.6.5. Cid digestion of solid samples

Acid digestion of solid samples taken from pre-treatment and SRB columns were performed to evaluate metal concentration in solid phase after treatment of synthetic AMD. Solid samples from inlet and exit sections of both columns were analysed (inlet section: solid material in the first 20 cm of column height near the bottom inlet of feed; exit section: solid material in the last 20 cm of column height near the upper output of treated effluent). Solid samples were dried in a oven (60°C for three days) and grounded (< 74 μm). Four representative sub-samples (0.1 g) of each column section were considered. Each sub-sample was placed in a Teflon recipient and digested by using 4 mL of an oxidising mixture of concentrated acids (HNO_3:HCl=3:1) and 6 mL HF, in a microwave oven (800 W, 4 minutes; 400 W, 4 minutes; 800 W, 4 minutes; 20 minutes of ventilation) (Barbaro et al., 1995). After complete digestion of solid samples, 5.6 g HBO_3 was added to avoid silica evaporation and each liquid sample was diluted to 100 mL with deionised water. Metal and metalloid concentrations in solution were determined by an Inductively Coupled Plasma Mass Spectrometer (ICP-MS).

2.6.6. Sequential extractions upon solid samples

Solid samples extracted from the inlet and exit sections of the II column after the treatment of 70 PV were used for sequential extraction tests. Solid samples were dried in a oven (60 °C for three days) and grounded (< 74 μm). Two representative sub-samples (2.5 g) of each column section were used in the following sequential extraction procedure (Campanella et al., 1995)

I Step: 2.5 g soil sample is treated by 45 mL of ammonium acetate 1 M at pH 5 by acetic acid under stirring for 24 hours at room temperature; suspension is then centrifuged at 3000 rpm for 20 minutes, diluted at 100 mL by deionised water and analysed by ICP.

II Step: The residual solid of previous step is treated by 22.5 mL of hydroxylammonium chloride 1 M and 22.5 mL of acetic acid (25%). After 24 h stirring at room temperature a solid-liquid separation is performed by centrifugation as before and the metal-bearing solution is diluted (to 100 mL) and analysed by ICP-MS.

III Step: The residual solid of previous step is treated by 12.5 mL of HCl 0.1 M and stirred for 24 h at room temperature. As in previous steps a solid-liquid separation is performed and the solution diluted to 25 mL is analysed for metal concentrations.

IV Step: The residual solid of previous step is treated by 12.5 mL of NaOH 0.5 M under stirring for 24 h at room temperature: for soil samples with large organic content this treatment should be repeated different times until a clear solution is obtained. All the solutions separated from the solids are then dried by an IR lamp at 60 °C and then digested by using 4 mL of HNO_3 (65%) and 2 mL HF (40%) in a microwave oven (250 W, 1 min; 0 W, 2 min; 250 W, 5 min; 400 W, 5 min; 600 W, 5 min). The acid solution is then diluted to 25 mL and analysed for metals by ICP-MS.

V Step: The residual solid of previous step is added to 12.5 mL of HNO_3 8 M and digested for 3 h at 80 °C. The solution is then diluted to 25 mL and analysed by ICP-MS for metal concentrations.

The residual solid of the fifth step is finally digested as described in section 2.6.5 to determine the metals in the mineralogical matrix.

3. RESULTS AND DISCUSSION

3.1. Preliminary Batch Tests of Chemical Precipitation

Limestone efficiency in iron removal was preliminary tested by batch tests, which also denoted the overall rate of the process. Replicated data showed a fast iron removal leading to 80% abatement within the first hour (Figure 2).

Simultaneously, a steep increase of pH was observed from 2.95 ± 0.05 to 4.4 ± 0.4 (mean values ± maximum dispersion by replicated data at the beginning of the experiment and after 1 hour). After the first hour, iron removal remained quite constant, while pH continued to raise up to 6.34 ± 0.04 (24 h) (Figure 2).

Limestone addition determined the increase of pH by carbonate dissolution and neutralisation of AMD acidity (equations 4 and 5). pH increase caused iron precipitation as a mixture of amorphous oxy and hydroxyl oxides (Kalin et al., 2006; Komnitsas et al., 2004) leading to 83% iron removal within 24 hours.

3.2. Preliminary Batch Test of Bioprecipitation

The efficacy of the SRB wild strain in sulphate removal was preliminary tested in batch tests using liquid medium. Biomass activity was evaluated by lactate and sulphate consumption (Figure 3).

Experimental data showed 55 % abatement of sulphates in 10 days with a slow decrease during the first three days, followed by a steep reduction (4th day) and then a slow further removal. The source of organic carbon (lactate) presented a similar trend during time. Optical density used to evaluate cell concentration was characterised by a typical transient of biological growth, with a latent phase (till the 3th day), an exponential phase and a final stationary phase (Figure 3).

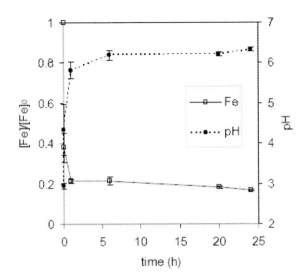

Figure 2. Preliminary batch tests of AMD treatment by chemical precipitation: iron and pH during time

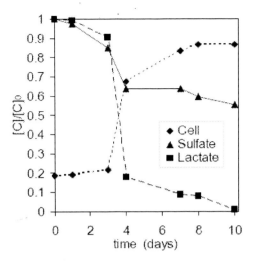

Figure 3. Preliminary batch tests of SRB growth: dimensionless concentration of biomass, lactate and sulphates during time

3.3. Column Test of Chemical Precipitation: Pre-Treatment of AMD

3.3.1. Overall performances of pre-treatment column

The efficiency of AMD pre-treatment was monitored during time for pH, iron and toxic metal concentrations in the different outputs of the column. The abatement of the different species (iron and toxic metals) were reported as dimensionless concentrations versus treated volume expressed as pore volume (PV).

pH and iron concentration in the 10th output of the column for increasing values of treated volume were reported in Figure 4.

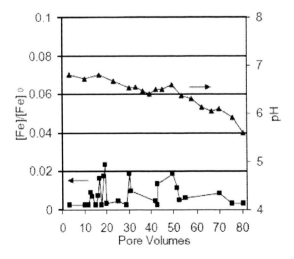

Figure 4. Column tests of chemical precipitation: iron concentration and pH in the 10th output for increasing volumes of treated AMD

It can be seen that during time a slight decrease of pH occurred due to a reduced buffering capacity of limestone. Nevertheless, a complete removal of iron was obtained (>99%) with residual concentrations ranging from 1 to 10 ppm.

The other monitored species in treated effluents were toxic elements added in synthetic AMD: arsenic, copper, zinc and manganese. Mean values of residual metals after pre-treatment (PV=70) were reported in Table 3. Arsenic and copper were completely removed by the chemical treatment, while zinc and manganese presented residual increasing concentration during the time of this experiment. Arsenic was not detected in anyone of the output stream monitored during this column experiment (instrument detection limit 0.01 ppm). Similarly, copper presented output concentrations about 0.1-0.2 ppm.

For increasing volumes of treated AMD, the formation of dark brown-reddish precipitates of iron oxides and hydroxides was observed within the column. Semi-quantitative information about the composition of such precipitates were obtained by fluorescence analysis of the finest size fraction of the filling material separated after dismantlement of pre-treatment column. This pulverulent and reddish size fraction was chosen as representative of precipitated solid mainly made up of iron oxides and hydroxides. This assumption was confirmed by fluorescence analysis (Table 4) showing that about 22% w/w of this solid was made up of iron, while native limestone presented very low concentration of this element (0.3%).

As a general trend, all toxic elements used for synthetic AMD formulation (Cu, As, Zn and Mn) presented higher concentrations in the precipitated solid than in the native limestone. On the other side typical elements of native limestone (Al, Si and K of aluminosilicate matrix) presented comparable concentrations in precipitate and limestone.

Semi-quantitative nature of fluorescence analysis (restricted only to a size fractionated sample of the column filling) did not allow to verify the material mass balance of such elements in solid and liquid phase. Nevertheless, interesting information can be obtained by enrichment factors evaluated as percent concentration in the reddish precipitate divided the percent concentration in native limestone. Such estimates from fluorescence analyses denoted that iron was 73 times more concentrated in the precipitated solid extracted from the column. In the same solid, arsenic presented a concentration 28 times larger than in native limestone. Enrichment factors for copper (14), zinc (5) and manganese (1.6) were in agreement with the trend of residual concentration observed in liquid effluents.

3.3.2. Mechanisms of abatement in pre-treatment column

Abatement of toxic metals in pre-treatment column can be related to different mechanisms and, in particular, precipitation of insoluble species and sorption onto both limestone and precipitated iron oxides. In fact, iron oxides and hydroxides present high sorbent properties due to large surface area rich of hydroxyl groups able to complex species in solution (Stumm, 1987).

Further insight into mechanisms of pollutant's removal can be then obtained by:

analysis of metal profiles along the column for increasing volumes of treated AMD,
simulation of metal speciation
acid digestion and selective extractions of metal concentration from solid samples taken
    from different sections of the columns.

Table 4. Element composition (%) of natural limestone and I column precipitate

Element	Limestone	I column Precipitate
Al	0.207	0.811
Si	1.459	0.403
S	0.122	0.519
K	0.229	0.223
Ca	34.470	15.090
Fe	0.310	22.520
Cu	0.080	1.075
Zn	0.308	1.535
Mn	0.220	0.348
As	0.004	0.109

As for species profiles along the column, pH showed an opposite trend with respect to iron and copper concentration (Figure 5): pH raised from the bottom to the top, while residual iron and copper concentrations decreased along the same direction.

The abatement of iron and copper only for sufficiently high pH environment denoted that chemical precipitation by pH raise is the main mechanism responsible for removal of both pollutants.

Conversely, the profiles of zinc along the column outputs denoted only a partial diminution of residual concentration from the bottom to the top of the column (Figure 6). Similar profiles were found for manganese (Figure 7), but in this case larger residual concentration were observed even during the early stage of pre-treatment.

The analysis of zinc and manganese profiles along the column for different volumes of treated AMD denoted the gradual diminution of abatement of these pollutants in the pre-treatment column (Figures 6 and 7). Such experimental finding denoted that sorption rather than precipitation is the main responsible mechanism operating in zinc and manganese removal.

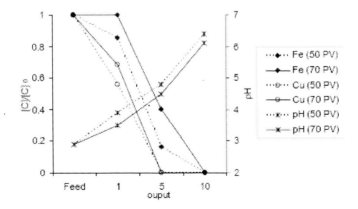

Figure 5. Column test of chemical precipitation: pH, iron, and copper profiles along the column outputs for increasing volumes of treated AMD

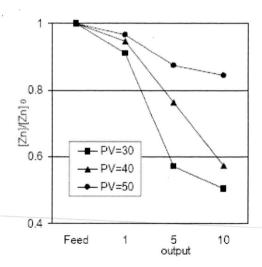

Figure 6. Column test of chemical precipitation: zinc profiles along the column outputs for increasing volumes of treated AMD

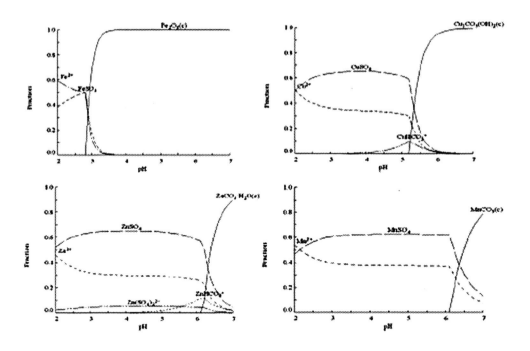

Figure 7. Column test of chemical precipitation: manganese profiles along the column outputs for increasing volumes of treated AMD

A qualitative distinction among precipitation and sorption for the different toxic species can be obtained by considering the chemical speciation for a solution such as the synthetic AMD used here (Figures 8). Speciation diagrams (as fraction of each species versus pH) were obtained by a dedicated software for chemical equilibria (Medusa, 2001) in the investigated conditions of AMD system (mean concentrations reported in Table 3).

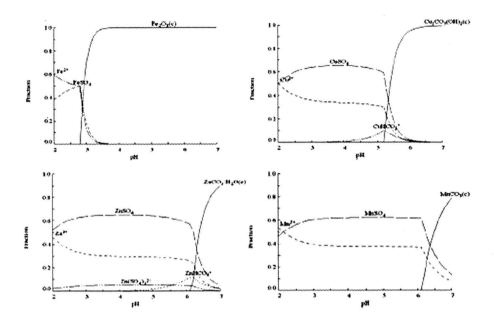

Figure 8. Speciation diagrams of iron, copper, zinc and manganese in AMD system (mean concentrations reported in Table 3; electric potential E_h=300mV)

Fraction diagrams showed that chemical pre-treatment by limestone (pH = 6) removed completely iron and copper as Fe_2O_3 (100%) and $Cu_2CO_3(OH)_2$ (95%), respectively. On the other side, both zinc and manganese remained in solution because of larger solubility of their hydroxides, oxides and carbonate salts. Partial abatement of these metals observed along the column length (Figures 6 and 7) can be due to sorption onto limestone gravel and iron hydroxide precipitates.

Arsenic speciation (nor reported here) denoted the complete solubility of this species. Nevertheless, arsenic removal during iron precipitation was already reported in the literature (Casiot et al., 2005; Gault et al., 2005). In particular, according to speciation diagrams in the operating conditions investigated in column tests, arsenic adsorption onto iron oxides and hydroxides can be considered as the predominating mechanisms in the removal of this metalloid.

Main mechanisms occurring in metal removal were further addressed by performing the acid digestion of the filling material extracted from the pre-treatment column (PV=80). Such analyses performed on the solid filling before complete exhaustion of column, can give information about the general efficiency of the process and also the specific way each metal was accumulated in solid phase by comparing pollutant content in the solid material of the bottom (inlet section) and upper (exit section) parts of the column. Table 5 reports the average values ± standard deviations obtained from the acid digestion of four distinct solid samples of the inlet section and the exit section of the pre-treatment column.

The analysis of these data showed that iron was mainly concentrated in the exit section of the column rather than in the inlet section. The same trend was observed for copper, while manganese and zinc presented similar concentrations in solid phase both in the inlet and in the exit sections.

Table 5. Acid digestion (ppm) for solid samples taken from the inlet and exit sections of the pre-treatment column (I) and the SRB column (II)

Metal	I column		II column	
	inlet section	exit section	inlet section	exit section
Fe	2500±50	5400±80	-	-
Cu	440±20	2050±50	-	-
As	28±2	53±4	-	-
Mn	325±20	290±20	870±40	4990±90
Zn	840±30	880±30	940±50	5010±60

The different behaviour of iron and copper with respect to manganese and zinc confirmed the main hypothesis about the mechanisms of their removal deduced both by residual concentrations in the liquid phase (Figures 5, 6 and 7) and by simulation of metal speciation (Figure 8). In fact, both iron and copper were completely removed in the pre-treatment column and speciation diagrams showed that these metals become insoluble for sufficiently high pH. pH gradient along the pre-treatment column (Figure 5) denoted a gradual decrease of H^+ concentration from the inlet to the exit section of the column. As a consequence, metal precipitation due to pH rise mainly occurred the column part where sufficiently high values of pH were reached (exit section). This was the case of iron and copper. On the other hand, metals which remained soluble in the investigated range of pH conditions, manganese and zinc, presented a quite uniform concentration in solid phase along the column, because their partial removal was mainly due to adsorption phenomena not specifically related to pH gradient in the range of conditions here considered. Finally, arsenic data from acid digestion denoted a significant increase in the solid phase of the exit section: this finding confirmed this metalloid can be removed both by sorption and coprecipitation with iron (this mechanism manly occurring in the exit section when iron precipitation prevailed due to pH rise).

3.4. Column Tests of Biological Precipitation

3.4.1. Overall performances of SRB column

After the start-up of the SRB inoculated column, treated effluents from the column of chemical precipitation were fed to this second column. In this second step of AMD purification, SRB reduced sulphates to sulphides, which determined metal precipitation as sulphides.

Sulphate abatement in the different outputs of the column denoted the improvement of sulphate reduction for increasing volumes of treated effluents (Figure 9).

Such effect was especially evident for the first output where the reduction of sulphate was mainly due to SRB metabolism (PV > 60). The non monotonous trends observed for the fifth output can be related to the simultaneous occurrence of sulphate sorption and sulphate reduction. In the early stage sulphate sorption onto solid matrices prevailed over bioreduction according to the well-known biosorption properties of humic substances (Dzombak et al., 1986; Tipping, 1993; Westall et al., 1998). Gradual saturation of sorption sites determined an increase in effluent concentration. Activation of SRB activity caused the further diminution especially evident in the fifth output.

Batch growth tests using the mixture of the II column as solid media for SRB growth confirmed the relevance of biosorption in sulphate removal. In fact, blank suspensions of the solid mixture without inoculum were characterised by significant sulphate abatement, even though no sulphide release was revealed (Figure 10).

After 22 days, inoculated suspensions (releasing H_2S) showed 83% sulphate abatement, while blank samples presented 58% abatement confirming that solid components used for II column filling are characterised by significant sorbing capacities. In particular sorption properties of the main component of II column filling (compost) denoted 53% abatement after 22 days showing its relevant contribution in biosorbing properties of the filling mixtures.

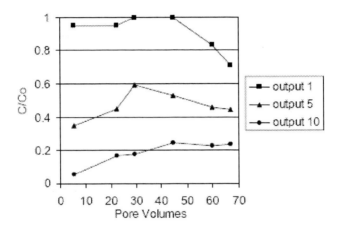

Figure 9. Column tests of biological precipitation: sulphate profiles in the different outputs for increasing volumes of treated AMD

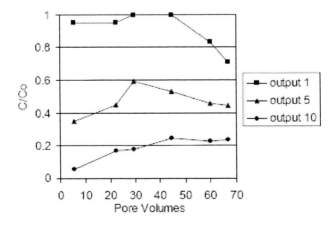

Figure 10. Sulphate abatement in batch tests using II column filling material inoculated by SRB (SRB), II column filling material without SRB inoculum (Blank), and compost used in II column filling material.

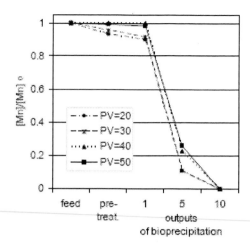

Figure 11. Column tests of biological precipitation: manganese profiles in the different outputs for increasing volumes of treated AMD.

Zinc was completely removed and can not be detected in any output of the SRB column (<0.01 ppm). Manganese profiles along the column denoted that sulphate reducing bacteria metabolism and simultaneous biosorption onto solid substrates completely remove this metal (Figure 11).

3.4.2. Mechanisms of abatement in SRB column

Further information about the removal of both these metals were obtained by comparing their concentrations in solid samples extracted from the inlet and exit sections of the SRB column before its exhaustion (after 60 PV).

Average values from acid digestions (Table 5) denoted that both zinc and manganese were mainly concentrated in the solid material taken from the upper part of the column (exit section).

A deeper insight into metal removal can be obtained by sequential extractions: these procedures are all based on the general principle of reacting a soil sample with chemical solutions characterised by increasing strength. In particular sequential extraction scheme used here was a modified Tessier procedure, which allows an operative speciation of metal in the solid phase according to the following scheme:

I Step: exchangeable metals;
II Step: complexed onto iron and manganese oxides;
III Step: weakly bound to organic matter;
IV Step: strongly bound to organic matter;
V Step: bound to the sulphide phase.

This sequential extraction scheme was chosen to isolate the contributions due to biosorption (III+IV steps) and bioprecipitation (IV step) in Mn and Zn abatement in comparison with other extraction procedures reported in the literature (Swash and Monhemius, 2005) that are characterised by a unique extractive step for metals bound to organics and metals precipitated as sulphides.

Table 6. Sequential extractions (expressed as % of extracted metals) for solid samples taken from the inlet and exit sections of the pre-treatment column (I) and the SRB column (II)

Sequential extractions						
Metal	Section	I Step	II Step	III+IV Steps	V steps	Residue
Mn	Inlet	23	46	30	1	1
Mn	Exit	33	31	31	4	1
Zn	Inlet	22	34	43	0	1
Zn	Exit	17	23	56	3	1

Nevertheless, it should be pointed out that only semi-quantitative comparison can be obtained by sequential extractions due to the numerous limits of this experimental procedure (low reproducibility, error propagations, strong influence of operating conditions, effective selectivity of the extracting reagents, re-adsorption of metals during extraction) (Pagnanelli et al., 2004). Experimental results of sequential extractions were reported in Table 6 as % of extracted metal in each step. This data denoted the predominant contribution of the biosorbing properties of biological matrices both in Mn and Zn removal and the localisation of SRB activities in the upper parts of the column. This second finding can be due to the migration of gaseous H_2S in the upper part of the column and consequently the localisation in such compartment of the precipitation of metal as sulphides.

4. CONCLUSION

In this study a combined chemical-biological treatment was tested for the decontamination of synthetic AMD containing Fe, Cu, As, Zn, and Mn.

Column tests denoted that this combined approach succeeded in attenuating AMD pollution. In particular, pre-treatment by natural limestone increased the pH allowing the complete removal of metals with low solubility as hydroxides and carbonates (Fe and Cu). A refining treatment in the second column (made up of natural organic matter inoculated by SRB) allowed the removal of those metals which can not be removed simply by increasing pH (Mn and Zn). Treated effluents from the second column were purified by these residual pollutants initially contained in AMD. Specific mechanisms of removal for each metal were investigated by considering profiles of pollutants along the column length, simulations of metal speciation in the investigated conditions, biosorption tests, acid digestion and sequential extractions from solid samples of column filling material.

The combination of such results led to the following main conclusions:

- iron was completely removed by precipitation in the pre-treatment column, whose active zone was only that with adequate pH conditions as evidenced by acid digestion of column sections;
- copper was completely precipitated in the first treatment column by the same chemical mechanism observed for iron;

manganese and zinc cannot be removed only for pH rise: dynamic trends of axial profiles showed a shifting front of saturation along the column and denoted that marginal removals in the first column were due only to sorption onto limestone and iron precipitates; both metals were completely removed in the second column of biological treatment by biosorption and bioprecipitation mechanisms;

arsenic was removed in the first column both by sorption and coprecipitation with iron being concentrated in the solid from the column section active for iron precipitation.

In conclusion, this work presents a methodological approach to developing a lab-study investigation able to isolate the different mechanisms simultaneously occurring in pollutant removal in biological treatment by SRB. Such preliminary quantification should be the basis for further kinetic modelling in order to determine the half-lives of the different pollutants specifically related to SRB metabolism, which is the only long-term active mechanism in full-scale PRB. Other mechanisms such as biosorption should be isolated and not accounted for in such kinetic parameters in order to avoid overestimates of SRB performances in long term run.

Experimental results reported here showed that biosorption is not a negligible mechanism in lab-scale study generally used for half-lives estimates. Column tests can not then be used as stand alone data for kinetic characterisation of such biological systems but should be accompanied by further investigation specifically aiming to mechanism isolation and quantification.

REFERENCES

Alvarez, M. T., Crespo, C. & Mattiasson, B. (2007). Precipitation of Zn(II), Cu(II) and Pb(II) at bench-scale using hydrogen sulphide from the utilization of volatile fatty acids *Chemosphere*, *66*, 1677-1683.

Barbaro, M., Passariello, B., Quaresima, S., Casciello, A. & Marabini, A. (1995). Analysis of Rare Earth Elements in Rock Samples by Inductively Coupled Plasma-Mass Spectrometry (ICP-MS). *Microchem, J.*, *51*, 312-318.

Baker, B. J. & Banfield, J. F. (2003). Microbial communities in acid mine drainage. *FEMS Microbiol. Ecol.*, *44*, 139-152.

Benner, S. G., Blowes, D. W., Gould, W. D., Herbert, R. B. & Ptacek C. J. (1999). Geochemistry of a Permeable Reactive Barrier for Metals and Acid Mine Drainage. *Environ. Sci. Technol.*, *33*, 2793-2799.

Campanella, L., D'Orazio, D., Petronio, B. M. & Pietrantonio, E. (1995). Proposal for a metal speciation study in sediments. *Analytica Chimica Acta,* *309*, 387-393.

Carrera, J., Alcolea, A., Bolzicco, J., Knudby, C. & Ayora, C. (2001). *An experimental geochemical barrier at Aznalcollar*. Thornton and Oswald Eds Proc. 3rd Internat. Conf. Groundwater Quality Sheffield UK, 18-21 June, 407-409.

Casiot, C., Lebrun, S., Morin, G., Bruneel, O., Personne´, J. C. & Elbaz-Poulichet, F. (2005). Sorption and redox processes controlling arsenic fate and transport in a stream impacted by acid mine drainage *Sci. Total Environ.*, *347*, 122-130.

Chang, I. S., Shin, P. K. & Kim, B. H. (2000). Biological treatment of acid mine drainage under sulphate-reducing conditions with solid waste materials as substrate. *Wat. Res., 34,* 1269-1277.

Christensen, B., Laake, M. & Lien T. (1996). Treatment of acid mine water by sulfate-reducing bacteria; results from a bench scale experiment. *Wat. Res., 30,* 1617-1624.

Cocos, I. A., Zagury, G. J., Clement, B. & Samson, R. (2002). Multiple factor design for reactive mixture selection for use in reactive walls in mine drainage treatment. *Wat. Res., 32,* 167-177.

Dzombak, D. A., Fish, W. & Morel F. M. M. (1986). Metal-humate interactions. 1. Discrete ligand and continuous distribution models. *Environ. Sci. Technol., 20,* 669-675.

Foucher, S., Battaglia-Brunet, F., Ignatiadis, I.& Morinet, D. (2001). Treatment by sulfate-reducing bacteria of chessy acid-mine drainage and metals recovery. *Chemical Engineering Science, 56,* 1639-1645.

Gault, A. G., Cooke, D. R., Townsend, A. T., Charnock, J. M. & Polya D. A. (2005). Mechanisms of arsenic attenuation in acid mine drainage from Mount Bischoff, western Tasmania *Sci. Total Environ., 345,* 219-228.

Gibert, O., de Pablo, J., Cortina, J. L. & Ayora C. (2002). Treatment of acid mine drainage by sulphate-reducing bacteria using permeable reactive barriers: A review from laboratory to full-scale Experiments. *Re/Views in Environmental Science and Bio/Technology, 1,* 327-333.

Gibert, O., de Pablo, J., Cortina, J. L. & Ayora, C. (2003). Evaluation of municipal compost/limestone/ iron mixtures as filling material for permeable reactive barriers for in-situ acid mine drainage treatment. *J. Chem. Technol. Biot., 78,* 489-496.

Gibert, O., de Pablo, J., Cortina, J. L. & Ayora, C. (2004). Chemical characterization of natural organic substrates for biological mitigation of acid mine drainage. *Water Res., 38,* 4186-4196.

Groudev, S. N., Spasova, I. I. & Georgiev, P. S. (2001). In situ bioremediation of soils contaminated with radioactive elements and toxic heavy metals. *Int. J. Miner. Process., 62,* 301-308.

Hammack, R. W. & Edenborn H. M. (1992). The removal of nickel from mine waters using bacterial sulfate reduction. *Appl. Microbiol. Biot., 37,* 674-678.

Herbert, R. B., Benner, S. G. & Blowes, D. W. (2000). Solid phase iron-sulfur geochemistry of a reactive barrier for treatment of mine drainage. *Applied Geochemistry, 15,* 1331-1343.

Jarvis, A. P., Moustafa, M., Orme, P. H. A. & Younger, P. L. (2006). Effective remediation of grossly polluted acidic, and metal-rich, spoil heap drainage using a novel, low-cost, permeable reactive barrier in Northumberland, UK. *Environ. Pollut., 143,* 261-268.

Johnson, D. B. (2006). Biohydrometallurgy and the environment: intimate and important interplay *Hydrometallurgy, 83,* 153-166.

Johnson, D. B. & Hallberg, K. B. (2005). Acid mine drainage remediation options: a review. *Sci. Total Environ., 338,* 3-14.

Johnson, D. B. & Hallberg K. B. (2005b). Biogeochemistry of the compost bioreactor components of a composite acid mine drainage passive remediation system *Science Total environment, 338,* 81-93.

Kalin, M., Fyson, A. & Wheeler, W. N. (2006). The chemistry of conventional and alternative treatment systems for the neutralization of acid mine drainage. *Sci. Total Environ.*, *366*, 395-408.

Komnitsas, K., Bartzas, G. & Paspaliaris, I. (2004). Efficiency of limestone and red mud barriers: laboratory column studies. *Miner. Eng.*, *17*, 183-194.

Ludwig, R. D., McGregor, R. G., Blowes, D. W., Benner, S. G. & Mountjoy, K. (2002). A permeable reactive barrier for treatment of heavy metals. *Ground Water*, *40*, 59-66.

Medusa Program by ©Ignasi Puigdomenech, 2001

Morrison, S. J., Matzler, D. R. & Dwyer B. P. (2002). Removal of As, Mn, Mo, Se, U, V and Zn from groundwater by zero-valent iron in a passive treatment cell: reaction progress modelling. *J. Contaminant Hydrology*, *56*, 99-116.

Neculita, C. M., Zagury, G. J. & Bussiere, B. (2007). Passive treatment of acid mine drainage in bioreactors using sulphate reducing bacteria: Critical review and research needs. *J. Environ. Qual.*, *36*, 1-16.

Pagnanelli, F., Moscardini, E., Giuliano, V. & Toro, L. (2004). Sequential extraction of heavy metals in river sediments of an abandoned pyrite mining area: pollution detection and affinity series. *Environ. Pollut.*, *132*, 189-201.

Pagnanelli, F., Luigi, M., Mainelli, S. & Toro, L. (2007). Use of natural materials for the inhibition of iron oxidizing bacteria involved in the generation of acid mine drainage. *Hydrometallurgy*, *87*, 27-35.

Peppas, A., Komnitsas, K. & Halikia, I. (2000). Use of organic covers for acid mine drainage control. *Miner. Eng.*, *13*, 563-574.

Postgate J. R. (1979). *The sulphate-reducing bacteria.* Cambridge University Press.

Sheoran, A. S. & Sheoran, V. (2006). Heavy metal removal mechanism of acid mine drainage in wetlands: A critical review. *Miner. Eng.*, *19*, 105-116.

Simon, M., Martın, F., Garcıa, I., Bouza, P., Dorronsoro, C. & Aguilar, J. (2005). Interaction of limestone grains and acidic solutions from the oxidation of pyrite tailings. *Environ. Pollut.*, *135*, 65-72.

Stumm, W. (1987). *Aquatic surface chemistry* John Wiley and Sons, Inc.

Swash, P. M. & Monhemius, A. J. (2005). Characteristics and stabilities of residues from the Wheal Jane constructed wetlands. *Science Total Environment*, *338*, 95-105.

Tipping, E. (1993). Modeling ion binding by humic acids. *Colloids Surf. A*, *73*, 117.

Tunay, O. & Kabdasli, N. I. (1994). Hydroxide precipitation of complexed metals. *Water Research*, *28*, 2117-2124.

Waybrant, K. R., Blowes D. W. & Ptacek C. J. (1998). Selection of Reactive Mixtures for Use in Permeable Reactive Walls for Treatment of Mine Drainage. *Environ. Sci. Technol.*, *32*, 1972-1979.

Waybrant, K. R., Ptacek, C. J. & Blowes D. W. (2002). Treatment of mine drainage using permeable reactive barrers: column experiments. *Environ. Sci. Technol.*, *36*, 1349-1356.

Westall, J. C., Jones, J. D., Turner, G. D. & Zachara, J. M. (1995). Models for Association of Metal Ions with Heterogeneous Environmental Sorbents. 1. Complexation of Co(II) by Leonardite Humic Acid as a Function of pH and $NaClO_4$ Concentration. *Environ. Sci. Technol.*, *29*, 951-959.

Whitehead, P. G., Hall, G., Neal, C. & Prior, H. (2005). Chemical behaviour of the Wheal Jane bioremediation system. *Science Total Environ, 338*, 41-51.

Wilkin, R. T. & McNeil, M. S. (2003). Laboratory evaluation of zero-valent iron to treat water impacted by acid mine drainage. *Chemosphere, 53*, 715-725.

INDEX

A

abatement, x, 54, 122, 144, 157, 240, 242, 244, 250, 251, 252, 253, 255, 256, 257, 258
abiotic, 241, 242
access, 110, 111, 112, 130, 141, 142
accessibility, 5
acclimatization, 49
accounting, 8, 65, 138, 167, 183, 228
acetate, 246, 249
acetic acid, 25, 249
acid mine drainage, 261
acidic, 261, 262
activated carbon, 27
activation energy, 79, 159
active treatment, 10, 11, 30, 48, 53, 190, 191, 194, 197, 199
adjustment, 25, 195, 244
adsorption, 6, 9, 11, 14, 19, 23, 27, 28, 30, 32, 33, 34, 53, 54, 60, 65, 67, 148, 207, 213, 242, 244, 255, 256, 259
aerobe, 131
aesthetics, 190
age, 21, 28, 34, 44, 73, 74, 94, 102
ageing, 80, 112
agencies, 2, 157
agents, 242
aggregates, 200, 204
aggregation, 200
agriculture, 28
algae, 28, 53, 58, 64
alkalinity, 242, 244
alloys, 103
alternatives, 15, 19, 140, 242, 244, 262
alters, 200
aluminium, viii, ix, 68, 72, 95, 106, 109, 139, 151, 171, 175, 176, 211, 215, 244

aluminosilicate, 252
aluminum, 2, 7, 8, 9, 11, 13, 14, 15, 16, 18, 23, 25, 26, 27, 28, 32, 37, 38, 39, 44, 46, 56, 58, 61, 129, 133, 191, 193, 194, 200, 207, 208, 221, 222
aluminum oxide, 191, 208
AMD, x, 239, 240, 241, 242, 243, 244, 245, 246, 247, 248, 250, 251, 252, 253, 254, 255, 256, 257, 258, 259
ammonia, 54, 190, 191, 192, 194, 195, 196, 198, 202, 205, 214
ammonium, 15, 40, 52, 194, 201, 249
amorphous, 250
anaerobic sludge, 24, 25
ANC, 171, 173, 174
angiosperm, 108
antimony, viii, 72, 73, 75, 76, 77, 90, 94, 99, 103, 106, 109, 176
applications, ix, 20, 26, 141, 147, 150, 206, 227, 228, 229, 231, 235, 236, 241, 242, 243, 244
aqueous solutions, 56, 58, 65, 80, 84, 86, 198
aquifers, 2, 40, 166
arsenic, viii, x, 2, 6, 9, 14, 16, 43, 46, 64, 72, 73, 90, 92, 94, 95, 96, 103, 105, 106, 109, 122, 158, 161, 175, 176, 178, 180, 183, 187, 188, 215, 239, 252, 255, 256, 260, 261
ash, 6, 11, 27, 31, 63, 102, 190, 194, 195, 196, 213, 217, 219, 225, 226, 241
assessment, 31, 55, 58, 61, 112, 116, 152, 164, 165, 166, 171, 184, 207, 220, 228, 236, 237, 241
atomic emission spectrometry, 10

B

bacteria, x, 4, 10, 11, 23, 25, 27, 28, 31, 34, 35, 36, 38, 47, 54, 56, 58, 61, 62, 65, 66, 79, 100, 110, 121, 123, 124, 125, 127, 129, 130, 131, 132, 137,

138, 141, 142, 145, 146, 152, 153, 159, 212, 239, 240, 241, 242, 244, 245, 246, 247, 258, 261, 262
bacterial, 244, 246, 261
bacterium, 145
banks, 100, 108
barium, 75, 76, 77, 176
barrier, x, 239, 242, 243, 244, 260, 261, 262
barriers, x, 36, 40, 41, 52, 63, 68, 239, 242, 243, 244, 261, 262
bauxite, 160
bicarbonate, 12, 23, 35, 36, 38, 43, 191, 193, 194, 208, 213
binding, 19, 200, 262
bioaccumulation, 34, 213
bioassay, 112
bioavailability, 40, 115, 228
biochemistry, 53
biodegradable, 242
biodegradable materials, 242
biodegradation, 245
biodiversity, 51, 108
bioindicators, 108, 112
biological, x, 239, 240, 241, 242, 243, 244, 245, 247, 250, 257, 258, 259, 260, 261
biological activity, 33, 120
biological media, 132
biological processes, 44, 197
biological systems, 260
biologically, 240, 242
biomass, 27, 28, 32, 34, 49, 50, 58, 59, 65, 66, 67, 68, 245, 251
biopolymer, 21
bioreactors, 261, 262
bioremediation, 67, 261, 263
biosorption, x, 239, 240, 244, 245, 256, 257, 258, 259, 260
biotechnological, 244
biotic, 3, 4, 9, 30, 31, 33, 132, 149, 158
bismuth, 90, 94, 99, 103
blocks, 99, 103, 105
bonding, 199
boreholes, viii, 40, 155, 162, 166, 169, 171, 172, 175, 176, 183, 207
breeding, 108, 110
buffer, 37
by-products, 5, 48, 49, 73

C

Ca^{2+}, 11, 12, 23, 123, 134, 135, 171, 180
cadmium, viii, 2, 4, 6, 9, 13, 16, 18, 20, 21, 22, 23, 26, 28, 43, 46, 60, 72, 73, 94, 95, 99, 106, 109, 122, 128, 158, 176, 179, 215
calcium, 2, 10, 11, 12, 13, 14, 15, 16, 21, 23, 34, 35, 37, 38, 46, 134, 139, 171, 174, 175, 191, 194, 195, 198, 208, 215, 246
calcium carbonate, 11, 14, 15, 21, 34, 46, 191, 208, 246
calibration, 248
campaigns, ix, 227
capacity, 252
carbon, 9, 12, 14, 17, 18, 19, 25, 32, 34, 36, 42, 44, 46, 49, 50, 59, 72, 124, 131, 132, 243, 244, 250
carbon dioxide, 12, 14, 17, 18, 25, 34, 72, 131
carbonates, 244, 259
carcinogen, 19
carrier, 23, 25, 29, 59
catalysis, 136
catalyst, 47
catalytic activity, 159
cation, 9, 23, 30, 34, 141, 169
cauterization, 167
cell, 37, 43, 46, 66, 246, 248, 250, 262
cell growth, 248
cellulose, 28, 36, 60, 243, 246
channels, 32, 38, 41, 50, 69, 73, 79, 95, 96, 142, 190, 197
charge density, 200
chemical, x, 239, 240, 244, 245, 246, 247, 248, 250, 251, 252, 253, 254, 255, 256, 258, 259
chemical bonds, 200
chemical composition, 245
chemical modeling, 143
chemical properties, 137, 195
chemical reactions, 72, 73, 130, 169
chemical stability, 13, 21, 49
chemicals, 241, 245
chemisorption, 27, 33
chemistry, x, 240, 262
chloride, 248, 249
chromatography, 10, 214
chromium, 2, 20, 21, 22, 23, 24, 29, 53, 106, 122, 176, 215
circulation, 94, 139
civil society, 165
civil war, 102
classes, 103
classification, 74, 89, 90, 105, 107, 167, 193
clay minerals, 135, 175
clays, 241
cleavages, 79
climate, 66, 120, 123
closure, 7, 68, 103, 150, 160, 205, 228, 240
CO2, 12, 13, 31, 64, 123, 124, 134, 135, 136, 170, 176, 214
coagulation, 13, 14, 25, 53, 194, 196, 202

coatings, 13, 18
cobalt, viii, 4, 20, 23, 25, 28, 58, 60, 72, 94, 95, 99, 103, 106, 109, 122, 215
colonization, 73
combined effect, 38, 200
combustion, 5, 6, 57, 88, 132, 194, 196
communication, 184
communities, 260
community, 60, 112, 160, 187, 213
complex interactions, 142
complexity, 83
compliance, 220
components, ix, 66, 85, 124, 130, 211, 217, 224, 242, 244, 246, 257, 261
composite, 261
composites, 141
composition, ix, 9, 53, 55, 80, 83, 85, 86, 94, 96, 112, 123, 128, 135, 140, 147, 151, 152, 154, 156, 163, 166, 170, 187, 189, 191, 208, 211, 215, 216, 217, 221, 224, 245, 248, 252, 253
compositions, 248
compost, 18, 29, 30, 35, 36, 37, 38, 39, 40, 41, 43, 44, 50, 57, 68, 242, 243, 246, 247, 257, 261
compounds, 19, 30, 34, 37, 42, 60, 123, 127, 131, 132, 136, 195, 200
compression, 203, 204
computation, 171
concentrates, 77, 93, 97, 150, 151
condensation, 27
conditioning, ix, 189, 199, 200, 208
conductivity, vii, 39, 41, 71, 92, 95, 100, 106, 108, 176
configuration, 18, 25, 39, 233, 244
connectivity, 8
conservation, 108, 110
consolidation, 196, 204
constructed wetlands, 30, 33, 53, 54, 56, 59, 60, 61, 197, 262
construction, vii, 5, 42, 53, 57, 207
consulting, 147
consumption, 8, 143, 156, 159, 250
contact time, 207, 219
contaminant, 5, 9, 10, 29, 30, 32, 36, 48, 133
contaminated soils, 109
contamination, 6, 19, 48, 63, 67, 79, 90, 94, 103, 107, 108, 118, 120, 153, 156, 160, 161, 176, 198, 228, 236, 237
control, viii, 5, 6, 55, 59, 62, 63, 64, 77, 92, 97, 119, 122, 140, 141, 142, 143, 146, 148, 152, 208, 241, 262
control measures, 143
convergence, 9
conversion, 124, 126, 130, 226

copyright, iv
correlation, 179, 231
correlation function, 231
cost-benefit analysis, 68
costs, viii, ix, 14, 15, 18, 24, 48, 49, 64, 119, 144, 189, 195, 204, 227, 228
cotton, 204
covering, 141, 165
critical value, 229, 235
crystal growth, 25
crystal structure, 138, 148
crystalline, 103, 127, 171, 195, 221
crystallinity, 15, 137
crystallization, 25, 29, 64, 83, 106
crystals, 137, 147, 198, 215
cultivation, 246
culture, 109, 148
cyanide, 50, 180
cycles, 73
cycling, 34, 51, 148, 149

D

data collection, 215
data set, 179
dating, 73, 236
death, 213
decay, 7, 180
decision making, 228
decomposition, 34, 51, 59
decontamination, x, 239, 259
defects, 79, 137, 138
deformation, 92
degradation, 36, 73, 121, 131, 144, 156, 178, 179, 188, 213, 221
degraded area, 73
dehydrate, 127
dehydration, 84
denitrifying, 35
density, x, 15, 16, 17, 38, 61, 76, 122, 148, 190, 194, 197, 198, 201, 208, 227, 229, 233, 235, 237, 250
Department of Energy, 67, 226
Department of the Interior, 54, 56, 57, 58, 61
deposition, 151, 185, 237, 244
deposits, vii, 2, 9, 49, 50, 62, 66, 71, 72, 74, 75, 76, 77, 79, 88, 96, 97, 101, 113, 114, 117, 131, 152, 157, 161, 162, 184, 185, 186, 187, 232, 235, 236
derivatives, 194
dermatitis, 161, 183
desorption, 30
destruction, 32, 91, 199, 206, 213
detection, 96, 139, 179, 252, 262
developed countries, viii, 155, 159

diffusion, 5, 27, 52, 130, 141, 142, 159, 241
digestion, 248, 252, 255, 256, 259
diluent, 29
discharges, 7, 53, 68, 77, 99, 132, 136, 160, 198, 212, 213, 225
discrimination, 176
disequilibrium, 130
dismantlement, x, 240, 245, 252
dispersion, 73, 77, 79, 90, 91, 92, 107, 200, 250
dissolved oxygen, 8, 37, 39, 47, 67, 112, 121, 140, 141, 151
distilled water, 245, 248
distribution, 56, 73, 90, 98, 105, 108, 114, 124, 144, 149, 199, 233, 235, 261
diversity, 24, 83
dominance, 108, 178
dosage, 200
dosing, 12, 13, 16, 46, 49
drawing, 109, 204
drinking water, 180, 183
dry, 247
drying, 15, 141, 204, 205, 206, 208, 248
durability, 49
duration, x, 2, 219, 224, 240
dyes, 63
dykes, 167, 180
dynamics, 32, 34, 36, 144, 149

E

earth, vii, x, 56, 227, 231, 236, 237
ecology, 58, 144, 154
ecosystem, 56, 68, 213
editors, 144, 154
efficacy, 250
effluent, ix, 11, 14, 15, 17, 18, 19, 21, 22, 24, 25, 26, 32, 33, 34, 35, 37, 43, 44, 45, 48, 53, 57, 58, 59, 63, 66, 96, 100, 102, 105, 122, 129, 133, 135, 137, 146, 147, 162, 189, 190, 207, 220, 247, 248, 257, 259
electric conductivity, vii, 71, 80, 92, 95, 99, 100, 106
electric potential, 255
electrical conductivity, 175, 214, 215
electrochemistry, 150
electroflotation, 19
electron, 10, 23, 25, 46, 131, 132, 148, 151, 195, 242
electron microscopy, 151
electrons, 80
electroplating, 21, 50, 51
encapsulation, 52
endangered species, 108
endothermic, 208

energy, 5, 11, 21, 29, 48, 49, 50, 124, 126, 131, 132, 199
energy supply, 48
engineering, 56, 58, 120, 244
environment, 13, 18, 25, 32, 34, 43, 58, 59, 72, 75, 77, 78, 86, 87, 106, 107, 108, 112, 116, 120, 121, 123, 130, 131, 134, 140, 146, 148, 151, 153, 207, 213, 220, 221, 253, 261
environmental, x, 239, 240
environmental audit, 161, 163
environmental conditions, 108, 130, 223
environmental impact, viii, ix, 68, 72, 73, 74, 77, 78, 99, 103, 110, 112, 116, 121, 122, 155, 157, 186, 212, 227, 228, 229
environmental protection, 2, 160
Environmental Protection Agency (EPA), 19, 42, 43, 62, 65, 144, 153, 155, 160, 164, 180, 183, 184, 188
enzymes, 124
equilibrium, 3, 8, 28, 53, 83, 84, 85, 86, 109, 123, 146, 149, 152, 170, 175, 200
equipment, 98, 190, 194, 206
erosion, 6, 33, 90, 91, 94, 97
estimating, 235
ethanol, 25, 26
evaporation, 4, 6, 32, 60, 77, 78, 89, 99, 206, 249
evolution, 50, 53, 145, 146
exclusion, 130, 140, 145, 165
experimental condition, 224
exploitation, vii, viii, 71, 72, 75, 76, 77, 88, 93, 97, 99, 102, 103, 115, 228
exponential, 246, 250
exporter, 31
exposure, 2, 37, 67, 78, 120, 122, 130, 140, 142, 163, 176
extraction, 5, 19, 29, 52, 53, 58, 65, 105, 116, 169, 180, 212, 235, 249, 258, 259, 262

F

facies, 161
factor analysis, 188
factories, 102, 103
fatty acids, 260
ferric ion, 112, 132, 136, 142, 212
ferrite, 14, 57
ferrous ion, 212
films, 129
filters, 24, 46, 47, 204, 205, 246
filtration, 11, 14, 15, 19, 20, 21, 23, 26, 58, 191, 197, 199, 200, 204, 205, 207, 208, 242
financial support, 225
fire retardants, 7

floating, 31, 46, 108
flocculation, 25, 27, 196, 202
flood, 5, 8, 104
flooding, 6, 94, 129, 141, 241
flora and fauna, 213
flotation, 5, 11, 14, 19, 20, 27, 49, 60, 63, 68, 69
flow, 244
fluctuations, 7, 9, 42, 43
flue gas, 194, 196
fluidized bed, 17, 18, 194, 196
fluorescence, 226, 248, 252
food, 73, 213
foreign exchange, 159
formula, 81, 171, 213, 231, 234
fouling, 21, 49, 53
fractures, 5, 43, 79, 129
fragments, 136
freezing, 32
freshwater, 2, 58, 185
fuel, 196
funding, 225
fungi, 27, 28, 47, 159

G

gallium, 29, 59
gases, 137
generation, viii, ix, x, 2, 5, 6, 8, 33, 37, 38, 51, 52, 60, 62, 72, 73, 78, 79, 87, 105, 119, 120, 121, 122, 123, 125, 130, 132, 136, 137, 139, 140, 142, 143, 149, 150, 154, 189, 194, 227, 228, 229, 232, 236, 240, 241, 244, 262
geochemical, 260
geochemistry, 261
geology, viii, 77, 99, 110, 118, 140, 142, 152, 155, 166, 178, 179, 184, 185, 188, 235
geometrical parameters, 235
germanium, 103
glycerine, 248
gold, viii, 23, 29, 50, 58, 72, 76, 88, 90, 92, 94, 99, 124, 232, 233
grades, 76, 77, 88, 93, 102
grading, 163, 165, 233
grains, 18, 132, 137, 139, 190, 262
graphite, 215
gravity, 37, 97, 203, 204, 206, 208
grids, 229
groundwater, 2, 6, 8, 10, 40, 41, 42, 50, 52, 55, 56, 60, 63, 64, 107, 150, 156, 160, 166, 169, 170, 171, 172, 173, 175, 176, 183, 185, 212, 262
groups, 27, 30, 90, 103, 200, 207, 252
growth, 24, 34, 39, 46, 52, 62, 108, 130, 131, 203, 242, 244, 245, 246, 248, 250, 251, 257

growth rate, 108
Guangdong, 69
guidelines, ix, 49, 62, 111, 156, 157, 160

H

habitat, 46, 108, 112, 132, 190, 213
hardness, 122, 162, 170, 171, 193
hazards, 15, 122, 213
HDPE, 214
health, 67, 120, 122, 147, 228, 235
health care, 235
heat, 126, 128, 131, 139, 145, 147
heat release, 131
heat transfer, 147
heating, 197
height, 203, 230, 247, 248
heterogeneity, ix, 105, 227, 228, 235, 236
high density polyethylene, 214
host, vii, 8, 9, 71, 73, 79, 80, 90, 91, 94, 95, 99, 102, 103, 117
hot spots, 228, 235
humate, 261
humic acid, 262
humic substances, 256
humidity, 80, 83, 84, 106, 111, 129, 153
hybrid, 24
hydrogen, 2, 3, 4, 6, 8, 10, 11, 12, 23, 24, 25, 26, 28, 36, 39, 133, 141, 169, 199, 212, 223, 260
hydrogen gas, 10
hydrogen peroxide, 6, 12, 23, 141
hydrolysis, 3, 4, 23, 37, 124, 136, 169
Hydrometallurgy, 261, 262
hydron, 79
hydrophyte, 108
hydrothermal activity, 232
hydrothermal system, 72
hydroxide, 10, 11, 12, 13, 14, 15, 16, 18, 19, 24, 26, 27, 31, 33, 37, 38, 41, 47, 49, 50, 83, 84, 86, 124, 126, 130, 132, 136, 159, 193, 194, 195, 196, 197, 198, 199, 207, 242, 255
hydroxides, 252, 255, 259
hydroxyl, 250, 252
hydroxyl groups, 252
hypothesis, 170, 256

I

Iberian Pyrite Belt, v, vii, 71, 72, 73, 74, 75, 76, 77, 81, 92, 96, 108, 110, 111, 112, 113, 114, 115, 116, 117, 118
images, 107, 228, 236

immersion, 130, 205
immobilization, 67
Impact Assessment, ix, 156, 160, 184, 187
impacts, 160, 166, 187
implementation, 228, 235
impurities, 179
inclusion, 237
indicators, 143
indigenous, 110, 112
indium, 103
industry, viii, 28, 53, 57, 62, 77, 98, 108, 119, 120, 143, 147, 196, 205, 212
inefficiency, 195, 199
infinite, 230
infrastructure, 94, 97, 98, 190
inhibition, 18, 62, 121, 141, 241, 262
initiation, 140
inoculation, 49, 247
inoculum, 246, 257
insight, 252, 258
instability, 94
intensity, 248
interaction, 44, 78, 92, 95, 135, 150, 151, 215, 261, 262
interface, 79, 144
interference, 62
intrusions, viii, 155, 165
ion-exchange, 53, 62
ionization, 84, 200
ions, 3, 4, 10, 11, 12, 21, 22, 23, 27, 28, 38, 40, 50, 55, 58, 59, 60, 61, 63, 64, 65, 66, 67, 68, 69, 79, 80, 84, 86, 106, 108, 128, 133, 146, 166, 169, 178, 179, 190, 212, 213, 214, 215, 221, 223, 224, 226
iris, 35
isolation, 142, 245, 260
isotope, 66, 153

K

K^+, 23, 82, 171, 175
kerosene, 19, 29
kidney, 176
kinetic model, 30, 260
kinetic parameters, 260
kinetics, 14, 18, 29, 31, 47, 52, 62, 112, 118, 151, 154, 180, 207

L

labor, 206
lakes, 9, 123, 143, 144, 146, 148, 150, 153, 156, 213

lamella, 17
laminar, 213
land, 5, 11, 15, 29, 30, 46, 47, 48, 49, 50, 140, 145, 204, 206
land use, 140
landfills, 49, 77, 79, 102
landscape, 73, 74, 90, 99, 108
laws, 114, 150
layering, 150
leaching, vii, ix, 5, 6, 8, 15, 34, 51, 60, 61, 77, 89, 90, 102, 104, 105, 144, 145, 146, 147, 148, 150, 151, 154, 156, 157, 158, 160, 161, 163, 166, 180, 183, 207, 208
lead, 246
Leonardite, 262
lesions, 183
liberation, 120
lifespan, 6, 43
ligand, 18, 52, 261
liquid phase, 17, 22, 27, 252, 256
literature, 244, 246, 255, 258
living conditions, 148
longevity, 2, 6, 29, 32, 36, 37, 44, 64, 68
long-term, 241, 260
low temperatures, 46
Luo, 60

M

macromolecules, 20
magnesium, 2, 11, 13, 14, 15, 23, 35, 139, 171, 174
maintenance, 2, 5, 11, 25, 29, 32, 37, 48, 97
majority, 10, 16, 73, 121
manure, 36, 39, 40, 42, 242, 243, 246, 247
mapping, x, 77, 103, 107, 115, 116, 227
mass spectrometry, 214
matrix, 28, 249, 252
measures, 63, 122, 212
media, 28, 32, 40, 43, 47, 48, 57, 84, 86, 204, 205, 206, 228, 244, 245, 246, 257
median, 172
Mediterranean climate, 74
membranes, 6, 21, 29, 52, 59, 63
mental disorder, 161
mercury, 2, 18, 52, 56, 73, 90, 94, 99, 106, 161, 176, 180, 183, 187, 215, 226
metabolism, 124, 148, 256, 258, 260
metal extraction, 19, 88
metal hydroxides, 14, 30, 37, 38, 169, 193, 195, 198
metal oxides, 27
metal recovery, 19, 25, 27
metalloids, viii, 72, 92, 103, 109, 161
meter, 205

methodology, 229, 236
Mg^{2+}, 23, 134, 171
$MgSO_4$, 246
Microbial, 260
microbial communities, 32, 34, 36
microbial community, 67
microcosms, 56
microcrystalline, 10
microscope, 195
microwave, 249
migration, 6, 49, 159, 213, 259
milligrams, 123
mine operators, 164
mine soil, 147, 188
mine tailings, 241
mineral precipitations, viii, 72, 109
minerals, x, 239, 240
mines, 240
misleading, x, 165, 240
mixing, 5, 9, 16, 17, 94, 200, 202, 248
mobility, 40, 63, 65, 73, 112, 223, 224
model, 10, 67, 83, 86, 106, 109, 145, 146, 150, 205, 207, 225, 231, 233, 234, 235, 236, 237
modeling, 149, 235
modelling, 9, 59, 116, 236, 260, 262
models, 53, 87, 152, 235, 237, 261
moisture, 139, 199, 201
moisture content, 201
molar ratios, 172
mole, 123, 124, 126, 134, 135
molecular oxygen, 4
molecular weight, 200, 201
molecules, 16, 199
molybdenum, viii, 72, 94, 99, 109, 158
Moses, 125, 127, 151
mountains, 74
movement, 42, 91, 97, 129, 141

N

Na^+, 23, 169, 171
Na2SO4, 246, 248
NaCl, 248
Native Americans, 73
NCV, 165
necrosis, 108
neglect, 244, 245
net neutralization potential, x, 167, 227
network, 77, 97, 98, 102, 109, 235
neutralization, 262
nickel (Ni), viii, 2, 4, 6, 9, 13, 14, 19, 20, 21, 22, 23, 25, 26, 28, 41, 50, 54, 58, 60, 67, 72, 106, 109, 122, 138, 150, 176, 180, 215, 236, 241, 243, 261

nitrates, 40
nitric acid, 248
nitrogen, 25, 35, 36, 52, 169
nitrogen compounds, 169
nitrogen gas, 25
N-N, 35
nodules, 74
NPR, 162, 163, 165, 168, 169
nucleation, 25
nutrients, 34, 36, 39, 108, 159

O

obligate, 131
observations, 132, 202, 221, 245
olive, 246
operating system, 198
optimization, 17, 49, 235, 237
order, x, 10, 30, 48, 72, 105, 135, 136, 140, 176, 179, 190, 194, 196, 199, 200, 207, 208, 213, 214, 215, 228, 229, 235, 240, 241, 244, 260
ores, 2, 7, 39, 65, 72, 76, 79, 99, 115, 119, 120, 128, 154, 161, 163, 164, 165, 166, 167, 176, 228, 232, 240
organic, x, 240, 241, 242, 243, 244, 246, 249, 250, 258, 259, 261, 262
organic compounds, 20
organic materials, 30, 35, 46, 241
organic matter, 23, 30, 34, 36, 38, 39, 40, 43, 46, 65, 67, 136, 137, 159, 180, 244, 258, 259
osmosis, 20, 21, 22, 49, 62, 66, 69
oxidation products, 127, 129, 156, 178
oxidation rate, 3, 4, 47, 79, 130, 132, 138, 149
oxidative reaction, 139
oxides, viii, 2, 27, 30, 32, 33, 46, 47, 72, 75, 76, 77, 78, 81, 83, 84, 86, 88, 98, 99, 105, 107, 109, 127, 136, 145, 191, 196, 198, 208, 244, 250, 252, 255, 258
oxygen, x, 2, 4, 5, 6, 33, 35, 36, 37, 40, 72, 79, 80, 86, 87, 119, 120, 123, 124, 125, 126, 127, 129, 130, 132, 137, 138, 139, 140, 141, 142, 145, 153, 156, 158, 159, 193, 197, 212, 220, 239, 240, 241
oxyhydroxides, 244
ozone, 12

P

parameters, 2, 4, 14, 16, 24, 29, 62, 86, 87, 200, 228, 235, 260
particles, 15, 19, 27, 30, 133, 137, 139, 196, 198, 199, 200, 204, 206, 207, 215
Pasco, 65

passive, vii, ix, x, 1, 2, 5, 6, 11, 24, 29, 32, 33, 37, 38, 39, 46, 47, 48, 50, 54, 55, 57, 60, 61, 62, 63, 64, 68, 69, 189, 190, 191, 194, 196, 197, 199, 206, 226, 239, 242, 261, 262
passive treatment, vii, ix, 1, 2, 6, 29, 32, 33, 37, 46, 48, 54, 55, 57, 60, 61, 62, 63, 68, 189, 190, 191, 196, 197, 206, 226, 262
pathways, 83
Pb, 241, 242, 260
PCA, 179
peat, 27, 28, 36, 39, 56, 242
percolation, 15, 105, 166
performance, vii, x, 1, 16, 21, 24, 31, 32, 33, 34, 36, 37, 38, 39, 42, 44, 47, 48, 53, 57, 58, 60, 62, 65, 66, 67, 68, 69, 199, 200, 205, 239, 240
permeability, 6, 37, 40, 41, 137, 139, 141, 206, 235, 243
permeable membrane, 20
permeation, 59
permission, 15, 20, 22, 26, 31, 41, 42, 43, 45
peroxide, 6, 12, 141
perseverance, 236
phase diagram, 80
phosphates, 93, 200, 241
phosphorus, 207
photodegradation, 30
photosynthesis, 29, 46
physical properties, 123, 138, 190, 194, 198, 208
physiology, 131
phytoremediation, 63, 109
planning, viii, 122, 228, 235
plants, 31, 33, 34, 48, 51, 53, 57, 60, 63, 73, 83, 84, 102, 103, 104, 117
plaque, 30
plasma, 10
platform, 42, 74
platinum, 176
pollutants, 7, 239, 242, 245, 252, 253, 259, 260
pollution, vii, viii, 55, 63, 68, 69, 111, 115, 119, 121, 122, 127, 142, 143, 147, 152, 156, 160, 176, 184, 213, 226, 241, 242, 244, 259, 262
poly(vinyl chloride), 214
polyacrylamide, 201, 202
polymer, 17, 200, 201, 202
polymers, 200, 201, 202
pomace, 246
population density, 159
pore, 247, 251
porosity, 38, 94, 137
porous media, 147
positive correlation, 179
potassium, 18, 23, 90, 139, 221
power, 29, 90, 98, 196, 214, 231

power plants, 196
prediction, viii, 9, 139, 142, 152, 154, 228, 229, 235, 236, 237
pressure, 5, 15, 20, 21, 37, 50, 62, 66, 69, 117, 130, 137, 139, 142, 152, 199, 202, 204, 205, 208
prevention, viii, 52, 61, 63, 111, 123, 142, 143, 146, 155, 157, 160, 186, 241
prices, 48, 88, 97, 102
Prince William Sound, 149
Principal Components Analysis, 117
probability, 235
probe, 179, 214
procedures, 246, 258
production, x, 5, 8, 23, 25, 34, 36, 38, 40, 75, 77, 93, 97, 120, 127, 130, 132, 133, 138, 139, 140, 141, 142, 144, 147, 149, 225, 239, 241, 242, 246
productivity, 46, 213
program, 58, 92, 97, 142
project, ix, 57, 58, 67, 75, 77, 97, 109, 111, 116, 156, 160, 161, 164, 165, 228
propagation, 125, 140, 212
properties, 28, 52, 60, 117, 147, 152, 189, 190, 191, 192, 193, 194, 195, 196, 197, 198, 200, 208, 252, 256, 257, 259
protective coating, 141
protons, 240
prototype, 56
public, 242
public awareness, 213
pulse, 229, 230
pumping, 247
purification, 57, 256
purity, 37
PVC, 47, 214

Q

quartz, viii, 90, 94, 155, 162, 163, 165, 167, 179, 180, 217

R

radiation, 215, 248
radicals, 200
radius, 231
rain, 7, 90, 91, 95
rainfall, 123
range, ix, 3, 4, 10, 13, 16, 22, 23, 24, 30, 31, 34, 49, 62, 122, 132, 189, 190, 193, 196, 199, 208, 220, 231, 232, 233, 234, 256
rare earth elements, 154
reactants, 126, 129, 142

reaction mechanism, 116
reaction rate, 132, 137, 139
reactions, ix, 4, 5, 9, 12, 33, 56, 67, 73, 84, 100, 105, 123, 124, 125, 129, 131, 132, 133, 134, 135, 136, 138, 139, 143, 151, 158, 159, 167, 169, 175, 180, 211, 212, 214, 217, 219, 220, 221, 222, 223, 224, 240, 241
reactivity, 11, 17, 40, 137, 138, 139, 244
reagent, 245
reagents, 10, 11, 12, 13, 14, 18, 22, 48, 49, 60, 194, 196, 197, 208, 259
reconstruction, 230
recovery, 1, 25, 26, 29, 48, 49, 50, 57, 58, 60, 61, 65, 108, 261
recycling, 14, 15, 25, 63, 112
red mud, 262
redox, 260
reduction, 242, 244, 250, 256, 261
refining, 259
reforms, 159
regeneration, 125, 127, 148, 240
region, 7, 72, 74, 77, 88, 90, 92, 97, 103, 108, 110, 190, 237
regulators, 147
regulatory framework, 157, 160
rehabilitation, viii, 51, 72, 75, 92, 109, 235
rehabilitation program, 92
rejection, 22
relationship, 80, 232
relevance, 116, 148, 257
reliability, 48, 228, 235
relief, 90, 92, 97
remediation, vii, x, 1, 10, 26, 40, 52, 55, 57, 60, 62, 64, 65, 121, 122, 123, 140, 142, 148, 188, 225, 226, 239, 261
remote sensing, 107
replication, 229
reproduction, 213
resale, 14
research, 244, 245, 262
reserves, ix, 88, 163, 227, 228, 233
residues, 103, 262
resins, 14, 23
resistance, 21, 190, 197, 200
resources, 76, 147, 176, 183, 188, 206
respect, x, 159, 166, 240, 242, 253, 256
respiration, 34
respiratory, 213
retention, 19, 24, 25, 28, 29, 36, 41, 56, 57, 59, 193, 196, 197, 207, 208
rhodium, 215
risk, viii, ix, 18, 44, 73, 108, 120, 140, 143, 150, 227, 228, 235, 237

risk assessment, 108, 150, 228, 237
rivers, 241
rodents, 36
room temperature, 249
runoff, 6, 43, 79, 106, 127

S

safety, 23, 94, 235
salinity, 156
salt, 7, 56
salts, 9, 21, 23, 84, 100, 106, 129, 139, 255
sample, 246, 247, 248, 249, 252, 258
sampling, ix, x, 92, 99, 214, 227, 228, 229, 230, 231, 234, 235, 236, 237, 246
sand, 243
SAPS, 36, 43, 44, 45, 50, 58, 190, 197
saturation, 80, 99, 175, 256, 260
savings, 26
sawdust, 28, 39, 40, 242
scale system, vii, 1
scaling, 21
screening, 164
search, 48
sediment, 30, 39, 40, 46, 78, 92, 111, 117, 141, 144, 147, 159, 213
sedimentation, 30, 31, 43, 72, 193, 194, 198, 206, 213
sediments, 33, 46, 58, 87, 92, 94, 96, 97, 101, 103, 104, 106, 108, 112, 117, 120, 132, 137, 149, 154, 159, 185, 207, 221, 260, 262
seed, 15
selecting, 190
selectivity, 23, 259
selenium, viii, 72, 99, 109, 158
semiconductor, 150
separation, 11, 14, 15, 17, 19, 23, 60, 89, 203, 249
sequencing, 24
series, 240, 262
settlements, 115
severity, 94
sewage, 28, 34, 35, 40, 51, 53, 57, 65, 205
SGP, 117
shape, 16, 102, 235
shear, 162, 165, 166
side effects, 200
signals, 229, 230
silica, 6, 90, 92, 191, 196, 206, 208, 246, 249
silicon, 14, 139, 194
silver, 2, 72, 76, 90, 94, 99, 103, 106, 232, 233
simulation, 147, 149, 228, 236, 252, 256
simulations, 259

SiO2, ix, 27, 135, 178, 192, 211, 213, 216, 217, 224, 226
sites, 240, 256
skeleton, 205
skin, 21, 183
slag, 38, 47, 64, 76, 90, 97, 99, 103, 104, 105, 106, 113, 114, 116
sludge, vii, ix, 14, 15, 16, 17, 23, 24, 25, 26, 27, 28, 40, 49, 50, 51, 54, 57, 58, 61, 64, 65, 66, 68, 111, 112, 121, 122, 189, 190, 191, 192, 193, 194, 195, 196, 197, 198, 199, 200, 201, 202, 203, 204, 205, 206, 207, 208, 225, 243
sludge disposal, 189, 193, 195, 199, 207
smoothing, 235
sodium, 11, 14, 15, 16, 18, 19, 23, 25, 139, 141, 190, 194, 195, 214, 246, 248
sodium hydroxide, 11, 14, 15, 16, 23, 25, 190, 194, 195, 214
soil, 7, 32, 36, 54, 60, 65, 72, 117, 119, 145, 149, 198, 228, 235, 236, 241, 249, 258
soils, 261
solid phase, 72, 80, 83, 87, 141, 162, 244, 245, 248, 255, 256, 258
solid waste, 196, 261
solidification, 141, 241
solubility, 11, 14, 33, 42, 84, 86, 120, 130, 132, 133, 134, 200, 220, 223, 241, 255, 259
solutions, 248, 249, 258, 262
sorption, x, 5, 9, 30, 35, 39, 59, 65, 223, 239, 252, 253, 254, 255, 256, 257, 260
space, x, 21, 130, 136, 142, 196, 197, 200, 227, 229, 233, 236
speciation, 188, 244, 245, 252, 254, 255, 256, 258, 259, 260
species, 2, 9, 24, 34, 49, 72, 73, 81, 82, 85, 86, 108, 124, 126, 130, 148, 180, 194, 208, 213, 221, 223, 251, 252, 253, 254, 255
specific surface, 19, 79, 137, 207
spectroscopy, 115, 217
spectrum, 229, 230, 231
speed, 12, 142, 202, 205, 206
stability, 2, 15, 49, 50, 56, 68, 81, 86, 87, 106, 107, 113, 149, 151
stabilization, 108, 132, 200
standard deviation, 191, 192, 245, 246, 255
statistics, 90, 176, 182, 236
steel, 38, 64, 204
stoichiometry, 136, 138, 175
storage, 122, 140, 141, 198, 200, 206, 207, 241
stormwater, 66
strain, 132, 250
strategies, viii, 5, 55, 121, 241, 242
strategy, 5, 141, 142, 164, 188, 228

streams, 242
strength, 10, 57, 84, 85, 86, 200, 203, 258
stress, 62, 108
substrates, 36, 39, 43, 46, 54, 55, 223, 258, 261
subsurface flow, 68
succession, 74
successive approximations, 134
sulfate, 261
sulfur, 57, 72, 73, 80, 82, 83, 84, 90, 94, 97, 102, 103, 104, 105, 106, 123, 137, 138, 146, 147, 148, 149, 153, 238, 261
sulfur dioxide, 80
sulfuric acid, 145
sulphate, x, 239, 240, 242, 243, 244, 246, 250, 256, 257, 258, 261, 262
sulphur, x, 8, 10, 24, 25, 26, 100, 120, 124, 127, 128, 131, 132, 134, 135, 156, 157, 159, 160, 162, 167, 220, 225, 228, 235, 239, 241
supply, 17, 18, 34, 36, 49, 50, 125, 133
surface area, 34, 47, 48, 57, 137, 139, 159, 169, 205, 213, 252
surface chemistry, 262
surface energy, 79
surface properties, 190, 208
surfactant, 29, 51, 52, 66
surfactants, 241
surplus, 17, 162
survey, 170, 187
susceptibility, 185
suspensions, 148, 246, 257
sustainability, vii, 5
sustainable development, 109, 114, 143, 228, 235
symptoms, 143
synthesis, 217
synthetic, x, 239, 245, 246, 247, 248, 252, 254, 259
systems, x, 239, 242, 244, 247, 262

T

tanks, 14, 15, 17, 76, 89, 102, 105, 203
TCE, 242
Teflon, 249
temperature, 7, 10, 30, 42, 72, 79, 80, 83, 84, 85, 106, 114, 130, 139, 142, 148, 150, 153, 158, 207, 249
tension, 160
territory, 73, 74
testing, 202, 248
textbooks, 176
thallium, 94, 176
theoretical, x, 240, 245
thermal energy, 199
thermodynamic properties, 118, 151

Third World, 184
threat, 2
threshold, x, 227, 229
thresholds, 62
time, 246, 247, 250, 251, 252
tin, 16, 72
titanium, 23
toxic, x, 239, 240, 241, 244, 251, 252, 254, 261
toxic metals, x, 5, 24, 65, 67, 239, 240, 241, 251, 252
toxicity, 6, 25, 28, 51, 62, 65, 183, 213
trace elements, ix, 9, 36, 54, 72, 73, 78, 79, 91, 92, 100, 101, 106, 107, 108, 118, 122, 138, 139, 147, 179, 180, 188, 226, 227, 228
training, 228, 236
transformation, 84, 106, 130
transformations, 53
transition, 132, 203
transport, 10, 28, 52, 59, 77, 90, 129, 136, 143, 150, 156, 207, 260
transport processes, 28
transportation, vii, ix, 142, 189, 204
treatment methods, 5, 93, 142, 192, 194
trend, 250, 252, 253, 255
trends, ix, 65, 123, 146, 166, 185, 211, 217, 222, 224, 256, 260
tungsten, 73
turbulence, 46, 213

U

U.S. Geological Survey, 152, 185, 187, 226
uncertainty, 73, 228, 235
uniform, 32, 191, 256
uranium, 2, 28, 56, 145, 146, 150, 176
US Department of Commerce, 67
USDA, 188

V

vanadium, 106
variance, 90, 229, 231
variations, 7, 18, 42, 51, 54, 59, 100, 115, 138, 144, 147, 166, 188, 200
vegetation, 30, 32, 33, 36, 46
vein, 76, 77, 165, 168, 185, 236

velocity, 30, 213
ventilation, 249
village, 102, 232
viscosity, 190, 197, 198, 208
vision, 127

W

waste disposal, 99, 149, 187, 228, 229, 235, 236
waste disposal sites, 149, 228, 236
waste management, 58, 77, 137, 139, 147
waste treatment, 53
wastes, 241
wastewater, 19, 20, 21, 50, 53, 55, 59, 60, 62, 63, 66, 68, 160, 204, 205, 207, 208, 212, 225, 226, 242
wastewaters, 243, 244
water, x, 239, 240, 247, 248, 249, 261, 263
water evaporation, 104
water quality, 4, 7, 8, 32, 43, 52, 53, 65, 115, 121, 131, 144, 150, 153, 182, 188
water resources, 143, 176
water supplies, 185
watershed, 144, 188
waterways, 156
wavelengths, 248
weakness, 160
weathering, 240
wells, 162, 166, 172, 175, 176, 184, 228
wetlands, 29, 30, 31, 32, 33, 34, 35, 36, 37, 38, 44, 46, 47, 48, 49, 50, 51, 52, 53, 56, 57, 59, 63, 64, 65, 66, 67, 68, 137, 190, 262
wood, 41, 42, 44, 89, 146
wool, 204

X

XPS, 130
XRD, 188, 215, 217, 225

Z

zeolites, 27, 51, 63, 67, 225, 226
zirconium, 23